MACHINE LEARNING FOR FUTURE FIBER-OPTIC COMMUNICATION SYSTEMS

Edited by

ALAN PAK TAO LAU
FAISAL NADEEM KHAN

ELSEVIER

ACADEMIC PRESS
An imprint of Elsevier

Academic Press is an imprint of Elsevier
125 London Wall, London EC2Y 5AS, United Kingdom
525 B Street, Suite 1650, San Diego, CA 92101, United States
50 Hampshire Street, 5th Floor, Cambridge, MA 02139, United States
The Boulevard, Langford Lane, Kidlington, Oxford OX5 1GB, United Kingdom

Notices

Knowledge and best practice in this field are constantly changing. As new research and experience broaden our
understanding, changes in research methods, professional practices, or medical treatment may become necessary.

Practitioners and researchers must always rely on their own experience and knowledge in evaluating and using any
information, methods, compounds, or experiments described herein. In using such information or methods they
should be mindful of their own safety and the safety of others, including parties for whom they have a professional
responsibility.

To the fullest extent of the law, neither the Publisher nor the authors, contributors, or editors, assume any liability
for any injury and/or damage to persons or property as a matter of products liability, negligence or otherwise, or
from any use or operation of any methods, products, instructions, or ideas contained in the material herein.

Library of Congress Cataloging-in-Publication Data
A catalog record for this book is available from the Library of Congress

British Library Cataloguing-in-Publication Data
A catalogue record for this book is available from the British Library

ISBN: 978-0-323-85227-2

For information on all Academic Press publications
visit our website at https://www.elsevier.com/books-and-journals

Publisher: Mara Conner
Acquisitions Editor: Tim Pitts
Editorial Project Manager: Mica Ella Ortega
Production Project Manager: Surya Narayanan Jayachandran
Designer: Christian J. Bilbow

Typeset by VTeX

MACHINE LEARNING FOR FUTURE FIBER-OPTIC COMMUNICATION SYSTEMS

Dedicated to my family for their unwavering love, support, and encouragement. Without their monumental help, this book would never have come to fruition.

— Faisal Nadeem Khan

To my parents for their unconditional love and support throughout my life. To my wife Sandy and my son Sean, who give me meaning and purpose in life.

— Alan Pak Tao Lau

Contents

Contributors

Polina Bayvel
University College London, London, United Kingdom

Matthew Brand
Mitsubishi Electric Research Laboratories (MERL), Cambridge, MA, United States

Francesco Da Ros
Department of Photonic Engineering, Technical University of Denmark, Kgs. Lyngby, Denmark

Camille Delezoide
Smart Optical Fabric and Devices Lab, Nokia Bell Labs, Nozay, France

Qirui Fan
Department of Electrical Engineering, The Hong Kong Polytechnic University, Kowloon, Hong Kong SAR, China

Marija Furdek
Department of Electrical Engineering, Chalmers University of Technology, Gothenburg, Sweden

Christian Häger
Chalmers University of Technology, Gothenburg, Sweden

Takeshi Hoshida
Fujitsu Limited, Kawasaki, Japan

Luyao Huang
Shanghai Jiao Tong University, Shanghai, China

Memedhe Ibrahimi
Department of Electronics, Information and Bioengineering, Politecnico di Milano, Milan, Italy

Ognjen Jovanovic
Department of Photonic Engineering, Technical University of Denmark, Kgs. Lyngby, Denmark

Boris Karanov
Eindhoven University of Technology, Eindhoven, The Netherlands

Faisal Nadeem Khan
Tsinghua Shenzhen International Graduate School, Tsinghua University, Shenzhen, China

Toshiaki Koike-Akino
Mitsubishi Electric Research Laboratories (MERL), Cambridge, MA, United States

Keisuke Kojima
Mitsubishi Electric Research Laboratories (MERL), Cambridge, MA, United States

Alan Pak Tao Lau
Department of Electrical Engineering, The Hong Kong Polytechnic University, Kowloon, Hong Kong SAR, China

Patricia Layec
Smart Optical Fabric and Devices Lab, Nokia Bell Labs, Nozay, France

Zhengxuan Li
Shanghai Jiao Tong University, Shanghai, China

Chao Lu
Department of Electronic and Information Engineering, The Hong Kong Polytechnic University, Kowloon, Hong Kong SAR, China

Carlos Natalino
Department of Electrical Engineering, Chalmers University of Technology, Gothenburg, Sweden

Petros Ramantanis
Smart Optical Fabric and Devices Lab, Nokia Bell Labs, Nozay, France

Cristina Rottondi
Department of Electronics and Telecommunications, Politecnico di Torino, Torino, Italy

Marc Ruiz
Universitat Politècnica de Catalunya, Barcelona, Spain

Laurent Schmalen
Karlsruhe Institute of Technology, Karlsruhe, Germany

Behnam Shariati
Fraunhofer HHI, Berlin, Germany

Yingheng Tang
Mitsubishi Electric Research Laboratories (MERL), Cambridge, MA, United States
Electrical and Computer Engineering Dept., Purdue University, West Lafeyette, IN, United States

Takahito Tanimura
Fujitsu Limited, Kawasaki, Japan

Massimo Tornatore
Department of Electronics, Information and Bioengineering, Politecnico di Milano, Milan, Italy

Alba P. Vela
Universitat Politècnica de Catalunya, Barcelona, Spain

Luis Velasco
Universitat Politècnica de Catalunya, Barcelona, Spain

Ye Wang
Mitsubishi Electric Research Laboratories (MERL), Cambridge, MA, United States

Wanting Xu
Shanghai Jiao Tong University, Shanghai, China

Yongxin Xu
Shanghai Jiao Tong University, Shanghai, China

Metodi Yankov
Department of Photonic Engineering, Technical University of Denmark, Kgs. Lyngby, Denmark

Lilin Yi
Shanghai Jiao Tong University, Shanghai, China

Shaoliang Zhang
Acacia Communication Inc., Maynard, MA, United States

Darko Zibar
Department of Photonic Engineering, Technical University of Denmark, Kgs. Lyngby, Denmark

Preface

Traditionally, optical communications and networking are relatively static in time and rely on precise and well-understood mathematical models that work exceptionally well in many practical scenarios. However, next-generation fiber-optic communication networks are expected to support extremely high data rates and low latencies as well as enable dynamicity, flexibility, reliability, compatibility, and efficiency in order to meet the heterogeneous quality-of-service (QoS) requirements of numerous emerging applications such as cloud computing, Internet-of-things (IoT), 6G wireless, etc. Consequently, the researchers are often forced to push boundaries to extents where the underlying mathematics/physics of the problems is too difficult to be described explicitly or the numerical procedures involved are computationally too complex and impractical. Fortunately, given the unprecedented availability of large amounts of data in today's hyper-connected world, fast computing resources, and a new wave of research advances in the past decade, the data-driven machine learning (ML) methods have appeared as a new powerful tool since they do not necessitate rigid predefined models and can instead self-learn and uncover meaningful patterns/structures in the big data to provide useful results. This can not only speed up the design cycle of future optical networks but also reduce the complexity and cost of implementation.

ML is being hailed as a new direction of innovation to transform future optical communication systems. Signal processing paradigms based on ML are being considered to solve certain critical problems in optical networks that cannot be easily tackled using conventional approaches. Over the past few years, serious efforts have been made by both industry and academia to exploit ML methods in various aspects of fiber-optic communications and networking. Examples of such works include ML-based: design of various photonic components and sub-systems, end-to-end learning of complete fiber-optic communication system, adaptive allocation of network resources, compensation of nonlinear distortions introduced by optical fiber and other devices, prediction of light-paths' quality and optical performance monitoring, proactive fault detection/prevention and root cause analysis, cross-layer optimizations for software-defined networks (SDNs), intrusions detection and network security management, etc.

We are privileged to be able to invite leading researchers around the globe, working on ML applications in key areas of optical communications and networking, to contribute to different chapters of this book. Their works reflect on state-of-the-art researches and industrial practices and give knowledge and insights into the role ML-based mechanisms will play in the future realization of intelligent optical network infrastructures that can manage and monitor themselves, diagnose and resolve their problems in real time, and deliver efficient and intelligent services to the end users.

The book is organized into 11 chapters. The first chapter details the reasons behind current popularity of ML in fiber-optic networks and sheds light on the kind of problems where ML can play a crucial role. It also discusses mathematical foundations of several fundamental ML techniques such as artificial neural networks, support vector machines, K-means clustering, expectation-maximization algorithm, principal component analysis, independent component analysis, reinforcement learning, etc., as well as more advanced deep learning (DL) methods like deep neural networks, convolutional neural networks, recurrent neural networks, generative adversarial networks, etc., from communication theory and signal processing perspectives. Chapter 2 discusses ML-based techniques for intra-channel fiber nonlinearity compensation in long-haul fiber-optic communication systems while Chapter 3 covers the applications of conventional ML as well as DL methods in the design and optimization of digital signal processing modules for short-reach systems. Chapter 4 focuses on ML-based equalizers for mitigating nonlinear distortions introduced by optical fibers and other devices in passive optical networks. In Chapter 5, end-to-end learning and optimization of complete optical communication system using various ML models are discussed. Chapter 6 focuses on DL-based optical performance monitoring modules for achieving fully-automated network operation while Chapter 7 discusses ML-enabled quality-of-transmission estimation methods and their integration into optimization tools for adaptive allocation of network resources. In Chapter 8, ML-assisted optical spectrum analysis solutions for detecting and identifying various network devices failures are presented. Chapter 9 discusses constructive roles of ML and data analytics for the design and operation of low-margin optical networks while Chapter 10 focuses on the application of ML techniques for automating physical-layer security management. Finally, Chapter 11 addresses how DL models can be effectively incorporated into the design and optimization processes of various photonic devices to achieve desired target performance.

On the whole, with the up-to-date coverage and exhaustive treatment of several important topics in fiber-optic communications and networking, this book can effectively serve as a valuable reference for researchers, engineers, and students interested in the fields of ML-based signal processing and networking.

Faisal Nadeem Khan
Shenzhen, China, July 2021

Alan Pak Tao Lau
Hong Kong SAR, China, July 2021

Acknowledgments

Alan Pak Tao Lau would like to acknowledge the support of The Hong Kong Polytechnic University under project YBY3 and Hong Kong Government Research Grants Council General Research Fund (GRF) under project PolyU 15220120.

Faisal Nadeem Khan would like to acknowledge the support of Tsinghua Shenzhen International Graduate School and Tsinghua–Berkeley Shenzhen Institute under Scientific Research Startup Fund.

## CHAPTER ONE

Introduction to machine learning techniques: An optical communication's perspective

Faisal Nadeem Khan[a], Qirui Fan[b], Chao Lu[c], and Alan Pak Tao Lau[b]

[a]Tsinghua Shenzhen International Graduate School, Tsinghua University, Shenzhen, China
[b]Department of Electrical Engineering, The Hong Kong Polytechnic University, Kowloon, Hong Kong SAR, China
[c]Department of Electronic and Information Engineering, The Hong Kong Polytechnic University, Kowloon, Hong Kong SAR, China

1.1. Introduction

Artificial intelligence (AI) makes use of computers/machines to perform cognitive tasks, i.e., the ones requiring knowledge, perception, learning, reasoning, understanding and other similar cognitive abilities. An AI system is expected to do three things: (i) store knowledge, (ii) apply the stored knowledge to solve problems, and (iii) acquire new knowledge via experience. The three key components of an AI system include knowledge representation, machine learning (ML), and automated reasoning. ML is a branch of AI which is based on the idea that patterns and trends in a given data set can be learned automatically through algorithms. The learned patterns and structures can then be used to make decisions or predictions on some other data in the system of interest [1].

ML is not a new field as ML–related algorithms exist at least since the 1970s. However, tremendous increase in computational power over the last decade, recent groundbreaking developments in theory and algorithms surrounding ML, and easy access to an overabundance of all types of data worldwide (thanks to three decades of Internet growth) have all contributed to the advent of modern deep learning (DL) technology, a class of advanced ML approaches that displays superior performance in an ever-expanding range of domains. In the near future, ML is expected to power numerous aspects of modern society such as web searches, computer translation, content filtering on social media networks, healthcare, finance, and laws [2].

ML is an interdisciplinary field which shares common threads with the fields of statistics, optimization, information theory, and game theory. Most ML algorithms perform one of the following two types of pattern recognition tasks as shown in Fig. 1.1. In the first type, the algorithm tries to find some functional description of given data with the aim of predicting values for new inputs, i.e., *regression problem*. The second type attempts to find suitable decision boundaries to distinguish different data classes, i.e., *classification problem* [3], which is more commonly referred to as *clustering problem* in ML literature. ML techniques are well known for performing exceptionally well in scenarios in which it is too hard to explicitly describe the problem's underlying physics and mathematics.

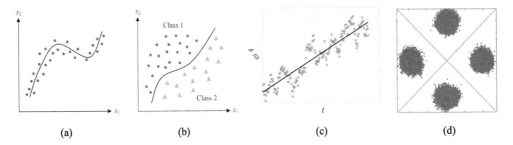

(a)   (b)   (c)   (d)

Figure 1.1 Given a data set, ML attempts to solve two main types of problems: (a) functional description of given data and (b) classification of data by deriving appropriate decision boundaries. (c) Laser frequency offset and phase estimation for quadrature phase-shift keying (QPSK) systems by raising the signal phase ϕ to the 4th power and performing regression to estimate the slope and intercept. (d) Decision boundaries for a received QPSK signal distribution.

Optical communication researchers are no strangers to regressions and classifications. Over the last decade, coherent detection and digital signal processing (DSP) techniques have been the cornerstone of optical transceivers in fiber-optic communication systems. Advanced modulation formats such as 16 quadrature amplitude modulation (16-QAM) and above together with DSP-based estimation and compensation of various transmission impairments such as laser phase noise have become the key drivers of innovation. In this context, parameter estimation and symbol detection are naturally regression and classification problems, respectively, as demonstrated by examples in Fig. 1.1(c) and (d). Currently, most of these parameter estimation and decision rules are derived from probability theory and adequate understanding of the problem's underlying physics. As high-capacity optical transmission links are increasingly being limited by transmission impairments such as fiber nonlinearity, explicit statistical characterizations of inputs/outputs become difficult. An example of 16-QAM multi-span dispersion-free transmissions in the presence of fiber nonlinearity and inline amplifier noise is shown in Fig. 1.2(a). The maximum likelihood decision boundaries in this case are curved and virtually impossible to derive analytically. Consequently, there has been an increasing amount of research on the application of ML techniques for fiber nonlinearity compensation (NLC). Another related area where ML flourishes is short-reach direct detection systems that are affected by chromatic dispersion (CD), laser chirp and other transceiver components imperfections, which render the overall communication system hard to analyze.

Optical performance monitoring (OPM) is another area with an increasing amount of ML-related research. OPM is the acquisition of real-time information about different channel impairments ubiquitously across the network to ensure reliable network operation and/or improve network capacity. Often, OPM is cost-limited so that one can

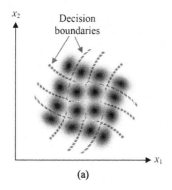 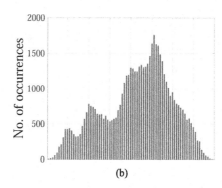

Figure 1.2 (a) Probability distribution and corresponding optimal decision boundaries for received 16-QAM symbols in the presence of fiber nonlinearity are hard to characterize analytically. (b) Probability distribution of received 64-QAM signal amplitudes. The distribution can be used to monitor optical signal-to-noise ratio (OSNR) and identify modulation format. However, this task will be extremely difficult if one relies on analytical modeling.

only employ simple hardware components and obtain partial signal features to monitor different channel parameters such as OSNR, optical power, CD, etc. [4][5]. In this case, the mapping between input and output parameters is intractable from underlying physics/mathematics, which in turn warrants ML. An example of OSNR monitoring using received signal amplitudes distribution is shown in Fig. 1.2(b).

Besides physical layer-related developments, optical network architectures and operations are also undergoing major paradigm shifts under the software-defined networking (SDN) framework and are increasingly becoming complex, transparent and dynamic in nature [6]. One of the key features of SDNs is that they can assemble large amounts of data and perform so-called big data analysis to estimate the network states as shown in Fig. 1.3. This in turn can enable (i) adaptive provisioning of resources such as wavelength, modulation format, routing path, etc., according to dynamic traffic patterns and (ii) advance discovery of potential components faults so that preventative maintenance can be performed to avoid major network disruptions. The data accumulated in SDNs can span from physical layer (e.g., OSNR of a certain channel) to network layer (e.g., client-side speed demand) and obviously have no underlying physics to explain their interrelationships. Extracting patterns from such cross-layer parameters naturally demands the use of data-driven algorithms such as ML.

This chapter is intended for the researchers in optical communications with a basic background in probability theory, communication theory and standard DSP techniques used in fiber-optic communications such as matched filters, maximum likelihood/maximum a posteriori (MAP) detection, equalization, adaptive filtering, etc. In this regard, a large class of ML techniques such as Kalman filtering, Bayesian learning, hidden Markov models (HMMs), etc., are actually standard statistical signal processing methods,

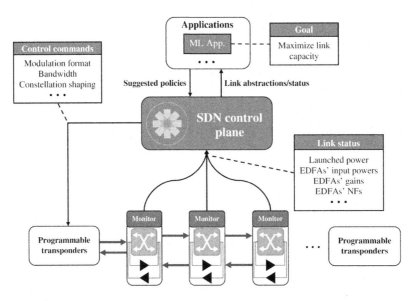

Figure 1.3 Dynamic network resources allocation and link capacity maximization via cross-layer optimization in SDNs.

and hence will not be covered here. We will first introduce supervised ML techniques such as artificial neural networks (ANNs), support vector machines (SVMs) and *K*-nearest neighbors (KNN) from communication theory and signal processing perspectives. This will be followed by popular unsupervised ML methods like *K*-means clustering, expectation-maximization (EM) algorithm, principal component analysis (PCA) and independent component analysis (ICA). Next, we will address reinforcement learning (RL) approach. Finally, more recent DL techniques such as deep neural networks (DNNs), convolutional neural networks (CNNs), recurrent neural networks (RNNs) and generative adversarial networks (GANs) will be discussed. The analytical derivations presented in this chapter are slightly different from those in standard introductory ML text to better align with the fields of communications and signal processing. By discussing ML through the language of communications and DSP, we hope to provide a more intuitive understanding of ML, its relation to optical communications and networking, and why/where/how it can play a unique role in specific areas of optical communications and networking.

The rest of the chapter is organized as follows. In Section 1.2, we will illustrate the fundamental conditions that warrant the use of a neural network and discuss the technical details of ANN, SVM and KNN algorithms. Section 1.3 will describe a range of basic unsupervised ML techniques while Section 1.4 will briefly discuss RL approach. Section 1.5 will be devoted to more recent DL algorithms. Section 1.6 will describe the future role of ML in optical communications and networking. Links for online resources and codes for standard ML algorithms will be provided in Section 1.7. Section 1.8 will conclude the chapter.

1.2. Supervised learning

What are the conditions that need ML for classification? Fig. 1.4 shows three scenarios with 2-dimensional (2D) data $\mathbf{x} = [x_1 \ x_2]^T$ and their respective class labels depicted as 'o' and '×' in the figure. In the first case, classifying the data is straightforward: the decision rule is to see whether $\sigma(x_1 - c)$ or $\sigma(x_2 - c)$ is greater or less than 0 where $\sigma(\cdot)$ is the decision function as shown. The second case is slightly more complicated as the decision boundary is a slanted straight line. However, a simple rotation and shifting of the input, i.e., $\mathbf{W}\mathbf{x} + \mathbf{b}$ will map one class of data to below zero and the other class above. Here, the rotation and shifting are described by matrix $\mathbf{W}$ and vector $\mathbf{b}$, respectively. This is followed by the decision function $\sigma(\mathbf{W}\mathbf{x} + \mathbf{b})$. The third case is even more complicated. The region for the 'green' (mid gray in print version) class depends on the outputs of the 'red' (dark gray in print version) and 'blue' (light gray in print version) decision boundaries. Therefore, one will need to implement an extra decision step to label the 'green' region. The graphical representation of this 'decision of decisions' algorithm is the simplest form of an ANN [8]. The intermediate decision output units are known as hidden neurons and they form the hidden layer.

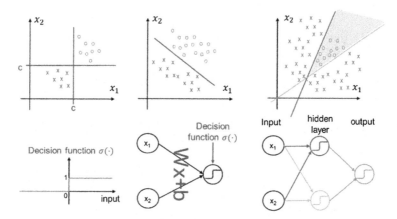

Figure 1.4 The complexity of classification problems depends on how the different classes of data are distributed across the variable space [7].

1.2.1 Artificial neural networks (ANNs)

Let $\big\{ (\mathbf{x}(1), \mathbf{y}(1)), (\mathbf{x}(2), \mathbf{y}(2)), \ldots (\mathbf{x}(L), \mathbf{y}(L)) \big\}$ be a set of L input-output pairs of M and K dimensional column vectors. ANNs are information processing systems comprising of an input layer, one or more hidden layers, and an output layer. The structure of a single hidden layer ANN with M input, H hidden and K output neurons is shown in Fig. 1.5. Neurons in two adjacent layers are interconnected where each connection has a variable weight assigned. Such ANN architecture is the simplest and most commonly-used one [8]. The number of neurons M in the input layer is determined by the dimension of the input data vectors $\mathbf{x}(l)$. The hidden layer enables the modeling of complex relationships between the input and output parameters of an ANN. There are no fixed rules for choosing the optimum number of neurons for a given hidden layer and the optimum number of hidden layers in an ANN. Typically, the selection is made via experimentation, experience and other prior knowledge of the problem. These are known as the *hyperparameters* of an ANN. For regression problems, the dimension K of the vectors $\mathbf{y}(l)$ depends on the actual problem nature. For classification problems, K typically equals to the number of class labels such that if a data point $\mathbf{x}(l)$ belongs to class k, $\mathbf{y}(l) = [0\ 0\cdots0\ 1\ 0\cdots0\ 0]^{\mathrm{T}}$ where the '1' is located at the kth position. This is called *one-hot encoding*. The ANN output $\mathbf{o}(l)$ will naturally have the same dimension as $\mathbf{y}(l)$ and the mapping between input $\mathbf{x}(l)$ and $\mathbf{o}(l)$ can be expressed as

$$
\begin{aligned}
\mathbf{o}(l) &= \sigma_2\left(\mathbf{r}(l)\right) \\
&= \sigma_2\left(\mathbf{W}_2\mathbf{u}(l) + \mathbf{b}_2\right) \\
&= \sigma_2\left(\mathbf{W}_2\sigma_1\left(\mathbf{q}(l)\right) + \mathbf{b}_2\right) \\
&= \sigma_2\left(\mathbf{W}_2\sigma_1\left(\mathbf{W}_1\mathbf{x}(l) + \mathbf{b}_1\right) + \mathbf{b}_2\right)
\end{aligned}
\tag{1.1}
$$

where $\sigma_{1(2)}(\cdot)$ are the *activation functions* for the hidden and output layer neurons, respectively. $\mathbf{W}_1$ and $\mathbf{W}_2$ are matrices containing the weights of connections between the input and hidden layer neurons and between the hidden and output layer neurons, respectively, while $\mathbf{b}_1$ and $\mathbf{b}_2$ are the bias vectors for the hidden and output layer neurons, respectively. For a vector $\mathbf{z} = [z_1\ z_2\ \cdots z_K]$ of length K, $\sigma_1(\cdot)$ is typically an element-wise nonlinear function such as the sigmoid function

$$
\sigma_1(\mathbf{z}) = \left[\frac{1}{1 + e^{-z_1}}\quad \frac{1}{1 + e^{-z_2}}\quad \cdots\quad \frac{1}{1 + e^{-z_K}}\right].
\tag{1.2}
$$

As for the output layer neurons, $\sigma_2(\cdot)$ is typically chosen to be a linear function for regression problems. In classification problems, one will normalize the output vector $\mathbf{o}(l)$ using the *softmax* function, i.e.,

$$
\mathbf{o}(l) = \mathit{softmax}\left(\mathbf{W}_2\ \mathbf{u}(l) + \mathbf{b}_2\right)
\tag{1.3}
$$

where

$$
\mathit{softmax}(\mathbf{z}) = \frac{1}{\sum_{k=1}^{K} e^{z_k}}\left[e^{z_1}\ e^{z_2}\cdots e^{z_K}\right].
\tag{1.4}
$$

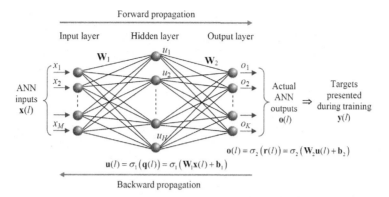

Figure 1.5 Structure of a single hidden layer ANN with input vector **x**(l), target vector **y**(l) and actual output vector **o**(l).

The softmax operation ensures that the ANN outputs conform to a probability distribution for reasons we will discuss below.

To train the ANN is to optimize all the parameters $\theta = \{\mathbf{W}_1, \mathbf{W}_2, \mathbf{b}_1, \mathbf{b}_2\}$ such that the difference between the actual ANN outputs **o** and the target outputs **y** is minimized. One commonly-used objective function (also called *loss function* in ML literature) to optimize is the mean square error (MSE)

$$E = \frac{1}{L} \sum_{l=1}^{L} E(l) = \frac{1}{L} \sum_{l=1}^{L} \left\| \mathbf{o}(l) - \mathbf{y}(l) \right\|^2. \tag{1.5}$$

Like most optimization procedures in practice, gradient descent is used instead of full analytical optimization. In this case, the parameter estimates for $n+1^{\text{th}}$ iteration are given by

$$\theta^{(n+1)} = \theta^{(n)} - \alpha \left. \frac{\partial E}{\partial \theta} \right|_{\theta^{(n)}} \tag{1.6}$$

where the step size α is known as the *learning rate*. Note that for computational efficiency, one can use a single input-output pair instead of all the L pairs for each iteration in Eq. (1.6). This is known as stochastic gradient descent (SGD) which is the standard optimization method used in common adaptive DSP such as constant modulus algorithm (CMA) and least mean squares (LMS) algorithm. As a trade-off between computational efficiency and accuracy, one can use a *mini-batch* of data $\{(\mathbf{x}(nP+1), \mathbf{y}(nP+1)), (\mathbf{x}(nP+2), \mathbf{y}(nP+2)) \ldots (\mathbf{x}(nP+P), \mathbf{y}(nP+P))\}$ of size P for the nth iteration instead. This can reduce the stochastic nature of SGD and improve accuracy. When all the data set has been used, the update algorithm will have completed one *epoch*. However, it is often the case that one epoch equivalent of updates is not enough for all the parameters to converge to their optimal values. Therefore, one can reuse the data set and the algorithm goes through the 2nd epoch for further parameter updates. There is no fixed rule to determine the number of epochs required for convergence [9].

The update algorithm is comprised of following main steps: (i) *Model initialization:* All the ANN weights and biases are randomly initialized, e.g., by drawing random numbers from a normal distribution with zero mean and unit variance; (ii) *Forward propagation:* In this step, the inputs $\mathbf{x}$ are passed through the network to generate the outputs $\mathbf{o}$ using Eq. (1.1). The input can be a single data point, a mini-batch or the complete set of L inputs. This step is named so because the computation flow is in the natural forward direction, i.e., starting from the input, passing through the network, and going to the output; (iii) *Backward propagation and weights/biases update:* For simplicity, let us assume SGD using 1 input-output pair $(\mathbf{x}(n), \mathbf{y}(n))$ for the $n+1^{\text{th}}$ iteration, sigmoid activation function for the hidden layer neurons and linear activation function for the output layer neurons such that $\mathbf{o}(n) = \mathbf{W}_2\,\mathbf{u}(n) + \mathbf{b}_2$. The parameters $\mathbf{W}_2, \mathbf{b}_2$ will be updated first followed by $\mathbf{W}_1, \mathbf{b}_1$. Since $E(n) = \|\mathbf{o}(n) - \mathbf{y}(n)\|^2$ and $\frac{\partial E(n)}{\partial \mathbf{o}(n)} = 2(\mathbf{o}(n) - \mathbf{y}(n))$, the corresponding update equations are

$$\mathbf{W}_2^{(n+1)} = \mathbf{W}_2^{(n)} - 2\alpha \sum_{k=1}^{K} \frac{\partial o_k(n)}{\partial \mathbf{W}_2} \left(o_k(n) - y_k(n) \right)$$

$$\mathbf{b}_2^{(n+1)} = \mathbf{b}_2^{(n)} - 2\alpha \frac{\partial \mathbf{o}(n)}{\partial \mathbf{b}_2} \left(\mathbf{o}(n) - \mathbf{y}(n) \right) \tag{1.7}$$

where $o_k(n)$ and $y_k(n)$ denote the kth element of vectors $\mathbf{o}(n)$ and $\mathbf{y}(n)$, respectively. In this case, $\frac{\partial \mathbf{o}(n)}{\partial \mathbf{b}_2}$ is the Jacobian matrix in which the jth row and m^{th} column is the derivative of the mth element of $\mathbf{o}(n)$ with respect to the jth element of $\mathbf{b}_2$. Also, the jth row and m^{th} column of the matrix $\frac{\partial o_k(n)}{\partial \mathbf{W}_2}$ denotes the derivative of $o_k(n)$ with respect to the jth row and m^{th} column of $\mathbf{W}_2$. Interested readers are referred to [10] for an overview of matrix calculus. Since $\mathbf{o}(n) = \mathbf{W}_2\,\mathbf{u}(n) + \mathbf{b}_2$, $\frac{\partial \mathbf{o}(n)}{\partial \mathbf{b}_2}$ is simply the identity matrix. For $\frac{\partial o_k(n)}{\partial \mathbf{W}_2}$, its kth row is equal to $\mathbf{u}(n)^{\text{T}}$ (where $(\cdot)^{\text{T}}$ denotes transpose) and is zero otherwise. Eq. (1.7) can be simplified as

$$\mathbf{W}_2^{(n+1)} = \mathbf{W}_2^{(n)} - 2\alpha \left(\mathbf{o}(n) - \mathbf{y}(n) \right) \mathbf{u}(n)^{\text{T}}$$

$$\mathbf{b}_2^{(n+1)} = \mathbf{b}_2^{(n)} - 2\alpha \left(\mathbf{o}(n) - \mathbf{y}(n) \right). \tag{1.8}$$

With the updated $\mathbf{W}_2^{(n+1)}$ and $\mathbf{b}_2^{(n+1)}$, one can calculate

$$\mathbf{W}_1^{(n+1)} = \mathbf{W}_1^{(n)} - 2\alpha \sum_{k=1}^{K} \frac{\partial o_k(n)}{\partial \mathbf{W}_1} \left(o_k(n) - y_k(n) \right)$$

$$\mathbf{b}_1^{(n+1)} = \mathbf{b}_1^{(n)} - 2\alpha \frac{\partial \mathbf{o}(n)}{\partial \mathbf{b}_1} \left(\mathbf{o}(n) - \mathbf{y}(n) \right). \tag{1.9}$$

Since the derivative of the sigmoid function is given by $\sigma_1'(\mathbf{z}) = \sigma_1(\mathbf{z}) \circ (\mathbf{1} - \sigma_1(\mathbf{z}))$ where $\circ$ denotes element-wise multiplication and $\mathbf{1}$ denotes a column vector of 1's with the same length as $\mathbf{z}$,

$$\frac{\partial \mathbf{o}(n)}{\partial \mathbf{b}_1} = \frac{\partial \mathbf{q}(n)}{\partial \mathbf{b}_1} \frac{\partial \mathbf{u}(n)}{\partial \mathbf{q}(n)} \frac{\partial \mathbf{o}(n)}{\partial \mathbf{u}(n)}$$

$$= \text{diag}\{\mathbf{u}(n) \circ (\mathbf{1} - \mathbf{u}(n))\} \cdot \left(\mathbf{W}_2^{(n+1)}\right)^{\mathrm{T}} \tag{1.10}$$

where $\text{diag}\{\mathbf{z}\}$ denotes a diagonal matrix with diagonal vector $\mathbf{z}$. Next,

$$\frac{\partial o_k(n)}{\partial \mathbf{W}_1} = \sum_j \frac{\partial o_k(n)}{\partial u_j(n)} \frac{\partial u_j(n)}{\partial q_j(n)} \frac{\partial q_j(n)}{\partial \mathbf{W}_1}$$

$$= \sum_j w_{2,k,j}^{(n+1)} u_j(n) (1 - u_j(n)) \frac{\partial q_j(n)}{\partial \mathbf{W}_1} \tag{1.11}$$

where $w_{2,k,j}^{(n+1)}$ is the kth row and jth column entry of $\mathbf{W}_2^{(n+1)}$. For $\frac{\partial q_j(n)}{\partial \mathbf{W}_1}$, its jth row is $\mathbf{x}(n)^{\mathrm{T}}$ and is zero otherwise. Eq. (1.11) can be simplified as

$$\frac{\partial o_k(n)}{\partial \mathbf{W}_1} = \left(\left(\mathbf{w}_{2,k}^{(n+1)}\right)^{\mathrm{T}} \circ \mathbf{u}(n) \circ (\mathbf{1} - \mathbf{u}(n))\right) \mathbf{x}(n)^{\mathrm{T}} \tag{1.12}$$

where $\mathbf{w}_{2,k}^{(n+1)}$ is the kth row of $\mathbf{W}_2^{(n+1)}$. Since the parameters are updated group by group starting from the output layer back to the input layer, this algorithm is called back-propagation (BP) algorithm (Not to be confused with the digital back-propagation (DBP) algorithm for fiber NLC). The weights and biases are continuously updated until convergence.

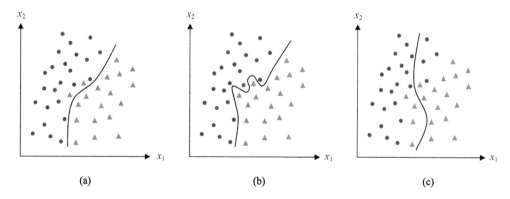

Figure 1.6 Example illustrating ANN learning processes with (a) no over-fitting or under-fitting, (b) over-fitting, and (c) under-fitting.

For the learning and performance evaluation of an ANN, the data sets are typically divided into three groups: training, validation and testing. The training data set is used to train the ANN. Clearly, a larger training data set is better since the more data an ANN sees, the more likely it is that it has encountered examples of all possible types of input. However, the learning time also increases with the training data size. There is no fixed rule for determining the minimum amount of training data needed since it often depends on the given problem. A rule of thumb typically used is that the size of the training data should be at least 10 times the total number of weights [1]. The purpose of the validation data set is to keep a check on how well the ANN is doing as it learns since during training there is an inherent danger of *over-fitting* (or *over-training*). In this case, instead of finding the underlying general decision boundaries as shown in Fig. 1.6(a), the ANN tends to perfectly fit the training data (including any noise components of them) as shown in Fig. 1.6(b). This in turn makes the ANN customized for a few data points and reduces its generalization capability, i.e., its ability to make predictions about new inputs which it has never seen before. The over-fitting problem can be avoided by constantly examining ANN's error performance during the course of training against an independent validation data set and enforcing an early termination of the training process if the validation data set gives large errors. Typically, the size of the validation data set is just a fraction ($\sim 1/3$) of that of training data set. Finally, the testing data set evaluates the performance of the trained ANN. Note that an ANN may also be subjected to *under-fitting* problem which occurs when it is under-trained and thus unable to perform at an acceptable level as shown in Fig. 1.6(c). Under-fitting can again lead to poor ANN generalization. The reasons for under-fitting include insufficient training time or number of iterations, inappropriate choice of activation functions, and/or insufficient number of hidden neurons used.

It should be noted that given an adequate number of hidden neurons, proper non-linearities, and appropriate training, an ANN with one hidden layer has great expressive power and can approximate any continuous function in principle. This is called the *universal approximation theorem* [11]. One can intuitively appreciate this characteristic by considering the classification problem in Fig. 1.7. Since each hidden neuron can be represented as a straight-line decision boundary, any arbitrary curved boundary can be approximated by a collection of hidden neurons in a single hidden layer ANN. This important property of an ANN enables it to be applied in many diverse applications.

1.2.2 Choice of activation functions

The choice of activation functions has a significant effect on the training dynamics and final ANN performance. Historically, sigmoid and hyperbolic tangent have been the most commonly-used nonlinear activation functions for hidden layer neurons. However, the rectified linear unit (ReLU) activation function has become the default choice among ML community in recent years. The above-mentioned three functions are given

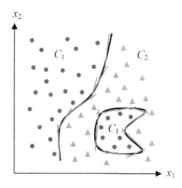

Figure 1.7 Decision boundaries for appropriate data classification obtained using an ANN.

by

$$\text{Sigmoid: } \sigma(z) = \frac{1}{1 + e^{-z}}$$
$$\text{Hyperbolic tangent: } \sigma(z) = \frac{e^z - e^{-z}}{e^z + e^{-z}} \tag{1.13}$$
$$\text{Rectified linear unit: } \sigma(z) = \max(0, z)$$

and their plots are shown in Fig. 1.8. Sigmoid and hyperbolic tangent are both differentiable. However, a major problem with these functions is that their gradients tend to zero as $|z|$ becomes large and thus the activation output gets saturated. In this case, the weights and biases updates for a certain layer will be minimal, which in turn will slow down the weights and biases updates for all the preceding layers. This is known as *vanishing gradient problem* and is particularly an issue when training ANNs with large number of hidden layers. To circumvent this problem, ReLU was proposed since its gradient does not vanish as z increases. Note that although ReLU is not differentiable at $z = 0$, it is not a problem in practice since the probability of having an entry exactly equal to 0 is generally very low. Also, as the ReLU function and its derivative are 0 for $z < 0$, around 50% of hidden neurons' outputs will be 0, i.e., only half of total neurons will be active when the ANN weights and biases are randomly initialized. It has been found that such sparsity of activation not only reduces computational complexity (and thus training time) but also leads to better ANN performance [12]. Note that while using ReLU activation function, the ANN weights and biases are often initialized using the method proposed by He *et al.* [13]. On the other hand, Xavier initialization technique [14] is more commonly employed for the hyperbolic tangent activation function. These heuristics-based approaches initialize the weights and biases by drawing random numbers from a truncated normal distribution (instead of standard normal distribution) with variance which depends on the size of the previous ANN layer.

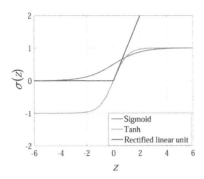

Figure 1.8 Common activation functions used in ANNs.

1.2.3 Choice of loss functions

The choice of loss function E has a considerable effect on the performance of an ANN. The MSE is a common choice in adaptive signal processing and other DSP in telecommunications. For regression problems, MSE works well in general and is also easy to compute. On the other hand, for classification problems, cross–entropy loss function defined as

$$E = -\frac{1}{L}\sum_{l=1}^{L}\sum_{k=1}^{K} y_k(l)\log\left(o_k(l)\right) \tag{1.14}$$

is often used instead of MSE [11]. The cross-entropy function can be interpreted by viewing the softmax output $\mathbf{o}(l)$ and the class label with one-hot encoding $\mathbf{y}(l)$ as probability distributions. In this case, $\mathbf{y}(l)$ has zero entropy and one can subtract the zero-entropy term from Eq. (1.14) to obtain

$$E = -\frac{1}{L}\sum_{l=1}^{L}\sum_{k=1}^{K} y_k(l)\log\left(o_k(l)\right) + \underbrace{\frac{1}{L}\sum_{l=1}^{L}\sum_{k=1}^{K} y_k(l)\log\left(y_k(l)\right)}_{=0}$$

$$= \frac{1}{L}\sum_{l=1}^{L}\sum_{k=1}^{K} y_k(l)\log\left(\frac{y_k(l)}{o_k(l)}\right) \tag{1.15}$$

which is simply the Kullback-Leibler (KL) divergence between the distributions $\mathbf{o}(l)$ and $\mathbf{y}(l)$ averaged over all input-output pairs. Therefore, the cross-entropy is in fact a measure of the similarity between ANN outputs and the class labels. The cross-entropy function also leads to simple gradient updates as the logarithm cancels out the exponential operation inherent in the softmax calculation, thus leading to faster ANN training. Appendix 1.A shows the derivation of BP algorithm for the single hidden layer ANN in

Fig. 1.5 with cross-entropy loss function and softmax activation function for the output layer neurons.

In many applications, a common approach to prevent over-fitting is to reduce the magnitude of the weights as large weights produce high curvatures which make the decision boundaries overly complicated. This can be achieved by including an extra regularization term in the loss function, i.e.,

$$E' = E + \lambda \|\mathbf{W}\|^2 \tag{1.16}$$

where $\|\mathbf{W}\|^2$ is the sum of squared element-wise weights. The parameter λ, called regularization coefficient, defines the relative importance of the training error E and the regularization term. The regularization term thus discourages weights from reaching large values and this often results in significant improvement in ANN's generalization ability [15].

1.2.4 Support vector machines (SVMs)

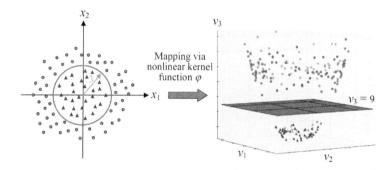

Figure 1.9 Example showing how a linearly inseparable problem (in the original 2D data space) can undergo a nonlinear transformation and becomes a linearly separable one in the 3-dimensional (3D) feature space.

In many classification tasks, it often happens that the two data categories are not easily separable with straight lines or planes in the original variable space. SVM is an ML technique that preprocesses the input data $\mathbf{x}(i)$ and transforms it into (sometimes) a higher-dimensional space $\mathbf{v}(i) = \varphi(\mathbf{x}(i))$, called *feature space*, where the data belonging to two different classes can be separated easily by a simple straight plane decision boundary or *hyperplane* [16]. An example is shown in Fig. 1.9 where one class of data lies within a circle of radius 3 and the other class lies outside. When transformed into the feature space $\mathbf{v} = (v_1, v_2, v_3) = (x_1, x_2, x_1^2 + x_2^2)$, the two data classes can be separated simply by the hyperplane $v_3 = 9$.

Let us first focus on finding the right decision hyperplane after the transformation into feature space as shown in Fig. 1.10(a). The right hyperplane should have the largest (and also equal) distance from the borderline points of the two data classes. This is graphically illustrated in Fig. 1.10(b). Had the data points been generated from two probability density functions (PDFs), finding a hyperplane with maximal margin from the borderline points is conceptually analogous to finding a maximum likelihood decision boundary. The borderline points, represented as solid dot and triangle in Fig. 1.10(b), are referred to as *support vectors* and are often most informative for the classification task.

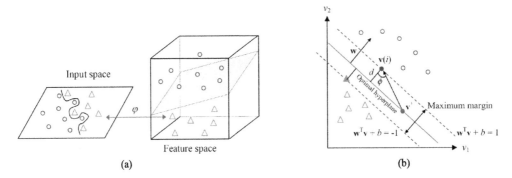

Figure 1.10 (a) Mapping from input space to a higher-dimensional feature space using a nonlinear kernel function φ. (b) Separation of two data classes in the feature space through an optimal hyperplane.

More technically, in the feature space, a general hyperplane is defined as $\mathbf{w}^{\mathrm{T}}\mathbf{v}+b=0$. If it classifies all the data points correctly, all the violet (dark gray in print version) points will lie in the region $\mathbf{w}^{\mathrm{T}}\mathbf{v}+b>0$ and the orange (light gray in print version) points will lie in the region $\mathbf{w}^{\mathrm{T}}\mathbf{v}+b<0$. We seek to find a hyperplane $\mathbf{w}^{\mathrm{T}}\mathbf{v}+b=0$ that maximizes the margin d as shown in Fig. 1.10(b). Without loss of generality, let the point $\mathbf{v}(i)$ reside on the hyperplane $\mathbf{w}^{\mathrm{T}}\mathbf{v}+b=1$ and is closest to the hyperplane $\mathbf{w}^{\mathrm{T}}\mathbf{v}+b=0$ on which $\mathbf{v}^{+}$ resides. Since the vectors $\mathbf{v}(i)-\mathbf{v}^{+}$, $\mathbf{w}$ and the angle ϕ are related by $cos\phi = \mathbf{w}^{\mathrm{T}}\left(\mathbf{v}(i)-\mathbf{v}^{+}\right)/\left(\|\mathbf{w}\|\,\|\mathbf{v}(i)-\mathbf{v}^{+}\|\right)$, the margin d is given as

$$
\begin{aligned}
d &= \left\|\mathbf{v}(i)-\mathbf{v}^{+}\right\| cos\phi \\
&= \left\|\mathbf{v}(i)-\mathbf{v}^{+}\right\| \cdot \frac{\mathbf{w}^{\mathrm{T}}\left(\mathbf{v}(i)-\mathbf{v}^{+}\right)}{\|\mathbf{w}\|\,\|\mathbf{v}(i)-\mathbf{v}^{+}\|} \\
&= \frac{\mathbf{w}^{\mathrm{T}}\left(\mathbf{v}(i)-\mathbf{v}^{+}\right)}{\|\mathbf{w}\|} = \frac{\mathbf{w}^{\mathrm{T}}\mathbf{v}(i)-\mathbf{w}^{\mathrm{T}}\mathbf{v}^{+}}{\|\mathbf{w}\|} \\
&= \frac{\mathbf{w}^{\mathrm{T}}\mathbf{v}(i)+b}{\|\mathbf{w}\|} = \frac{1}{\|\mathbf{w}\|}.
\end{aligned}
\tag{1.17}
$$

Therefore, we seek to find $\mathbf{w}, b$ that maximize $1/\|\mathbf{w}\|$ subject to the fact that all the data points are classified correctly. To characterize the constraints more mathematically, one can first assign the violet class label to 1 and orange class label to -1. In this case, if we have correct decisions for all the data points, the product $y(i)\left(\mathbf{w}^T\mathbf{v}(i)+b\right)$ will always be greater than 1 for all i. The optimization problem then becomes

$$\underset{\mathbf{w},b}{\operatorname{argmax}} \frac{1}{\|\mathbf{w}\|}$$

$$\text{subject to } y(l)\left(\mathbf{w}^T\mathbf{v}(l)+b\right)\geq 1, \quad l=1,2,\dots,L \tag{1.18}$$

and thus standard convex programming software packages such as CVXOPT [17] can be used to solve Eq. (1.18).

Let us come back to the task of choosing the nonlinear function $\varphi(\cdot)$ that maps the original input space $\mathbf{x}$ to feature space $\mathbf{v}$. For SVM, one would instead find a kernel function $K\left(\mathbf{x}(i),\mathbf{x}(j)\right)=\varphi\left(\mathbf{x}(i)\right)\cdot\varphi\left(\mathbf{x}(j)\right)=\mathbf{v}(i)^T\mathbf{v}(j)$ that maps to the inner product. Typical kernel functions include:

- Polynomials: $K\left(\mathbf{x}(i),\mathbf{x}(j)\right)=\left(\mathbf{x}(i)^T\mathbf{x}(j)+a\right)^b$ for some scalars a,b
- Gaussian radial basis function: $K\left(\mathbf{x}(i),\mathbf{x}(j)\right)=\exp\left(-a\|\mathbf{x}(i)-\mathbf{x}(j)\|^2\right)$ for some scalar a
- Hyperbolic tangent: $K\left(\mathbf{x}(i),\mathbf{x}(j)\right)=\tanh\left(a\mathbf{x}(i)^T\mathbf{x}(j)+b\right)$ for some scalars a,b.

The choice of a kernel function is often determined by the designer's knowledge of the problem domain [3].

Note that a larger separation margin typically results in better generalization of the SVM classifier. SVMs often demonstrate better generalization performance than conventional ANNs in various pattern recognition applications. Furthermore, multiple SVMs can be applied to the same data set to realize non-binary classifications such as detecting 16-QAM signals [18][19][20].

It should be noted that ANNs and SVMs can be seen as two complementary approaches for solving classification problems. While an ANN derives curved decision boundaries in the input variable space, the SVM performs nonlinear transformations of the input variables followed by determining a simple decision boundary or hyperplane as shown in Fig. 1.11.

1.2.5 *K*-nearest neighbors (KNN)

KNN is one of the simplest and most intuitive supervised ML algorithm [21]. Let $\left\{\left(\mathbf{x}(1),\mathbf{y}(1)\right),\left(\mathbf{x}(2),\mathbf{y}(2)\right),\dots\left(\mathbf{x}(L),\mathbf{y}(L)\right)\right\}$ be a learning set of observed data with L input feature vectors $\mathbf{x}$ of dimensionality M and their corresponding labels $\mathbf{y}$ of dimensionality N. Now, suppose we have a new sample feature vector $\mathbf{x}'$ and we want to predict its class label $\mathbf{y}'$. The idea in KNN algorithm is to identify K (where K is user-defined) feature vectors $\mathbf{x}(k)$, $k=1,2,\dots,K$ of the learning data set in the M-dimensional feature space that are nearest (in terms of some distance measure such as Euclidean distance) to the given sample vector $\mathbf{x}'$ [1]. The class label that most number

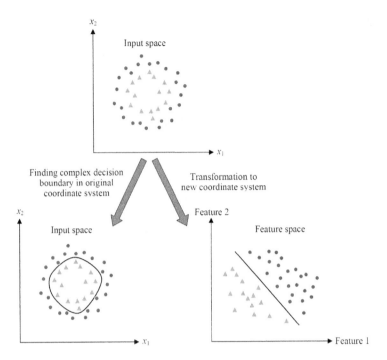

Figure 1.11 Example showing how an ANN determines a curved decision boundary in the original input space while an SVM obtains a simple decision boundary in the transformed feature space.

of these neighboring feature vectors $\mathbf{x}(k)$ belong to, also referred to as *plurality voting*, is then taken as the predicted class label $\mathbf{y}'$ as shown in Fig. 1.12.

Mathematically, let K_c represent the number of samples among the group of K nearest neighbors which belong to a particular class c such that

$$\sum_{c=1}^{N} K_c = K \tag{1.19}$$

then the predicted class c of the given feature vector $\mathbf{x}'$ can be obtained as

$$c = \underset{N}{\operatorname{argmax}}(K_c) \tag{1.20}$$

On the other hand, for a regression problem, label $\mathbf{y}'$ is estimated by simply taking the average of labels $\mathbf{y}(k)$ of the K feature vectors that are closest to $\mathbf{x}'$, i.e.,

$$\mathbf{y}' = \frac{1}{K} \sum_{k=1}^{K} \mathbf{y}(k) \tag{1.21}$$

In standard KNN algorithm, all K neighbors are treated equally during the plurality voting or averaging process. However, in situations where the radius enclosing the set of neighbors is large enough, one may consider assigning weights w_k, $k = 1, 2, ..., K$ to the neighbors depending on their distances from the given feature vector $\mathbf{x}'$. Such a variant is referred to as weighted-KNN algorithm [21]. A popular weighting scheme is to use the inverse squared distance, i.e.,

$$w_k = \frac{1}{d(\mathbf{x}(k), \mathbf{x}')^2}, \quad k = 1, 2, ..., K \tag{1.22}$$

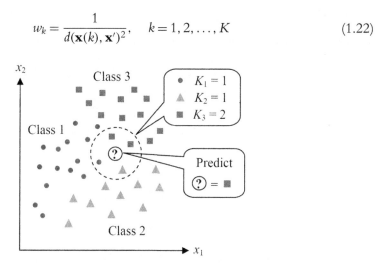

Figure 1.12 Illustration of classification performed using KNN algorithm for a 3-class problem with $K = 4$.

It is interesting to note that unlike ANN and SVM, the KNN algorithm does not have an explicit learning process and the training phase consists literally of just storing the labeled training examples in the system's memory. Moreover, since KNN approach defers the processing of stored training examples until making actual predictions, it is also known as *lazy learning* algorithm.

Compared to other supervised learning techniques, the basic KNN algorithm has fewer hyperparameters, i.e., K and the distance metric. The choice of K typically depends upon the data. A smaller value of K can make the results more sensitive to the noise associated with the individual data points. On the other hand, a larger K value though reduces the effect of the noise, it ends up looking at data points that are not really neighbors and are far away from the query point $\mathbf{x}'$, thereby degrading estimation accuracy. A good K value is often determined through cross–validation. A typical rule of thumb is that $K < \sqrt{L}$, where L is the size of observed data.

Although a powerful predictive approach, KNN has some practical limitations such as it is slow and computationally expensive [22]. This is due to the fact that in order to determine the nearest neighbor of a single query point $\mathbf{x}'$, one needs to compute the distances to all L training examples and hence the computational time/complexity for a large training set can be prohibitive. Several solutions have been proposed to make KNN substantially more efficient during prediction including: (i) using more sophisticated data structures like search trees to speed up the identification of a nearest neighbor, (ii) editing the training data set to eliminate the redundant data points, also called *condensing*, and (iii) working in a reduced dimension space by applying dimensionality reduction techniques like PCA to shorten the time for computing distances.

1.3. Unsupervised learning

The ANN and SVM are examples of *supervised learning* approach in which the class labels $\mathbf{y}$ of the training data are known. Based on this data, the ML algorithm generalizes to react accurately to new data to the best possible extent. Supervised learning can be considered as a closed-loop feedback system as the error between the ML algorithm's actual outputs and the targets is used as a feedback signal to guide the learning process.

In *unsupervised learning*, the ML algorithm is not provided with correct labels of the training data. Rather, it learns to identify similarities between various inputs with the aim to either categorize together those inputs which have something in common or to determine some better representation/description of the original input data. It is referred to as "unsupervised" because the ML algorithm is not told what the output should be rather it has to come up with it itself [23]. One example of unsupervised learning is data clustering as shown in Fig. 1.13.

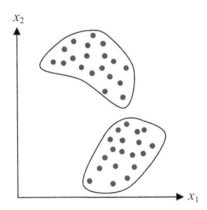

Figure 1.13 Data clustering based on unsupervised learning.

Unsupervised learning is becoming more and more important because in many real circumstances it is practically not possible to obtain labeled training data. In such scenarios, an unsupervised learning algorithm can be applied to discover some similarities between different inputs for itself. Unsupervised learning is typically used in tasks such as clustering, vector quantization, dimensionality reduction, and features extraction. It is also often employed as a preprocessing tool for extracting useful (in some particular context) features of the raw data before supervised learning algorithms can be applied. We hereby provide a review of few key unsupervised learning techniques.

1.3.1 K-means clustering

Let $\{\mathbf{x}(1), \mathbf{x}(2), \ldots \mathbf{x}(L)\}$ be the set of data points which is to be split into K clusters $C_1, C_2, \ldots C_K$. K-means clustering is an iterative unsupervised learning algorithm which aims to partition L observations into K clusters such that the sum of squared errors for data points within a group is minimized [15]. An example of this algorithm is graphically shown in Fig. 1.14. The algorithm initializes by randomly picking K locations $\boldsymbol{\mu}(j), j = 1, 2, \ldots, K$ as cluster centers. This is followed by two iterative steps. In the first step, each data point $\mathbf{x}(i)$ is assigned to the cluster C_k with the minimum Euclidean distance, i.e.,

$$C_k = \{\mathbf{x}(i) : \|\mathbf{x}(i) - \boldsymbol{\mu}(k)\| < \|\mathbf{x}(i) - \boldsymbol{\mu}(j)\| \ \forall \ j \in \{1, 2, \ldots, K\}\backslash\{k\}\} \qquad (1.23)$$

In the second step, the new center of each cluster C_k is calculated by averaging out the locations of data points that are assigned to cluster C_k, i.e.,

$$\boldsymbol{\mu}(k) = \sum_{\mathbf{x}(i) \in C_k} \mathbf{x}(i) \qquad (1.24)$$

The two steps are repeated iteratively until the cluster centers converge. Several variants of K-means algorithm have been proposed over the years to improve its computational efficiency as well as to achieve smaller errors. These include fuzzy K-means, hierarchical K-means, K-means++, K-medians, K-medoids, etc.

1.3.2 Expectation-maximization (EM) algorithm

One drawback of K-means algorithm is that it requires the use of hard decision boundaries whereby a data point can only be assigned to one cluster even though it might lie somewhere midway between two or more clusters. The EM algorithm is an improved clustering technique which assigns a probability to the data point belonging to each cluster rather than forcing it to belong to one particular cluster during each iteration [23]. The algorithm assumes that a given data distribution can be modeled as a superposition of K jointly Gaussian probability distributions with distinct means and covariance matrices $\boldsymbol{\mu}(k)$, $\boldsymbol{\Sigma}(k)$ (also referred to as *Gaussian mixture models*). The EM algorithm is a two-step iterative procedure comprising of expectation (E) and maximization (M) steps [3]. The E step computes the a posteriori probability of the class label given each data point using the current means and covariance matrices of the Gaussians, i.e.,

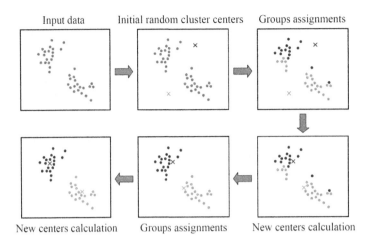

Input data   Initial random cluster centers   Groups assignments

New centers calculation   Groups assignments   New centers calculation

Figure 1.14 Example to illustrate initialization and two iterations of *K*-means algorithm. The data points are shown as dots and cluster centers are depicted as crosses.

$$
\begin{aligned}
p_{ij} &= p\left(C_j|\mathbf{x}(i)\right) \\
&= \frac{p\left(\mathbf{x}(i)|C_j\right)p\left(C_j\right)}{\sum_{k=1}^{K}p\left(\mathbf{x}(i)|C_k\right)p\left(C_k\right)} \\
&= \frac{\mathbf{N}\left(\mathbf{x}(i)|\boldsymbol{\mu}(k),\boldsymbol{\Sigma}(k)\right)}{\sum_{k=1}^{K}\mathbf{N}\left(\mathbf{x}(i)|\boldsymbol{\mu}(k),\boldsymbol{\Sigma}(k)\right)}
\end{aligned}
\tag{1.25}
$$

where $\mathbf{N}\left(\mathbf{x}(i)|\boldsymbol{\mu}(k),\boldsymbol{\Sigma}(k)\right)$ is the Gaussian PDF with mean and covariance matrix $\boldsymbol{\mu}(k),\boldsymbol{\Sigma}(k)$. Note that we have inherently assumed equal probability $p(C_j)$ of each class, which is a valid assumption for most communication signals. In scenarios where this assumption is not valid, e.g., the one involving probabilistic constellation shaping (PCS), the actual non-uniform probabilities $p\left(C_j\right)$ of individual symbols shall instead be used in Eq. (1.25). The M step attempts to update the means and covariance matrices according to the updated soft-labeling of the data points, i.e.,

$$
\begin{aligned}
\boldsymbol{\mu}(j) &= \frac{\sum_{i=1}^{L}p_{ij}\mathbf{x}(i)}{\sum_{i=1}^{L}p_{ij}} \\
\boldsymbol{\Sigma}(k) &= \sum_{i=1}^{L}p_{ij}(\mathbf{x}(i)-\boldsymbol{\mu}(j))(\mathbf{x}(i)-\boldsymbol{\mu}(j))^{\mathrm{T}}
\end{aligned}
\tag{1.26}
$$

A graphical illustration of EM algorithm and its convergence process is shown in Fig. 1.15. Fig. 1.15(a) shows the original data points in green (light gray in print version) which are to be split into two clusters by applying EM algorithm. The two Gaussian probability distributions are initialized with random means and unit covariance matrices and are depicted using red (mid gray in print version) and blue (dark gray in print

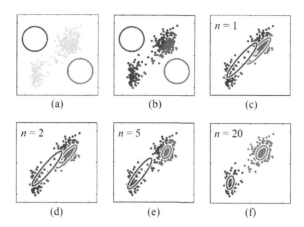

Figure 1.15 Example showing the concept of EM algorithm. (a) Original data points and initialization. Results after (b) first E step; (c) first M step; (d) 2 complete EM iterations; (e) 5 complete EM iterations; and (f) 20 complete EM iterations [3].

version) circles. The results after first E step are shown in Fig. 1.15(b) where the posterior probabilities in Eq. (1.25) are expressed by the proportion of red and blue colors for each data point. Fig. 1.15(c) depicts the results after first M step where the means and covariance matrices of the red and blue Gaussian distributions are updated using Eq. (1.26), which in turn uses the posterior probabilities computed by Eq. (1.25). This completes the 1st iteration of the EM algorithm. Fig. 1.15(d) to (f) show the results after 2, 5 and 20 complete EM iterations, respectively, where the convergence of the algorithm and consequently effective splitting of the data points into two clusters can be clearly observed.

1.3.3 Principal component analysis (PCA)

PCA is an unsupervised learning technique for features extraction and data representation [24][25]. It is often used as a preprocessing tool in many pattern recognition applications for the extraction of limited but most critical data features. The central idea behind PCA is to project the original high-dimensional data onto a lower-dimensional feature space that retains most of the information in the original data as shown in Fig. 1.16. The reduced dimensionality feature space is spanned by a small (but most significant) set of orthonormal eigenvectors, called principal components (PCs). The first PC points in the direction along which the original data has the greatest variability and each successive PC in turn accounts for as much of the remaining variability as possible. Geometrically, we can think of PCA as a rotation of the axes of the original coordinate system to a new set of orthogonal axes which are ordered based on the amount of variation of the original data they account for, thus achieving dimensionality reduction.

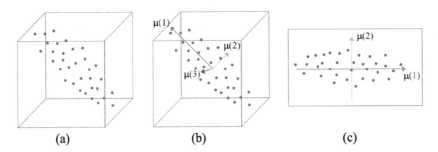

Figure 1.16 Example to illustrate the concept of PCA. (a) Data points in the original 3D data space; (b) Three PCs ordered according to the variability in original data; (c) Projection of data points onto a plane defined by the first two PCs while discarding the third one.

More technically, consider a data set $\{\mathbf{x}(1), \mathbf{x}(2), \ldots \mathbf{x}(L)\}$ with L data vectors of M dimensions. We will first compute the mean vector $\overline{\mathbf{x}} = \frac{1}{L}\sum_{i=1}^{L}\mathbf{x}(i)$ and the covariance matrix $\boldsymbol{\Sigma}$ can then be estimated as

$$\boldsymbol{\Sigma} \approx \frac{1}{L}\sum_{i=1}^{L}(\mathbf{x}(i) - \overline{\mathbf{x}})(\mathbf{x}(i) - \overline{\mathbf{x}})^{\mathrm{T}} \tag{1.27}$$

where $\boldsymbol{\Sigma}$ can have up to M eigenvectors $\boldsymbol{\mu}(i)$ and corresponding eigenvalues λ_i. We then sort the eigenvalues in terms of their magnitude from large to small and choose the first S (where $S \ll M$) corresponding eigenvectors such that

$$\sum_{i=1}^{S}\lambda_i / \sum_{i=1}^{M}\lambda_i > R \tag{1.28}$$

where R is typically above 0.9 [25]. Note that, as compared to the original M-dimensional data space, the chosen eigenvectors span only an S-dimensional subspace that in a way captures most of the data information. One can understand such procedure intuitively by noting that for a covariance matrix, finding the eigenvectors with large eigenvalues corresponds to finding linear combinations or particular directions of the input space that give large variances, which is exactly what we want to capture. A data vector $\mathbf{x}$ can then be approximated as a weighted-sum of the chosen eigenvectors in this subspace, i.e.,

$$\mathbf{x} \approx \sum_{i=1}^{S}w_i\boldsymbol{\mu}(i) \tag{1.29}$$

where $\boldsymbol{\mu}(i), i = 1, 2, \ldots, S$ are the chosen orthogonal eigenvectors such that

$$\boldsymbol{\mu}^{\mathrm{T}}(m)\boldsymbol{\mu}(l) = \left\{ \begin{array}{l} 1 \ if \ l = m \\ 0 \ if \ l \neq m \end{array} \right\} \tag{1.30}$$

Multiplying both sides of Eq. (1.29) with $\boldsymbol{\mu}^T(k)$ and then using Eq. (1.30), we get

$$w_k = \boldsymbol{\mu}^T(k)\,\mathbf{x}, \quad k = 1, 2, \ldots, S \tag{1.31}$$

The vector $\mathbf{w} = [w_1\ w_2 \ldots w_S]^T$ of weights describing the contribution of each chosen eigenvector $\boldsymbol{\mu}(k)$ in representing $\mathbf{x}$ can then be considered as a feature vector of $\mathbf{x}$.

1.3.4 Independent component analysis (ICA)

Another interesting technique for features extraction and data representation is ICA. Unlike PCA which uses orthogonal and uncorrelated components, the components in ICA are instead required to be statistically independent [1]. In other words, ICA seeks those directions in the feature space that are most independent from each other. Fig. 1.17 illustrates the conceptual difference between PCA and ICA. Finding the independent components (ICs) of the observed data can be useful in scenarios where we need to separate mutually independent but unknown source signals from their linear mixtures with no information about the mixing coefficients. An example is the task of polarization demultiplexing at the receiver using DSP. For a data set $\{\mathbf{x}(1), \mathbf{x}(2), \ldots, \mathbf{x}(L)\}$, one seeks to identify a collection of basis vectors $\mathbf{v}(1), \mathbf{v}(2), \ldots \mathbf{v}(S)$ so that $\mathbf{x} \approx \sum_{k=1}^{S} w_k \mathbf{v}(k)$ and the empirical distributions of $w_k, k = 1, 2, \ldots, S$ across all the data $\mathbf{x}$ are statistically independent. This can be achieved by minimizing the mutual information between different w_k.

ICA is used as a preprocessing tool for extracting data features in many pattern recognition applications and is shown to outperform conventional PCA in many cases [26]. This is expected because unlike PCA which is derived from second-order statistics (i.e., covariance matrix) of the input data, ICA takes into account high-order statistics of the data as it considers complete probability distribution.

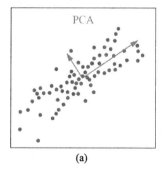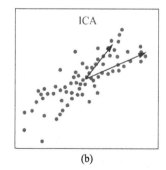

Figure 1.17 Example 2D data fitted using (a) PCs bases and (b) ICs bases. As shown, the orthogonal basis vectors in PCA may not be efficient while representing non-orthogonal density distributions. In contrast, ICA does not necessitate orthogonal basis vectors and can thus represent general types of densities more effectively.

We would like to highlight here that the dimensionality of the transformed space in ML techniques can be higher or lower than the original input space depending upon the nature of the problem at hand. If the objective of the transformation is to simply reduce the input data dimensionality (e.g., for decreasing the computational complexity of the learning system) then the dimensionality of the transformed space should be lower than that of original one. On the other hand, a transformation to a higher-dimensional space may be desirable if the data classes can be separated more easily by a classifier in the new space.

1.4. Reinforcement learning (RL)

RL is an experience-driven learning approach with the objective to develop autonomous algorithms that are able to learn optimal behaviors through interaction with their environments and improve their performance over time through trial and error. It uses the formal framework of Markov decision processes (MDP) to describe how a learning *agent* interacts with an unfamiliar, dynamic and stochastic *environment* in terms of *states*, *actions* and *rewards* [27]. As shown in Fig. 1.18, at time step t, an agent observes a state $\mathbf{s}(t)$ from its environment and then interacts with it by taking an action $\mathbf{a}(t)$ because of which the environment and the agent transition to a new state $\mathbf{s}(t+1)$ depending upon the current state and the action chosen. As the environment transitions to its new state, it provides a feedback signal called reward $r(t+1)$, which is then used by the agent to update its knowledge.

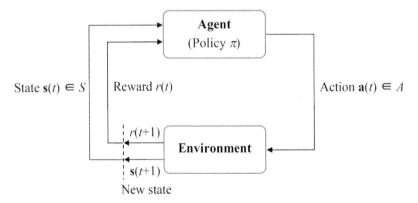

Figure 1.18 The RL cycle. In state $\mathbf{s}(t)$, the agent uses its policy π to take an action $\mathbf{a}(t)$ and receives the corresponding reward $r(t+1)$ from the environment ending up in new state $\mathbf{s}(t+1)$.

Mathematically, we can describe RL as an MDP consisting of
- A set S of possible states of the agent
- A set A of possible actions of the agent

- Transition probability distribution $p(\mathbf{s}(t+1)|\mathbf{s}(t), \mathbf{a}(t))$, i.e., distribution over next state given state–action pair
- An instantaneous reward function $\mathcal{R}(\mathbf{s}(t), \mathbf{a}(t), \mathbf{s}(t+1))$ after transition from $\mathbf{s}(t)$ to $\mathbf{s}(t+1)$ under action $\mathbf{a}(t)$
- A discount factor $\gamma \in [0, 1]$, where 0 signifies more emphasis on instantaneous rewards and vice versa

The behavior of an RL agent at a given time is defined either by a deterministic policy π which maps the perceived environment states to the actions to be taken when in those states, i.e., $\pi: S \to A$ or, alternatively, by a stochastic policy π that maps the states to a probability distribution over actions, i.e., $\pi: S \to p(A = \mathbf{a}|S)$. A policy lies at the core of an RL agent since it alone is sufficient to determine the agent's behavior. In general, it may be a lookup table, a simple function, or may involve an extensive search process. A rollout of a chosen policy accumulates rewards from the environment thus resulting in a return R given by

$$R = \sum_{t=0}^{\infty} \gamma^t r(t) \qquad (1.32)$$

The aim of an agent is to learn an optimal policy π_{opt} (i.e., a strategy) that maximizes the expected cumulative reward in the environment, i.e.,

$$\pi_{\text{opt}} = \underset{\pi}{\text{argmax}}\, \mathbb{E}\,[R|\pi] \qquad (1.33)$$

The optimal policy in RL problems can be determined by employing two main types of approaches: (i) methods based on value functions and (ii) methods based on direct policy search [28].

Value functions-based approach essentially relies on estimating the value (i.e., expected return) of being in a particular state. While the immediate desirability of an environmental state in RL is well indicated by a reward signal, the long-term desirability of states can be better described by also considering the states that are likely to follow and the corresponding rewards associated with those states. In this context, a *state-value function* $V^{\pi}(\mathbf{s})$ is defined as the expected return when starting in state $\mathbf{s}$ and successfully following the policy π. Simply put, $V^{\pi}(\mathbf{s})$ specifies how "good" it is to be in a given state and what exactly is good in the long run? Thus,

$$V^{\pi}(\mathbf{s}) = \mathbb{E}\,[R|\mathbf{s}, \pi] \qquad (1.34)$$

An optimal state-value function $V^{\pi}_{\text{opt}}(\mathbf{s})$ corresponding to an optimal policy π_{opt} in Eq. (1.33) can thus be written as

$$V^{\pi}_{\text{opt}}(\mathbf{s}) = \max_{\pi} V^{\pi}(\mathbf{s}) \; \forall \mathbf{s} \in S \qquad (1.35)$$

Although the function $V^{\pi}_{\text{opt}}(\mathbf{s})$ suffices to determine optimality, it is often useful to define action-values. Therefore, the *state-action value function* or *quality function* $Q^{\pi}(\mathbf{s}, \mathbf{a})$ is defined as the expected return from taking action $\mathbf{a}$ in state $\mathbf{s}$ and then following the policy π. In other words, $Q^{\pi}(\mathbf{s}, \mathbf{a})$ describes how good a state-action pair really is? Thus,

$$Q^{\pi}(\mathbf{s}, \mathbf{a}) = \mathbb{E}[R|\mathbf{s}, \mathbf{a}, \pi] \qquad (1.36)$$

and the optimal quality function $Q^{\pi}_{\text{opt}}(\mathbf{s}, \mathbf{a})$ corresponding to π_{opt} in Eq. (1.33) is then given as

$$Q^{\pi}_{\text{opt}}(\mathbf{s}, \mathbf{a}) = \max_{\pi} Q^{\pi}(\mathbf{s}, \mathbf{a}) \ \forall \mathbf{s} \in S \text{ and } \mathbf{a} \in A \qquad (1.37)$$

In RL problems, the optimal quality function $Q^{\pi}_{\text{opt}}(\mathbf{s}, \mathbf{a})$ can be learnt by exploiting the fact that it satisfies the Bellman optimality equation [29] which has the following recursive form

$$Q^{\pi}_{\text{opt}}(\mathbf{s}(t), \mathbf{a}(t)) = \mathbb{E}_{\mathbf{s}(t+1)}\left[r(t+1) + \gamma \max_{\mathbf{a}(t+1)} Q^{\pi}_{\text{opt}}(\mathbf{s}(t+1), \mathbf{a}(t+1))|\mathbf{s}(t), \mathbf{a}(t)\right] \qquad (1.38)$$

The knowledge of $Q^{\pi}_{\text{opt}}(\mathbf{s}, \mathbf{a})$ makes it possible to determine the optimal policy π_{opt} of the MDP by simply taking an action that maximizes $Q^{\pi}_{\text{opt}}(\mathbf{s}, \mathbf{a})$. Thus,

$$\pi_{\text{opt}} = \underset{\mathbf{a} \in A}{\operatorname{argmax}} \ Q^{\pi}_{\text{opt}}(\mathbf{s}, \mathbf{a}) \qquad (1.39)$$

Examples of most popular value functions-based RL algorithms include Q-learning, state–action–reward–state–action (SARSA), and deep Q-network (DQN).

The second approach for determining π_{opt} in RL relies on directly searching for an optimal policy in some subset of the policy space. The policy search methods can choose actions without actually consulting a value function. In this case, a parameterized policy $\pi(\mathbf{s}, \mathbf{a})$ is typically selected and its parameters θ are then updated in each step to maximize the expected return $\mathbb{E}[R|\theta]$ by employing either gradient-based or gradient-free stochastic optimization. Policy search methods are quite effective in high-dimensional or continuous action spaces and can also learn stochastic policies. However, they often tend to converge to local optima, can be inefficient to evaluate, and encounter high variance [30]. Monte-Carlo policy gradient, trust region policy optimization (TRPO), and hill-climbing are some well known policy search-based RL algorithms.

RL is particularly useful in solving interactive problems in which it is often impossible to attain examples of desired behavior which are not only correct but are also representative of all the possible situations in which the model may have to act ultimately. In an uncharted territory, an RL model should be able to learn from its own experiences instead of getting trained by an external supervisor with a training data set of labeled examples.

Due to their inherent self-learning and adaptability characteristics, RL algorithms have been considered for various tasks in optical networks including network self-configuration, adaptive resource allocation, etc. In these applications, the actions performed by the RL algorithms may include choosing spectrum or modulation format, rerouting data traffic, etc., while the reward may be the maximization of network throughput, minimization of latency or packet loss rate, etc. [31][32]. Currently, there are limited applications of RL in the physical layer of optical communication systems. This is partly because in most cases the reward (objective function) can be explicitly expressed as a continuous and differentiable function of the actions. An example is the CMA algorithm where the actions are the filter tap weights and the objective is to produce output signals with a desired amplitude. For these type of optimization problems, we simply refer to them as adaptive signal processing instead of RL.

1.5. Deep learning techniques

1.5.1 Deep learning vs. conventional machine learning

The recent emergence of DL technologies has taken ML research to a whole new level. DL algorithms have demonstrated comparable or better performance than humans in a lot of important tasks including image recognition, speech recognition, natural language processing, information retrieval, etc. [2][33]. Loosely speaking, DL systems consist of multiple layers of nonlinear processing units (thus deeper architectures) and may even contain complex structures such as feedback and memory. DL then refers to learning the parameters of these architectures for performing various pattern recognition tasks.

One way to interpret DL algorithms is that they automatically learn and extract higher-level features of data from lower-level ones as the input propagates through various layers of nonlinear processing units, resulting in a hierarchical representation of data. For example, while performing a complex human face recognition task using a DL-based multilayer ANN (called DNN), the first layer might learn to detect edges, the second layer can learn to recognize more complex shapes such as circles or squares which are built from the edges. The third layer may then recognize even more complex combinations and arrangements of shapes such as the location of two ovals and a triangle in between, which in turn starts to resemble parts of a human face with two eyes and a nose. Such an ability to automatically discover and learn features at increasingly high levels of abstraction empowers DL systems to learn complex relationships between inputs and outputs directly from the data instead of using human-crafted features.

As an example of this notion of hierarchical learning, Fig. 1.19 shows the use of a DNN as well as a conventional ANN on signal's eye-diagrams to monitor OSNR. In the first approach, the eye-diagrams are directly applied as images at the input of

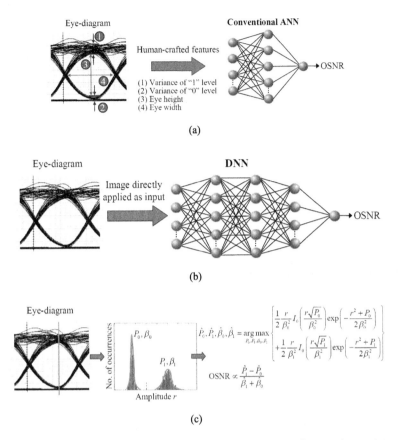

Figure 1.19 Example illustrating OSNR monitoring using eye-diagrams' features by applying (a) DNN, (b) conventional ANN, and (c) analytical modeling and parameters fitting.

the DNN, as shown in Fig. 1.19(a), and it is made to automatically learn and discover OSNR-sensitive features without any human intervention. The extracted features are subsequently exploited by DNN for OSNR monitoring. In contrast, with conventional ANNs, prior knowledge in optical communications is utilized in choosing suitable features for the task, e.g., the variances of "1" and "0" levels and eye-opening can be indicative of OSNR. Therefore, these useful features are manually extracted from the eye-diagrams and are then used as inputs to an ANN for the estimation of OSNR as shown in Fig. 1.19(b). For completeness, Fig. 1.19(c) shows an analytical and non-ML approach to determine OSNR by finding the powers and noise variances that best fit the noise distributions of "1" and "0" levels knowing that they follow Rician distribution. In this case, a specific mathematical formula or computational instruction is pre-coded into the program and there is nothing to learn from the input data. Fig. 1.20 compares the underpinning philosophies of the three different approaches discussed above. Note

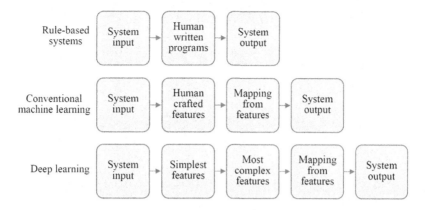

Figure 1.20 Conceptual differences between rule-based systems, conventional ML, and DL approaches for pattern recognition.

that, in principle, there is no hard rule on how many layers are needed for an ML model in a given problem. In practice, it is generally accepted that when more underlying physics/mathematics of the problem is used to identify and extract the suitable data features as inputs, the ML model tends to be simpler.

It should be noted that deep architectures are more efficient or more *expressive* than their shallow counterparts [34]. For example, it has been observed empirically that compared to a shallow neural network, a DNN requires much fewer number of neurons and weights (i.e., around 10 times less connections in speech recognition problems [35]) to achieve the same performance.

A major technical challenge in DL is that the conventional BP algorithm and gradient-based learning methods used for training shallow networks are inherently not effective for training networks with multiple layers due to the *vanishing gradient problem* [2]. In this case, different layers in the network learn at significantly different speeds during the training process, i.e., when the layers close to the output are learning well, the layers close to the input often get stuck. In the worst case, this may completely stop the network from further learning. Several solutions have been proposed to address the vanishing gradient problem in DL systems. These include: (i) choosing specific activation functions such as ReLU [12], as discussed earlier; (ii) pretraining of network one layer at a time in a greedy way and then fine-tuning the entire network through BP algorithm [36]; (iii) using some special architectures such as long short-term memory (LSTM) networks [37]; and (iv) applying network optimization approaches which avoid gradients (e.g., global search methods such as genetic algorithm). The choice of a given solution typically depends on the type of DL model being trained and the degree of computational complexity involved.

1.5.2 Deep neural networks (DNNs)

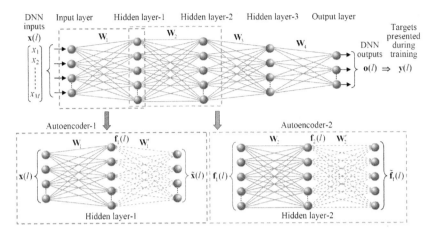

Figure 1.21 Schematic diagram of a three hidden layers DNN (top). Two autoencoders used for the pretraining of first two hidden layers of the DNN (bottom). The decoder parts in both autoencoders are shown in gray color with dotted weight lines.

Unlike shallow ANNs, DNNs contain multiple hidden layers between input and output layers. The structure of a simple three hidden layers DNN is shown in Fig. 1.21 (top). DNNs can be trained effectively using the BP algorithm. To avoid vanishing gradient problem during training of DNNs, following two approaches are typically adopted. In the first method, the ReLU activation function is simply used for the hidden layers neurons due to its non-saturating nature. In the second approach, a DNN is first pretrained one layer at a time and then the training process is fine-tuned using BP algorithm [36]. For pretraining of hidden layers of the DNNs, *autoencoders* are typically employed which are essentially feed-forward neural networks. Fig. 1.21 (bottom) shows two simple autoencoders used for the unsupervised pretraining of first two hidden layers of the DNN. First, hidden layer-1 of the DNN is pretrained in isolation using autoencoder-1 as shown in the figure. The first part of autoencoder-1 (called encoder) maps input vectors **x** to a hidden representation $\mathbf{f}_1$ while the second part (called decoder) reverses this mapping in order to synthesize the initial inputs **x**. Once autoencoder-1 learns these mappings successfully, hidden layer-1 is considered to be pretrained. The original input vectors **x** are then passed through the encoder of autoencoder-1 and the corresponding representations $\mathbf{f}_1$ (also called feature vectors) at the output of pretrained hidden layer-1 are obtained. Next, vectors $\mathbf{f}_1$ are utilized as inputs for the unsupervised pretraining of hidden layer-2 using autoencoder-2, as depicted in the figure. This procedure is repeated for the pretraining of hidden layer-3 and the corresponding feature vectors $\mathbf{f}_3$ are then used for the supervised pretraining of final output layer by setting the desired outputs **y** as targets. After isolated pretraining of hidden and output layers, the complete DNN is trained (i.e., fine-tuned) using BP algorithm with $\mathbf{x}$ and $\mathbf{y}$ as inputs and targets, respectively. By adopting this autoencoders-based hierarchical learning approach, the vanishing gradient problem can be successfully bypassed in DNNs.

1.5.3 Convolutional neural networks (CNNs)

CNNs are a type of neural network primarily used for pattern recognition within images though they have also been applied in a variety of other areas such as speech recognition, natural language processing, video analysis, etc. The structure of a typical CNN is shown in Fig. 1.22(a) comprising of a few alternating *convolutional* and *pooling* layers followed by an ANN-like structure towards the end of the network. The convolutional layer consists of neurons whose outputs only depend on the neighboring pixels of the input as opposed to fully-connected (FC) layers in typical ANNs as shown in Fig. 1.22(b). That is why it is called *local network*, *local connected network* or *local receptive field* in ML literature. The weights are also shared across the neurons in the same layer, i.e., each neuron undergoes the same computation $\mathbf{w}^{\mathrm{T}}(\cdot) + b$ but the input is a different part of the original image. This is followed by a decision-like nonlinear activation function and the output is called a *feature map* or *activation map*. For the same input image/layer, one can build multiple feature maps, where the features are learned via a training process. A parameter called *stride* defines how many pixels we slide the $\mathbf{w}^{\mathrm{T}}(\cdot) + b$ filter across the input image horizontally/vertically per output. The stride value determines the size of a feature map. Next, a *max-pooling* or *sub-sampling* layer operates over the feature maps by picking the largest value out of 4 neighboring neurons as shown in Fig. 1.22(c). Max-pooling is essentially nonlinear down-sampling with the objective to retain the largest identified features while reduce the dimensionality of the feature maps.

The $\mathbf{w}^{\mathrm{T}}(\cdot) + b$ operation essentially multiplies part of the input image with a 2D function $g(s_x, s_y)$ and sums the results as shown in Fig. 1.23. The sliding of $g(s_x, s_y)$ over all spatial locations is the same as *convolving* the input image with $g(-s_x, -s_y)$ (hence the name convolutional neural networks). Alternatively, one can also view the $\mathbf{w}^{\mathrm{T}}(\cdot) + b$ operation as *cross-correlating* $g(s_x, s_y)$ with the input image. Therefore, a high value will result if that part of the input image resembles $g(s_x, s_y)$. Together with the decision-like nonlinear activation function, the overall feature map indicates which location in the original image best resembles $g(s_x, s_y)$, which essentially tries to identify and locate a certain feature in the input image. With this insight, the interleaving convolutional and sub-sampling layers can be intuitively understood as identifying higher-level and more complex features of the input image.

The training of a CNN is performed using a modified BP algorithm which updates convolutional filters' weights and also takes the sub-sampling layers into account. Since a lot of weights are supposedly identical as the network is essentially performing the convolution operation, one will update those weights using the average of the corresponding gradients.

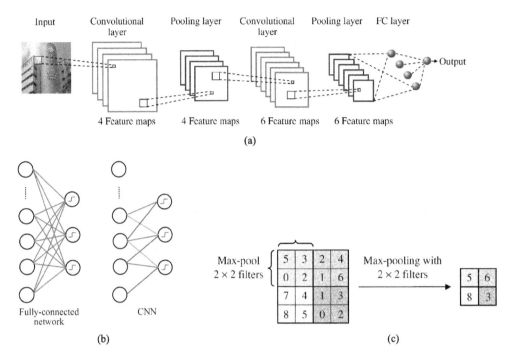

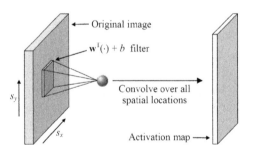

Figure 1.22 (a) A simple CNN architecture comprising of two sets of convolutional and pooling layers followed by an FC layer on top. (b) In a CNN, a node in the next layer is connected to a small subset of nodes in the previous layer. The weights (indicated by colors of the edges) are also shared among the nodes. (c) Nonlinear down-sampling of feature maps via a max-pooling layer.

Figure 1.23 Convolution followed by an activation function in a CNN. Viewing the $\mathbf{w}^T(\cdot) + b$ operation as *cross-correlating* a 2D function $g(s_x, s_y)$ with the input image, the overall feature map indicates which location in the original image best resembles $g(s_x, s_y)$.

1.5.4 Recurrent neural networks (RNNs)

In our discussion up to this point, different input–output pairs $(\mathbf{x}(i), \mathbf{y}(i))$ and $(\mathbf{x}(j), \mathbf{y}(j))$ in a data set are assumed to have no relation with each other. However, in a lot of real-world applications such as speech recognition, handwriting recognition, stock market performance prediction, inter-symbol interference (ISI) cancelation in communications, etc., the sequential data has important spatial/temporal dependence to be learned. An RNN is a type of neural network that performs pattern recognition for data sets with memory. RNNs have feedback connections, as shown in Fig. 1.24, and thus enable the information to be temporarily memorized in the networks [38]. This property allows RNNs to analyze sequential data by making use of their inherent memory.

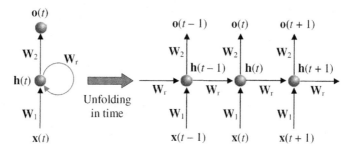

Figure 1.24 Schematic diagram of an RNN and the unfolding in time.

Consider an RNN as shown in Fig. 1.24 with an input $\mathbf{x}(t)$, an output $\mathbf{o}(t)$ and a hidden state $\mathbf{h}(t)$ representing the memory of the network, where the subscript t denotes time. The model parameters $\mathbf{W}_1$, $\mathbf{W}_2$ and $\mathbf{W}_r$ are input, output and recurrent weight matrices, respectively. An RNN can be unfolded in time into a multilayer network [39], as shown in Fig. 1.24. Note that unlike a feed-forward ANN which employs different parameters for each layer, the same parameters $\mathbf{W}_1$, $\mathbf{W}_2$, $\mathbf{W}_r$ are shared across all steps which reflects the fact that essentially same task is being performed at each step but with different inputs. This significantly reduces the number of parameters to be learned. The hidden state $\mathbf{h}(t)$ and output $\mathbf{o}(t)$ at time step t can be computed as

$$\mathbf{h}(t) = \sigma_1(\mathbf{W}_1\mathbf{x}(t) + \mathbf{W}_r\mathbf{h}(t-1) + \mathbf{b}_1) \tag{1.40}$$

$$\mathbf{o}(t) = \sigma_2\left(\mathbf{W}_2\mathbf{h}(t) + \mathbf{b}_2\right) \tag{1.41}$$

where $\mathbf{b}_1$ and $\mathbf{b}_2$ are the bias vectors while $\sigma_1(\cdot)$ and $\sigma_2(\cdot)$ are the activation functions for the hidden and output layer neurons, respectively. Given a data set $\{(\mathbf{x}(1), \mathbf{y}(1)), (\mathbf{x}(2), \mathbf{y}(2)), \ldots (\mathbf{x}(L), \mathbf{y}(L))\}$ of input-output pairs, the RNN is first unfolded in time to represent it as a multilayer network and then BP algorithm is applied on this graph, as shown in Fig. 1.25, to compute all the necessary matrix derivatives $\left\{\frac{\partial E}{\partial \mathbf{W}_1}, \frac{\partial E}{\partial \mathbf{W}_2}, \frac{\partial E}{\partial \mathbf{W}_r}, \frac{\partial E}{\partial \mathbf{b}_1}, \frac{\partial E}{\partial \mathbf{b}_2}\right\}$. The loss function can be cross-entropy or MSE. The matrix derivative $\frac{\partial E}{\partial \mathbf{W}_r}$ is a bit more complicated to calculate since $\mathbf{W}_r$ is shared across all hidden layers. In this case,

$$\frac{\partial E}{\partial \mathbf{W}_r} = \sum_{t=1}^{L} \frac{\partial E(t)}{\partial \mathbf{W}_r} = \sum_{t=1}^{L} \frac{\partial E(t)}{\partial \mathbf{h}(t)} \frac{\partial \mathbf{h}(t)}{\partial \mathbf{W}_r} = \sum_{t=1}^{L} \frac{\partial E(t)}{\partial \mathbf{h}(t)} \sum_{l=1}^{t} \frac{\partial \mathbf{h}(t)}{\partial \mathbf{h}(l)} \frac{\partial \mathbf{h}(l)}{\partial \mathbf{W}_r} \tag{1.42}$$

where most of the derivatives in Eq. (1.42) can be easily computed using Eqs. (1.40) and (1.41). The Jacobian $\frac{\partial \mathbf{h}(t)}{\partial \mathbf{h}(l)}$ is further decomposed into $\frac{\partial \mathbf{h}(t)}{\partial \mathbf{h}(t-1)}\frac{\partial \mathbf{h}(t-1)}{\partial \mathbf{h}(t-2)}\cdots\frac{\partial \mathbf{h}(l+1)}{\partial \mathbf{h}(l)}$ so that efficient updates naturally involve the flow of matrix derivatives from the last data point $(\mathbf{x}(L), \mathbf{y}(L))$ back to the first $(\mathbf{x}(1), \mathbf{y}(1))$. This algorithm is called back-propagation through time (BPTT) [40].

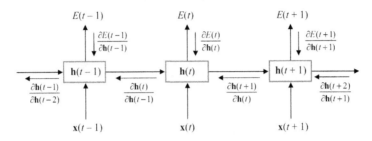

Figure 1.25 Flow of gradient signals in an RNN.

In the special case when the nonlinear activation function is absent, the RNN structure resembles a linear multiple-input multiple-output (MIMO) channel with memory 1 in communication systems. Optimizing the RNN parameters will thus be equivalent to estimating the channel memory given input and output signal waveforms followed by maximum likelihood sequence detection (MLSD) of additional received signals. Consequently, an RNN may be used as a suitable tool for channel characterization and data detection in *nonlinear* channels with memory such as long-haul transmission links with fiber Kerr nonlinearity or direct detection systems with CD, chirp or other component nonlinearities. Network traffic prediction may be another area where RNNs can play a useful role.

One major limitation of conventional RNNs in many practical applications is that they are not able to learn long-term dependencies in data (i.e., dependencies between events that are far apart) due to the so-called *exploding* and *vanishing gradient problems* encountered during their training. To overcome this issue, a special type of RNN architecture called long short-term memory (LSTM) network is designed which can model and learn temporal sequences and their long-range dependencies more accurately through better storing and accessing of information [37]. An LSTM network makes decision on whether to forget/delete or store the information based on the importance which it assigns to the information. The assigning of importance takes place through weights which are determined via a learning process. Simply put, an LSTM network learns over time which information is important and which is not. This allows LSTM network's short-term memory to last for longer periods of time as compared to conventional RNNs which in turn leads to improved sequence learning performance. More recently, attention and transformer models have shown further enhanced performance in efficiently capturing long-term dependencies in data sequences and thus, have become standards for natural language processing, computer vision, etc. [41].

1.5.5 Generative adversarial networks (GANs)

A GAN is a DL-based *generative* model that is capable of generating new data with similar characteristics as the input real data. A GAN comprises of two sub-networks called *generator* and *discriminator* networks which are trained together but have opposing goals, that is why the system as a whole is described as *adversarial* [42]. The generator and discriminator networks can be DNN, CNN, RNN, etc., in principle. The illustration of a GAN configuration is shown in Fig. 1.26.

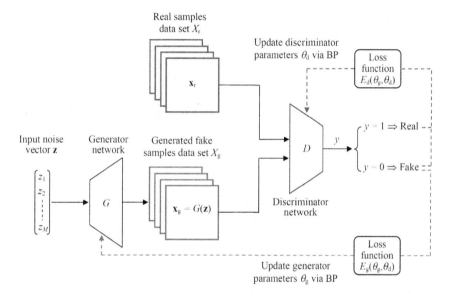

Figure 1.26 Schematic diagram of a basic GAN. The generator transforms a noise vector **z** sampled from a distribution $p_z(\mathbf{z})$ into a generated sample $\mathbf{x_g}$. The discriminator is a binary classifier that tries to distinguish between the generated and real samples $\mathbf{x_r}$.

We can represent generator and discriminator by nonlinear mapping functions $G(\mathbf{z}, \theta_g)$ and $D(\mathbf{x}, \theta_d)$, respectively, each of which is differentiable with respect to its inputs as well as its parameters $\theta_{g,d}$. Given a noise vector **z** sampled from a distribution $p_z(\mathbf{z})$ as input, the generator tries to generate synthetic but perceptually convincing samples $\mathbf{x_g} = G(\mathbf{z})$ that form a distribution $p_g(\mathbf{x})$. On the other hand, the inputs to the discriminator network are either samples $\mathbf{x_g}$ generated by the generator network or the real samples $\mathbf{x_r}$ drawn from a real data distribution $p_r(\mathbf{x})$ and it tries to classify the observations as "fake" or "real". The output of discriminator network, i.e., $y = D(G(\mathbf{z}))$ or $y = D(\mathbf{x_r})$, where $y \in [0, 1]$, is a single value that indicates the probability of the input being a fake (i.e., $y = 0$) or real (i.e., $y = 1$) sample.

To train a GAN, the generator and discriminator networks are trained simultaneously using SGD so that they may evolve together, where the generator learns to produce samples that can fool the discriminator while the discriminator learns to distinguish between the generated and real samples [43]. Both networks have associated loss functions $E_{g,d}$ defined in terms of both networks' parameters, i.e.,

$$E_g = \min_G \mathbb{E}_{\mathbf{x}_g \sim p_g(\mathbf{x})} \left[log \left(1 - D\left(\mathbf{x}_g\right)\right)\right] \tag{1.43}$$

$$E_d = \max_D \mathbb{E}_{\mathbf{x}_r \sim p_r(\mathbf{x})} \left[log D\left(\mathbf{x}_r\right)\right] + \mathbb{E}_{\mathbf{x}_g \sim p_g(\mathbf{x})} \left[log \left(1 - D\left(\mathbf{x}_g\right)\right)\right] \tag{1.44}$$

It is clear from Eq. (1.44) that the discriminator is essentially a binary classifier with a maximum log-likelihood objective. During training, the generator attempts to minimize $E_g(\theta_g, \theta_d)$ by controlling only θ_g while the discriminator tries to maximize $E_d(\theta_g, \theta_d)$ through controlling only θ_d. Since each sub-network's loss function also depends on other sub-network's parameters but it cannot control other's parameters, this scenario is not a conventional optimization problem. Instead, it can be better described as a *game* with Nash equilibrium as its solution, which in the current context is a tuple (θ_g, θ_d) that occurs when E_g is at minimum with respect to θ_g and E_d is at maximum with respect to θ_d.

During each training iteration, a batch of samples $\mathbf{x}_r$ is drawn from a data distribution $p_r(\mathbf{x})$ and a batch of noise vectors $\mathbf{z}$ is drawn from a distribution $p_z(\mathbf{z})$. Next, two gradient steps, i.e., one for updating θ_g to reduce $E_g(\theta_g, \theta_d)$ and the other for updating θ_d to increase $E_d(\theta_g, \theta_d)$, are made simultaneously. The learning process can thus be formulated as following optimization problem

$$\min_G \max_D V\left(D, G\right) = \mathbb{E}_{\mathbf{x}_r \sim p_r(\mathbf{x})} \left[log D\left(\mathbf{x}_r\right)\right] + \mathbb{E}_{\mathbf{z} \sim p_z(\mathbf{z})} \left[log \left(1 - D\left(G(\mathbf{z})\right)\right)\right] \tag{1.45}$$

where $V(D, G)$ is the value function which the generator attempts to minimize and the discriminator tries to maximize [44]. By minimizing $V(D, G)$, the generator attempts to minimize the probability that the discriminator will predict a 0 for fake samples. On the other hand, by maximizing $V(D, G)$, the discriminator aims to minimize the error of its predictions with respect to target values of 0 and 1 for fake and real samples, respectively. The Nash equilibrium of this minimax (i.e., adversarial) game is achieved at [42]

$$p_g\left(\mathbf{x}\right) = p_r\left(\mathbf{x}\right) \forall \mathbf{x}$$
$$\text{and } D\left(\mathbf{x}\right) = \frac{1}{2} \forall \mathbf{x} \tag{1.46}$$

The above-mentioned strategy ideally results in a generator network that produces convincingly realistic new data samples and a discriminator network that has learnt the characteristic feature representations of the input real data.

The ability to generate new data samples by learning the underlying distribution of the given training data and the generative process that creates them makes GANs extremely promising in applications suffering from data scarcity. In the context of fiber-optic communications, there are a number of practical situations where the acquisition of extensive data sets for ML models' training purpose is quite difficult. For example, obtaining large-scale data corresponding to various network failure scenarios is a challenging task due to low fault occurrence probability in modern fiber-optic networks. In such cases, GANs can be employed effectively to generate ample new data samples with similar statistical characteristics as the original limited real data, thus substantially mitigating the shortage of relevant training data for ML models used for failure prediction in optical networks.

1.6. Future role of ML in optical communications

The emergence of SDNs with their inherent programmability and access to enormous amount of network-related monitored data provides unprecedented opportunities for the application of ML methods in these networks. The vision of future intelligent optical networks integrates the programmability/automation functionalities of SDNs with data-analytics capabilities of ML technologies to realize self-aware, self-managing and self-healing network infrastructures. Over the past few years, we have seen an increasing amount of research on the application of ML techniques in various aspects of optical communications and networking. As ML is gradually becoming a common knowledge to the photonics community, we can envisage some potential significant developments in optical networks in the near future ushered in by the application of emerging ML technologies.

Looking to the future, we can foresee a vital role played by ML-based mechanisms across several diverse functional areas in optical networks, e.g., network planning and performance prediction, network maintenance and fault prevention, network resources allocation and management, etc. ML can also aid cross-layer optimization in future optical networks requiring big data analytics since it can inherently learn and uncover hidden patterns and unknown correlations in big data which can be extremely beneficial in solving complex network optimization problems. The ultimate objective of ML-driven next-generation optical networks will be to provide infrastructures which can monitor themselves, diagnose and resolve their problems, and provide intelligent and efficient services to the end users.

1.7. Online resources for ML algorithms

At present, there are numerous ML codes and examples readily available online and one seldom needs to write their own codes from the scratch. There are several off-the-shelf powerful frameworks available under open-source licenses such as TensorFlow, Pytorch, Caffe, etc. Matlab®, which is widely used in optical communications researches is not the most popular programming language among the ML community. Instead, Python is the preferred language for ML research partly because it is freely available, multi-platform, relatively easy to use/read, and has a huge number of libraries/modules available for a wide variety of tasks. We hereby report some useful resources including example Python codes using TensorFlow library to help interested readers get started with applying simple ML algorithms to their problems. More intuitive understanding of ANNs can be found at this visual playground [45]. The Python codes for most of the standard neural network architectures discussed in this chapter can be found in these Github repositories [46][47] with examples. For non-standard model design, Tensor-Flow also provides low-level programming interfaces for more custom and complex operations based on its symbolic building blocks, which are documented in detail in [48].

Modern ML framework is constantly evolving with the ongoing advances in computational paradigms like *dynamic computational graphs* and *differentiable programming* as well as with the availability of heterogeneous hardware (e.g., graphics processing unit (GPU), application-specific integrated circuits (ASICs), etc.) for acceleration. A stack of such driving technologies can facilitate compilation of a formulized hypothesis in the most automatic and efficient form, from software and hardware perspectives, thus giving rise to recent sophisticated and outstanding models [41][49]. Specifically, one of the most powerful techniques is *AutoDiff* that can be used to automatically evaluate derivatives of ML model's functions with respect to its parameters. Almost all modern ML libraries adopt AutoDiff along with well-abstracted programming interfaces to speed up the implementation of ML algorithms in a modular and structured way without necessarily going into fine details of the models. There are several works reported in the literature successfully incorporating above-mentioned tools as the researchers are becoming more aware of the power of ML ecosystem.

Recently, ML community has also developed some general-purpose platforms for other science and engineering disciplines with the same powerful underlying technology stacks that facilitate ML research. Notably, JAX [50] by DeepMind is one such platform that supports differentiation, vectorization, acceleration, etc., in GPU/tensor processing unit (TPU). Noticing JAX's advantages, we recently developed COMM-PLAX [51]—an open-source DSP framework for optical communications that allows synthesis and optimization of DSP modules as well as their merger with ML algorithms under the same coding platform. COMMPLAX aims to foster seamless integration of optical communications and ML disciplines.

1.8. Conclusions

In this chapter, we discussed how the rich body of ML techniques can be applied as a unique and powerful set of signal processing tools in fiber-optic communication systems. As optical networks become faster, more dynamic and more software-defined, we will see an increasing number of applications of ML and big data analytics in future networks to solve certain critical problems that cannot be easily tackled using conventional approaches. A basic knowledge and skills in ML will thus become necessary and beneficial for researchers in the field of optical communications and networking.

Appendix 1.A.

For cross-entropy loss function defined in Eq. (1.14), the derivative with respect to the output is given by

$$\frac{\partial E(n)}{\partial o_j(n)} = -\frac{y_j(n)}{o_j(n)}. \tag{1.47}$$

With softmax activation function for the output neurons,

$$
\begin{aligned}
\frac{\partial o_j(n)}{\partial r_k(n)} &= \frac{\left(\sum_{m=1}^{K} e^{r_m(n)}\right) e^{r_j(n)} \delta_{j,k} - e^{r_j(n)} \cdot e^{r_k(n)}}{\left(\sum_{m=1}^{K} e^{r_m(n)}\right)^2} \\
&= \frac{\left(\sum_{m=1}^{K} e^{r_m(n)}\right) e^{r_j(n)} \delta_{j,k} - e^{r_j(n)} \cdot e^{r_k(n)}}{\left(\sum_{k=1}^{K} e^{r_k(n)}\right)^2} \\
&= o_j(n)\,\delta_{j,k} - o_j(n)\,o_k(n)
\end{aligned}
\tag{1.48}
$$

where $\delta_{j,k} = 1$ when $j = k$ and 0 otherwise. Consequently,

$$
\begin{aligned}
\frac{\partial E(n)}{\partial r_k(n)} &= \sum_{j=1}^{K} \frac{\partial E(n)}{\partial o_j(n)} \frac{\partial o_j(n)}{\partial r_k(n)} \\
&= \sum_{j=1}^{K} -\frac{y_j(n)}{o_j(n)} \left(o_j(n)\,\delta_{j,k} - o_j(n)\,o_k(n)\right) \\
&= \sum_{j=1}^{K} -y_j(n)\left(\delta_{j,k} - o_k(n)\right) = o_k(n) - y_k(n)
\end{aligned}
\tag{1.49}
$$

as $\sum_{j=1}^{K} y_j(n) = 1$. Therefore,

$$\frac{\partial E(n)}{\partial \mathbf{r}(n)} = \mathbf{o}(n) - \mathbf{y}(n).$$

Now, since $\frac{\partial \mathbf{r}(n)}{\partial \mathbf{b}_2}, \frac{\partial \mathbf{r}(n)}{\partial \mathbf{b}_1}, \frac{\partial r_k(n)}{\partial \mathbf{W}_2}, \frac{\partial r_k(n)}{\partial \mathbf{W}_1}$ are the same as $\frac{\partial \mathbf{o}(n)}{\partial \mathbf{b}_2}, \frac{\partial \mathbf{o}(n)}{\partial \mathbf{b}_1}, \frac{\partial o_k(n)}{\partial \mathbf{W}_2}, \frac{\partial o_k(n)}{\partial \mathbf{W}_1}$ for MSE loss function and linear activation function for the output neurons (as $\mathbf{o}(n) = \mathbf{r}(n)$ for that case), it follows that the update equations Eq. (1.8) to Eq. (1.12) also hold for the ANNs with cross-entropy loss function and softmax activation function for the output neurons.

References

[1] S. Marsland, Machine Learning: An Algorithmic Perspective, second ed., CRC Press, Boca Raton, USA, 2015.
[2] Y. Bengio, A. Courville, P. Vincent, Representation learning: a review and new perspectives, IEEE Trans. Pattern Anal. Mach. Intell. 35 (8) (2013) 1798–1828.
[3] C.M. Bishop, Pattern Recognition and Machine Learning, Springer-Verlag, New York, USA, 2006.
[4] Z. Dong, F.N. Khan, Q. Sui, K. Zhong, C. Lu, A.P.T. Lau, Optical performance monitoring: a review of current and future technologies, J. Lightwave Technol. 34 (2) (2016) 525–543.
[5] F.N. Khan, Z. Dong, C. Lu, A.P.T. Lau, Optical performance monitoring for fiber-optic communication networks, in: X. Zhou, C. Xie (Eds.), Enabling Technologies for High Spectral-Efficiency Coherent Optical Communication Networks, John Wiley & Sons, Hoboken, USA, 2016 (Chapter 14).
[6] A.S. Thyagaturu, A. Mercian, M.P. McGarry, M. Reisslein, W. Kellerer, Software defined optical networks (SDONs): a comprehensive survey, IEEE Commun. Surv. Tutor. 18 (4) (2016) 2738–2786.
[7] F.N. Khan, Q. Fan, C. Lu, A.P.T. Lau, An optical communication's perspective on machine learning and its applications, J. Lightwave Technol. 37 (2) (2019) 493–516.
[8] R.A. Dunne, A Statistical Approach to Neural Networks for Pattern Recognition, John Wiley & Sons, Hoboken, USA, 2007.
[9] I. Kaastra, M. Boyd, Designing a neural network for forecasting financial and economic time series, Neurocomputing 10 (3) (1996) 215–236.
[10] C.D. Meyer, Matrix Analysis and Applied Linear Algebra, Society for Industrial and Applied Mathematics, Philadelphia, USA, 2000.
[11] R.O. Duda, P.E. Hart, D.G. Stork, Pattern Classification, second ed., John Wiley & Sons, New York, USA, 2007.
[12] X. Glorot, A. Bordes, Y. Bengio, Deep sparse rectifier neural networks, in: Proc. AISTATS, Fort Lauderdale, FL, USA, vol. 15, 2011, pp. 315–323.
[13] K. He, X. Zhang, S. Ren, J. Sun, Delving deep into rectifiers: surpassing human-level performance on ImageNet classification, in: Proc. ICCV, Santiago, Chile, 2015, pp. 1026–1034.
[14] X. Glorot, Y. Bengio, Understanding the difficulty of training deep feedforward neural networks, in: Proc. AISTATS, Chia Laguna Resort, Sardinia, Italy, 2010, pp. 249–256.
[15] A. Webb, Statistical Pattern Recognition, second ed., John Wiley & Sons, Chichester, UK, 2002.
[16] A. Statnikov, C.F. Aliferis, D.P. Hardin, I. Guyon, A Gentle Introduction to Support Vector Machines in Biomedicine, World Scientific, Singapore, 2011.
[17] M.S. Andersen, J. Dahl, L. Vandenberghe, CVXOPT: python software for convex optimization, available online at: https://cvxopt.org.
[18] M. Li, S. Yu, J. Yang, Z. Chen, Y. Han, W. Gu, Nonparameter nonlinear phase noise mitigation by using M-ary support vector machine for coherent optical systems, IEEE Photonics J. 5 (6) (2013) 7800312.

[19] D. Wang, M. Zhang, Z. Li, Y. Cui, J. Liu, Y. Yang, et al., Nonlinear decision boundary created by a machine learning-based classifier to mitigate nonlinear phase noise, in: Proc. ECOC, Valencia, Spain, 2015, Paper P.3.16.

[20] T. Nguyen, S. Mhatli, E. Giacoumidis, L.V. Compernolle, M. Wuilpart, P. Mégret, Fiber nonlinearity equalizer based on support vector classification for coherent optical OFDM, IEEE Photonics J. 8 (2) (2016) 7802009.

[21] K. Hechenbichler, K. Schliep, Weighted k-nearest-neighbor techniques and ordinal classification, Discussion Paper 399, Collaborative Research Center 386, University of Munich, Munich, Germany, 2004, available online at: https://epub.ub.uni-muenchen.de/1769/.

[22] T. Hastie, R. Tibshirani, J. Friedman, The Elements of Statistical Learning, second ed., Springer, New York, USA, 2009.

[23] M. Kirk, Thoughtful Machine Learning with Python, O'Reilly Media, Sebastopol, USA, 2017.

[24] I.T. Jolliffe, Principal Component Analysis, second ed., Springer-Verlag, New York, USA, 2002.

[25] J.E. Jackson, A User's Guide to Principal Components, John Wiley & Sons, Hoboken, USA, 2003.

[26] L.J. Cao, K.S. Chua, W.K. Chong, H.P. Lee, Q.M. Gu, A comparison of PCA, KPCA and ICA for dimensionality reduction in support vector machine, Neurocomputing 55 (1–2) (2003) 321–336.

[27] R.S. Sutton, A.G. Barto, Reinforcement Learning: An Introduction, second ed., MIT Press, Cambridge, USA, 2018.

[28] K. Arulkumaran, M.P. Deisenroth, M. Brundage, A.A. Bharath, Deep reinforcement learning: a brief survey, IEEE Signal Process. Mag. 34 (6) (2017) 26–38.

[29] R. Bellman, On the theory of dynamic programming, Proc. Natl. Acad. Sci. 38 (8) (1952) 716–719.

[30] P.-H. Su, P. Budzianowski, S. Ultes, M. Gasic, S. Young, Sample-efficient actor-critic reinforcement learning with supervised data for dialogue management, in: Proc. SigDial, Saarbrucken, Germany, 2017, pp. 147–157.

[31] Y.V. Kiran, T. Venkatesh, C.S. Murthy, A reinforcement learning framework for path selection and wavelength selection in optical burst switched networks, IEEE J. Sel. Areas Commun. 25 (9) (2007) 18–26.

[32] X. Chen, J. Guo, Z. Zhu, R. Proietti, A. Castro, S.J.B. Yoo, Deep-RMSA: a deep-reinforcement-learning routing, modulation and spectrum assignment agent for elastic optical networks, in: Proc. OFC, San Diego, CA, USA, 2018, Paper W4F.2.

[33] Y. Bengio, Learning deep architectures for AI, Found. Trends Mach. Learn. 2 (1) (2009) 1–127.

[34] Y. Bengio, O. Delalleau, On the expressive power of deep architectures, in: J. Kivinen, C. Szepesvári, E. Ukkonen, T. Zeugmann (Eds.), Algorithmic Learning Theory, Springer-Verlag, Heidelberg, Germany, 2011, pp. 18–36.

[35] L.J. Ba, R. Caurana, Do deep nets really need to be deep?, in: Proc. NIPS, Montreal, Canada, 2014, pp. 2654–2662.

[36] H. Larochelle, Y. Bengio, J. Louradour, P. Lamblin, Exploring strategies for training deep neural networks, J. Mach. Learn. Res. 10 (2009) 1–40.

[37] I. Goodfellow, Y. Bengio, A. Courville, Deep Learning, MIT Press, Massachusetts, USA, 2016.

[38] D.P. Mandic, J. Chambers, Recurrent Neural Networks for Prediction: Learning Algorithms, Architectures and Stability, John Wiley & Sons, Chichester, UK, 2001.

[39] R. Pascanu, C. Gulcehre, K. Cho, Y. Bengio, How to construct deep recurrent neural networks, in: Proc. ICLR, Banff, Canada, 2014.

[40] R. Pascanu, T. Mikolov, Y. Bengio, On the difficulty of training recurrent neural networks, in: Proc. ICML, Atlanta, GA, USA, 2013, pp. 1310–1318.

[41] A. Vaswani, N. Shazeer, N. Parmar, J. Uszkoreit, L. Jones, A.N. Gomez, L. Kaiser, I. Polosukhin, Attention is all you need, in: Proc. NIPS, Long Beach, CA, USA, 2017, pp. 5998–6008.

[42] I. Goodfellow, NIPS 2016 tutorial: generative adversarial networks, in: Proc. NIPS, Barcelona, Spain, 2016, available online at: https://arxiv.org/abs/1701.00160.

[43] X. Yi, E. Walia, P. Babyn, Generative adversarial network in medical imaging: a review, Med. Image Anal. 58 (2019) 101552.

[44] J.M. Wolterinka, K. Kamnitsasb, C. Ledigc, I. Išguma, Deep learning: generative adversarial networks and adversarial methods, in: S.K. Zhou, D. Rueckert, G. Fichtinger (Eds.), Handbook of Medical Image Computing and Computer Assisted Intervention, Academic Press, Cambridge, USA, 2019 (Chapter 23).

[45] D. Smilkov, S. Carter, TensorFlow — a neural network playground, available online at: http://playground.tensorflow.org.

[46] A. Damien, GitHub repository — TensorFlow tutorial and examples for beginners with latest APIs, available online at: https://github.com/aymericdamien/TensorFlow-Examples.

[47] M. Zhou, GitHub repository — TensorFlow tutorial from basic to hard, available online at: https://github.com/MorvanZhou/Tensorflow-Tutorial.

[48] TensorFlow, Guide for programming with the low-level TensorFlow APIs, available online at: https://www.tensorflow.org/programmers_guide/low_level_intro.

[49] T.B. Brown, N. Mann, N. Ryder, M. Subbiah, J. Kaplan, P. Dhariwal, et al., Language models are few-shot learners, in: Proc. NIPS, Vancouver, Canada, 2020, available online at: https://arxiv.org/abs/2005.14165.

[50] J. Bradbury, R. Frostig, P. Hawkins, M.J. Johnson, C. Leary, D. Maclaurin, et al., GitHub repository — JAX: composable transformations of Python + NumPy programs, available online at: http://github.com/google/jax.

[51] Q. Fan, C. Lu, A.P.T. Lau, GitHub repository — COMMPLAX: differentiable DSP for optical communication, available online at: https://github.com/remifan/commplax.

CHAPTER TWO

Machine learning for long-haul optical systems

Shaoliang Zhang[a] and Christian Häger[b]
[a]Acacia Communication Inc., Maynard, MA, United States
[b]Chalmers University of Technology, Gothenburg, Sweden

2.1. Introduction

The backbone of global Internet networks is nowadays supported by optical transmission systems to carry Terabits per second data rates. The transmission distance can range from a few meters in data centers to tens of thousands of kilometers in undersea cables. Thanks to the technological advance in high-speed optics and digital signal processing (DSP) algorithms, digital coherent transceivers play a key role in supporting capacity-demanding backbone networks to meet the ever-increasing Internet traffic [1]. Although advanced modulation formats are capable of increasing the spectral usage given the limited optical bandwidth in each fiber [2], the maximum achievable information rate of long-haul transmission systems is fundamentally limited by the fiber's deterministic Kerr nonlinear interaction between signal intensity and phase [3]. The evolution of the optical field in a nonlinear fiber is governed by nonlinear Schrödinger equation (NLSE) [4]

$$\frac{\partial u_{x/y}(t,z)}{\partial z} + j\frac{\beta_2}{2}\frac{\partial^2 u_{x/y}(t,z)}{\partial t^2} = j\frac{8}{9}\gamma\left[\left|u_{x/y}(t,z)\right|^2 + \left|u_{y/x}(t,z)\right|^2\right]u_{x/y}(t,z), \qquad (2.1)$$

where $u_{x/y}(t,z)$ is the optical field in the x and y polarization, respectively, β_2 is the group velocity dispersion, and γ is the nonlinear coefficient.

DSP-based nonlinearity compensation (NLC) algorithms have been developed to mitigate the nonlinear impairments by approximating NLSE using either single-step or multiple-step solution. The most popular digital backpropagation (DBP) algorithm is designed to propagate the received signals back to the transmitter side in a virtual link with the opposite sign of γ and β_2. As explained in more detail in Section 2.3.2 below, the signal propagation characterized by the NLSE is numerically approximated in a split-step Fourier method (SSFM), where the dispersion $\boldsymbol{D}$ and nonlinear phase rotation $\boldsymbol{N}$ are applied alternatively in a very short fiber length:

$$\frac{\partial u_{x/y}(t,z)}{\partial z} = (\boldsymbol{D} + \boldsymbol{N})\,u_{x/y}(t,z). \qquad (2.2)$$

Machine Learning for Future Fiber-Optic Communication Systems
https://doi.org/10.1016/B978-0-32-385227-2.00009-7
43

Although the NLC benefits using DBP are very attractive, many factors can reduce its performance gain, such as the number of steps per span (StPs). To trade-off between the performance and computation complexity, filtered DBP is proposed by emulating multiple spans in each DBP step where the intensity waveform is first filtered by a Gaussian low-pass filter (LPF) prior to de-rotating the signal phase. Compared to full-step DBP, at least 0.2 StPs and 0.6 StPs are found to be sufficient for quadrature phase shift keying (QPSK) [5] and 16 quadrature amplitude modulation (16QAM) [6], respectively.

The first-order linear perturbation of the NLSE has led to the single-step perturbation-based pre/post-distortion (PPD) algorithm for Gaussian [7] and root-raised cosine (RRC) pulses [8–10]. In the first-order perturbation theory, the solution to Eq. (2.1) consists of both linear $u_{0,x/y}(t, z)$ and nonlinear perturbation $\Delta u_{x/y}(t, z)$ terms [7,11]. Assuming much larger accumulated dispersion than symbol duration, the nonlinear perturbation terms for the symbol at $t = 0$ can be approximated as [12]

$$\Delta u_{x/y}(0, z) = \sum_{m,n} P_0^{3/2} \left(H_n H_{m+n}^* H_m + V_n V_{m+n}^* H_m \right) C_{m,n}, \tag{2.3}$$

where P_0, H_m and V_m, and $C_{m,n}$ are, respectively, the launch power, symbol sequences for the x and y polarization, and nonlinear perturbation coefficients, m and n are symbol indices with respect to the symbol of interest H_0 and V_0. The nonlinear perturbation coefficients $C_{m,n}$ can be analytically computed given the link parameters and signal pulse duration/shaping factors [7], whereas the triplets do not depend on the link and can be calculated directly from the received symbols. The advantage of the PPD algorithm is its capability of working with 1 sample per symbol and compensating nonlinearity in a single-step approach, whereas DBP and its variants typically work with 2 samples per symbol.

In this chapter, the application of machine learning (ML) is introduced for both single-step perturbation method in Section 2.2 and multi-step DBP method in Section 2.3. A short summary outlook of ML in long-haul systems is given in Section 2.4. A list of acronyms that are used in this chapter can be found in Table 2.1.

2.2. Application of machine learning in perturbation-based nonlinearity compensation

A deep neural network (DNN) model was applied to time-domain PPD at symbol rate to predict the received signal nonlinearity without prior knowledge of the link parameters [13]. The intra-channel cross-phase modulation (IXPM) and intra-channel four-wave mixing (IFWM) triplets are fed into a DNN to explore the correlation among these triplets. In this book chapter, a wide & deep neural network (NN) architecture [14] is used to improve the robustness of its application in the presence of signal nonlinear distortions from imperfect optical front-ends.

Table 2.1   List of acronyms.

ASE	amplified spontaneous emission
ASIC	application-specific integrated circuit
BER	bit error rate
CD(C)	chromatic dispersion (compensation)
CV	cross-validation
(L)DBP	(learned) digital backpropagation
(D)NN	(deep) neural network
DP	dual polarization
DSP	digital signal processing
FIR	finite impulse response
IFWM	intra-channel four-wave mixing
IXPM	intra-channel cross-phase modulation
MIMO	multiple-input multiple-output
ML	machine learning
MSE	mean square error
NLC	nonlinearity compensation
NLSE	nonlinear Schrödinger equation
PMD	polarization mode dispersion
PPD	perturbation-based pre/post-distortion
QAM	quadrature amplitude modulation
QPSK	quadrature phase shift keying
RRC	root-raised cosine
SMF	single-mode fiber
SP	single polarization
SPM	self-phase modulation
StPs	steps per span
SSFM	split-step Fourier method
WDM	wavelength division multiplexing

2.2.1 Wide & deep neural network

The advantage of the wide & deep NN architecture is its capability of learning both deep patterns via the deep path and simple rules through the short path. The triplets features are already studied in the literature to compensate for self-phase modulation (SPM). Therefore, these IXPM and IFWM features can be easily picked up by the NN to accurately characterize the signal nonlinear distortion. On the other hand, the received signal may suffer from nonlinear distortions caused by optical drivers and transimpedance amplifiers.

The wide & deep NN shown in Fig. 2.1 is built in TensorFlow 2 with an input layer consisting of $2 \times N_t$ triplets nodes for wide path and N_s neighbor symbols for deep path. The triplets features only flow through single hidden layer 1 while the neighbor symbols

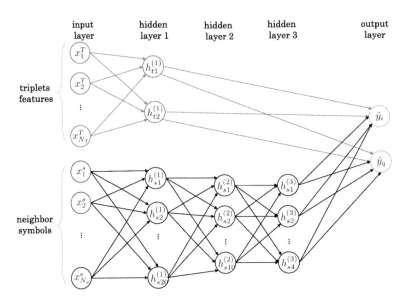

Figure 2.1 The architecture of the wide & deep NN to estimate the nonlinearity of the received signals. 2 hidden nodes for triplets inputs and 3 hidden layers for neighbor symbols inputs.

are fed into a DNN to exploit the relationship among them to improve the quality of the output signals. At the output layer, it takes the outputs from both the deep path and wide path and estimates the distortion in the received signal. The source code to create models in TensorFlow 2 is attached here for reference.

```
# wide path
input_A = keras.layers.Input(shape=X_train.shape[1:], name = "wide_input")
hiddenA = keras.layers.Dense(2, activation="relu")(input_A)
#   deep path
input_B = keras.layers.Input(shape=newX_neighbors.shape[1:], name = "deep_input")
hidden1 = keras.layers.Dense(20, activation="relu")(input_B)
dropOut1 = keras.layers.Dropout(0.5)(hidden1)
hidden2 = keras.layers.Dense(10, activation="relu")(dropOut1)
dropOut2 = keras.layers.Dropout(0.5)(hidden2)
hidden3 = keras.layers.Dense(4, activation="relu")(dropOut2)
# concatenate wide and deep path
concat = keras.layers.concatenate([hiddenA, hidden3])
# output layer
output = keras.layers.Dense(2, name = "output")(concat)
model = keras.Model(inputs=[input_A, input_B], outputs=[output])
```

In this particular example, 1 hidden layer is used for the input triplets features. On the other hand, 3 hidden layers with 20, 10 and 4 nodes, respectively, are applied to the neighbor symbols input features. Two dropout layers after the first and second hidden layer are inserted to mitigate overfitting. The built-in *RMSprop* learning algorithm with a learning rate of 0.002 and batch size of $B = 200$ is used to train the NN by transmitting known but randomly generated patterns, and searching for the best node tensor parameters that minimize the mean squared error (MSE) between the transmitted and received symbols after compensation, given by

$$\text{MSE} = \frac{1}{B} \sum_{i=1}^{B} \left| H_i - \left(\hat{H}_i - \hat{H}_{i,\text{NL}} \right) \right|^2, \tag{2.4}$$

where $\hat{H}_i$ and $\hat{H}_{i,\text{NL}}$, respectively, are the received symbols and estimated nonlinearity for x-polarization, H_i is the ith transmitted symbol, and $|\cdot|$ stands for absolute operation. The same notation is also used for V_i in y-polarization. Although the model is trained using x-polarization data, a similar performance improvement is observed for the y-polarization when applying the same model. The source code to set up the loss function is listed here.

```
import tensorflow.keras.backend as kb
# define MSE loss function
def custom_loss(y_actual, y_pred):
    custom_loss=kb.sum(kb.mean(kb.square(y_actual[:, :2] - y_actual[:, 2:]-y_pred), axis=0))
    return custom_loss

# set up model with loss and optimizer
model.compile(loss = custom_loss,
              optimizer=keras.optimizers.RMSprop(lr=2e-3))

# fit model
history = model.fit((X_train, X_neighbors), Y_train,
                    batch_size = 200,
                    epochs=10,
                    validation_data=((Xcv, Xcv_neighbors), ycv))
```

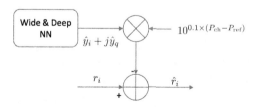

Figure 2.2 Block diagram of applying NLC to the received signals r_i.

Fig. 2.2 shows the block diagram to apply the predicted nonlinear distortions $\hat{y}_i + j\hat{y}_q$ to the received symbol r_i. Here P_{ref} and P_{ch} denotes the channel power of training and test data sets, respectively.

2.2.2 Data collection and pre-processing

The experimental data is collected from a digital coherent receiver after transmitting 32 Gbaud dual-polarization (DP) 16QAM signal over 35 spans of 80 km single-mode fiber (SMF) with 0.2 dB/km loss and 17 ps/nm/km chromatic dispersion (CD). The channel power pre-emphasis is swept from -5 dB up to 2 dB to investigate both linear and nonlinear performance. Note that 50% CD compensation (CDC) has been applied at the transmitter side to reduce the number of triplets needed for perturbation-based NLC algorithm [10].

About 200k symbols are used in the training datasets, and 50k are used as cross-validation (CV) datasets. Each of 5 test datasets contains 121k symbols, and all five test datasets are checked for the stability of the NLC improvement. To ensure data independence, the data pattern used in the training, CV and test datasets is measured to have maximum 0.6% normalized cross-correlation.

In order to more accurately characterize the fiber nonlinearity, signal nonlinearity has to be observed in the received training data. The launch channel power P_0 can be set beyond the optimum channel power to allow nonlinearity noise dominate over received amplified spontaneous emission (ASE) noises. In addition, de-noising averaging, such as averaging over multiple datasets, can be carried out for the fixed training data pattern to isolate the data-dependent nonlinearity from additive Gaussian ASE noises [13]. After cleaning up the ASE noises in the received training dataset, these data are ready to be used for computing the IXPM & IFWM triplets $\hat{H}_{n+k}\hat{H}^*_{m+n+k}\hat{H}_{m+k}$ and $\hat{V}_{n+k}\hat{V}^*_{m+n+k}\hat{H}_{m+k}$ as described in Eq. (2.3). Here, the subscripts m and n are selected from $-\lfloor L/2 \rfloor \leq m, n \leq \lceil L/2 \rceil - 1$, and stand for the symbol index with respect to the center symbols H_k and V_k in a symbol window length of L. Here $\lfloor \cdot \rfloor$ and $\lceil \cdot \rceil$ denote rounding downwards and upper towards the nearest integer.

Depending on the system parameters, such as fiber dispersion, signal baudrate and pulse shaping, the nonlinear perturbation coefficients $C_{m,n}$ can be analytically computed [7,10]. The complexity of non-Gaussian pulses involves triple summation to accurately derive these coefficients $C_{m,n}$ [10]. The number of triplets can be first selected by assuming either a Gaussian pulse or based on hyperbola characteristic of the nonlinear perturbation coefficients $C_{m,n}$ at the given m [13]. The latter can make the proposed wide & deep NN to become a system-agnostic algorithm without prior information of any system parameters. The iterative trimming of insignificant triplets can further reduce the complexity of the NN.

2.2.3 Training results

The channel power pre-emphasis is set to 2 dB higher than the nominal channel power in the training datasets. The number of triplets is $N_T = 1133$ and the number of neighbor symbols is $N_s - 1 = 6$, respectively. Fig. 2.3 plots the training loss trace of each batch and bit error rate (BER) of the CV datasets during the training stage. The training loss

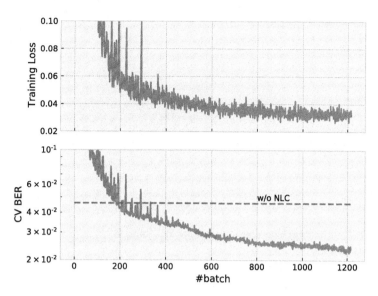

Figure 2.3 The trace of training loss and BER of CV datasets. The BER of CV datasets is ~4.6% without NLC.

can be reduced to ~2.5% after 1000 batches and the BER of the CV datasets is measured as low as ~2.1%, corresponding to ~1.64 dB Q-factor improvement. Note that the hyperparameters of the wide & deep NN, such as the number of neurons in each hidden layer and different activation function [15], can be further adjusted to improve the model accuracy and performance gain on the test datasets. Furthermore, it is more interesting to analyze the learned weights of these neurons and investigate how to better build a NN model to improve performance and/or reduce complexity.

The learned weights of neurons $h_{t_1}^{(1)}$ and $h_{t_2}^{(1)}$ associated with $N_T = 1133$ triplets inputs in the wide path shown in Fig. 2.1 are plotted in Fig. 2.4a, where the first half and second half of the total 2266 input triplets features are the real and imaginary part of IFWM & IXPM triplets in Eq. (2.3). A strong correlation between the learned weights $W_{t_1}^{(1)}$ and $W_{t_2}^{(1)}$ can be easily observed because of complex multiplication in Eq. (2.3). By overlapping $W_{t_2}^{(1)}$ of the real part of triplets, i.e., input features $x_0^T - x_{1132}^T$, with $W_{t_1}^{(1)}$ of the imaginary part of triplets, i.e., the features $x_{1133}^T - x_{2265}^T$ in Fig. 2.4b, their learned weights are almost the same to justify Eq. (2.3). The learned weights of other half are only flipped by sign. These learned features are contrary to the analytical derivation of the perturbation coefficients $C_{m,n}$ has only imaginary part in a 50% pre-CDC system [10]. If there is only imaginary part of $C_{m,n}$, the weights for both real and imaginary of triplets input should be similar. The mismatch may be caused by the inaccurate knowledge of the fiber parameters when applying 50% pre-CDC at the transmitter side. However, the learned weights are a good example of how applying ML models to

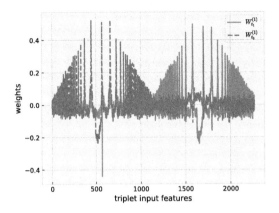

(a) The learned weights $W_{t_1}^{(1)}$ and $W_{t_2}^{(1)}$.

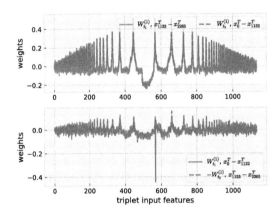

(b) The learned weights $W_{t_1}^{(1)}$ and $W_{t_2}^{(1)}$ after folding.

Figure 2.4 The learned weights of neurons $h_{t_1}^{(1)}$ and $h_{t_2}^{(1)}$ associated with triplets input features.

optical transmission field can complement DSP algorithms without the knowledge of the system setup. By setting the constraint of these correlations between $W_{t_1}^{(1)}$ and $W_{t_2}^{(1)}$, the number of trainable parameters can be reduced almost half to mitigate the common overfitting problem in ML algorithm.

Fig. 2.5 plots the learned weights of first hidden layers in the deep path, i.e., $h_{s_1}^{(1)}$ to $h_{s_{20}}^{(1)}$. Note that the input feature X_{s_3} is the current symbol to be decoded. Other than the fiber nonlinearity characterized by perturbation triplets terms, the learned weights of neurons suggest that correlation can be observed between its neighbor symbols which may be caused by nonlinearity in the optical front-ends.

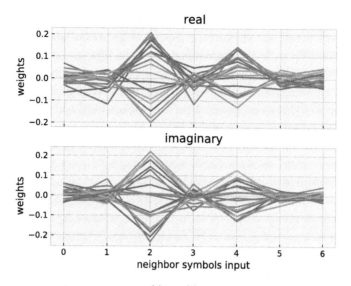

Figure 2.5 The learned weights of neurons $h_{s_1}^{(1)}$ to $h_{s_{20}}^{(1)}$ in the first hidden layers of deep path.

2.2.4 Results and discussion

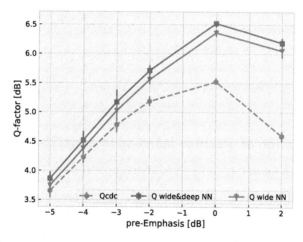

Figure 2.6 The Q-factor improvement using the wide & deep NN versus channel power pre-emphasis in the test datasets collected from a single-channel 32 Gbaud DP-16QAM over 2800 km.

Fig. 2.6 plots the performance gain using the wide & deep NN architecture compared with CDC and NN with the wide path only. At the training pre-emphasis = 2 dB, ~1.6 dB Q-factor improvement can be gained over CDC which demonstrates that the NN model can accurately predict the fiber nonlinearity based on the IFWM&IXPM triplets features. At the 0 dB pre-preemphasis, the wide & deep NN can still outperform CDC by ~0.8 dB. To investigate the performance gain from the neurons in the deep path, the NN is trained without the input features from neighbor symbols, i.e., X_s. In general, the deep path can contribute ~0.15 dB gain at all the channel powers thanks to the NN's self-learning characteristic to exploit the correlation of the symbols that are not fully characterized by the perturbation terms. The performance gain using NN-NLC is found in [13] to be better than conventional PPD at the same number of triplets.

Up to now all the training and test results are carried out in a single-channel case where the dominant impairment is SPM. It is very critical to extend this NN model to wavelength division multiplexing (WDM) systems to evaluate its robustness and performance gain. Although it is desirable to re-train the NN in different WDM configuration, the benefit of re-training the NN is found to be small compared to the model obtained in the single-channel case. The Q performance improvement over CDC is plotted in Fig. 2.7 versus the number of real multiplications per symbol for filtered DBP and Tx-side NN-NLC algorithm at both single- and WDM-channel cases. Note that the complexity of NN-NLC can be smaller because of clean symbols at the transmitter side. The complexity of filtered DBP is based on the analytical results in [6] by assuming FFT size = 4086. In general, Tx-side NN-NLC performs better than filtered-DBP only when the computation complexity is less than ~2000 real multiplications per symbol. The performance of Tx-side NN-NLC is able to match the performance of the filtered-DBP even at higher complexity up to 1 StpS. The Q-factor improvement both NN-NLC and filtered-DBP over CDC drops to 0.3 dB, 0.5 dB, respectively, at 37.5 GHz and 50 GHz WDM from ~0.9 dB at single channel. The mitigation of inter-channel nonlinearity needs to be addressed to see further improvement at affordable complexity.

2.3. Application of machine learning in digital backpropagation

2.3.1 Physics-based machine-learning models

Mathematically speaking, an NN like the one shown in Fig. 2.1 is a parameterized function, where the parameters (weights and biases) can be numerically optimized via some form of gradient descent. NNs are often celebrated as universal function approximators, which means that they can, at least in principle, learn to approximate any function to any desired degree of accuracy [16]. On the other hand, NNs are also often referred to as "black boxes" because they give little insight into their inner workings. In turn, this makes it difficult (although not impossible) to incorporate existing domain knowledge, e.g., to simplify the learning task.[1] Another shortcoming of NNs is that it can be

---

[1] The approach discussed in Section 2.2 can be seen as an example where domain knowledge about the propagation dynamics is used to pre-compute suitable input features for a NN equalizer.

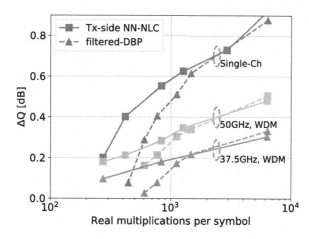

Figure 2.7 The comparison of Q-factor improvement between Tx-side NN-NLC and filtered DBP in single-channel, 50 GHz and 37.5 GHz WDM after 2800 km SMF transmission.

challenging to interpret their parameter configurations after the training, for example to gain additional insights into the studied problem when the trained network outperforms an established baseline method.

In the context of long-haul systems, NN equalizers can be regarded as the "ML counterpart" to the traditional approach of employing mathematical models and analytical techniques for compensating for transmission impairments. On the other hand, it is possible to unify NN-based and model-based approaches by exploiting existing domain knowledge for the construction of a suitable ML model. The general philosophy is to carefully extend and parameterize well-understood algorithms or physics models, rather than relying on black-box models like NNs. This is sometimes referred to as *model-based ML*.

A concrete example of model-based ML is the so-called neural belief propagation decoder for linear error-correcting codes proposed by Nachmani et al. in [17]. Here, the idea is to first "unfold" (or "unroll") the iterations of a standard belief propagation decoder and then attach tunable weights to each edge in the resulting feed-forward computation graph. The "unfolding" methodology is in fact quite general and can also be applied to other iterative algorithms in order to obtain computation graphs with multiple layers which can then be parameterized. An overview of other applications in the context of communication systems can be found in [18].

Model-based ML is gaining in popularity, particularly in the field of communications where decades of research efforts have led to a wide variety of sophisticated algorithms and methods. However, it should be mentioned that ML models obtained by parameterizing existing algorithms are not necessarily universal function approximators like NNs, thereby potentially restricting the class of functions that can be expressed and/or

Table 2.2 High-level comparison between ML approaches.

NN-based	model-based
• "generic" model structure that can fit many applications and tasks	• application-tailored structure based on known algorithms, physics, . . .
• universal function approximators	• not necessarily universal
• few guidelines for the model design (no. of layers, activation function, . . .)	• domain knowledge can be exploited for designing the model
• black boxes that can be difficult to "open" and interpret	• familiar model structure and building blocks can enable easy interpretability

"learned" by the model. A high-level comparison between NN-based and model-based ML can be found in Table 2.2.

In this section, we plan to review the model-based ML approach for nonlinear equalization in fiber-optic systems proposed in [19]. This approach first appeared in [20], where it is shown that the mathematical structure that results from applying spatial discretization methods to the NLSE is very similar to the structure of a conventional DNN with multiple layers. In a nutshell, the main idea is to obtain an ML model by properly parameterizing the resulting computation graph. The model can then be trained using conventional deep-learning tools. Compared to applying large NNs, such *physics-based* ML models are appealing because they incorporate important domain knowledge about the propagation dynamics directly into the model structure and obey well-established physical principles like symmetry and locality in order to reduce the number of trainable parameters. More details about the close relationship between physics and ML can be found for example in [21], where the authors argue that DNNs perform well because their functional form matches the hierarchical or Markovian structure that is present in most real-world data [21]. Indeed, applying spatial discretization methods to a differential equation can be seen as a practical example where such a structure arises, i.e., by decomposing the underlying physical process into a hierarchy of elementary steps. Similar observations were also made in [22], where the training of DNNs is augmented by penalizing non-physical solutions.

In the following, we describe the approach in [19] including some of its generalizations. We will start from simple models and then gradually introduce more complex elements into the design:

- In Section 2.3.2, we consider the simplest version of the model based on the standard NLSE, which is appropriate for single-polarization (SP) systems. In this case, the physics-based model alternates trainable finite impulse response (FIR) filters and fixed element-wise nonlinearities.

- In Section 2.3.3, we consider DP systems, where the propagation dynamics are described by a set of coupled NLSEs. The resulting ML model can be seen generalization of the previous case, where the FIR filters are replaced by 2×2 multiple-input multiple-output (MIMO) FIR filters.
- In Section 2.3.4, we consider the generalization to subband processing, where the model applies multiple filters in parallel. This structure effectively alternates filter banks and nonlinearities, which is very similar to the structure of convolutional NNs commonly used in image processing.

Lastly, we will discuss some applications of these models in Section 2.3.5.

2.3.2 Single-polarization systems

The starting point for channel modeling in long-haul systems is the NLSE, which can be derived from Maxwell's equations under some assumptions that are appropriate for SMFs [4]. The NLSE is a partial differential equation that defines the input–output relationship for optical baseband signals propagating through SMFs. For SP systems, let $u(t, z)$ denote the signal of interest, where we recall that $0 \leq z \leq L$ is the propagation distance, L is the total fiber length, and t denotes time. After simplifying Eq. (2.1) for SP systems, the NLSE is given by

$$\frac{\partial u(t, z)}{\partial z} = -j \frac{\beta_2}{2} \frac{\partial^2}{\partial t^2} u(t, z) + j\gamma |u(t, z)|^2 u(t, z). \tag{2.5}$$

As before, fiber loss and other impairments (e.g., noise, higher-order dispersion, etc.) have been neglected for simplicity.

In general, (2.5) does not admit a closed-form solution and must be solved using numerical methods. As mentioned in Section 2.1, one of the most popular methods is the SSFM which we describe in more detail in the following. Conceptually, we start by discretizing the spatial dimension and subdividing the entire fiber of length L into small segments of length δ, where $M = L/\delta$ is the total number of segments. It is then assumed that in each segment the effects stemming from the two terms on the right-hand side of (2.5) can be separated. More precisely, for $\gamma = 0$, (2.5) is linear and has the frequency-domain solution

$$U(f, z) = H(f, z)U(f, 0), \tag{2.6}$$

where $H(f, z) = \exp(j2\beta_2 \pi^2 f^2 z)$ is the frequency response of a (z-dependent) CD filter. Applying the inverse Fourier transform to (2.6) leads to

$$u(t, z) = h(t, z) * u(t, 0), \tag{2.7}$$

where $h(t, z)$ is the impulse response of the CD filter and $*$ denotes convolution. Note that $|H(f, z)| = 1$, i.e., CD manifests itself as a unit-gain all-pass filter. For $\beta_2 = 0$ on the other hand, one may verify that the solution is

$$u(t, z) = \sigma_z(u(t, 0)) \triangleq u(t, 0)e^{j\gamma z|u(t,0)|^2}, \tag{2.8}$$

where the nonlinear function $\sigma_z : \mathbb{C} \to \mathbb{C}$ has been defined implicitly. Thus, the nonlinear Kerr effect by itself induces a nonlinear phase shift of the signal that is proportional to the instantaneous signal power $|u(t, 0)|^2$. For the SSFM, it is then assumed that an approximate solution for each segment is given by sequentially applying (2.7) and (2.8). Denoting the input signal by $s(t) \triangleq u(t, 0)$ and the output signal by $r(t) \triangleq u(t, Z)$, the entire SSFM can then be written as

$$r(t) \approx \sigma_\delta(h(t, \delta) * \cdots * \sigma_\delta(h(t, \delta) * \sigma_\delta(h(t, \delta) * s(t)))). \tag{2.9}$$

It has been shown that the SSFM converges to the true solution $r(t)$ as $M \to \infty$ [4, p. 42]. The name of the method originates from the fact that the nonlinear phase shift operation and the linear filtering are commonly carried out in the time and frequency domain, respectively. Therefore, one forward and one inverse Fourier transform have to be performed per segment.

The mathematical structure that arises from the spatial discretization of the NLSE in (2.9) is very similar compared to that of a conventional DNN with multiple layers. In both cases, the input signal is processed by alternating linear steps and element-wise nonlinearities. In an NN, the trainable parameters are the weights and biases corresponding to the linear steps, whereas the nonlinearities are typically fixed (i.e., non-trainable) activation functions, e.g., a sigmoid or rectified linear unit. The idea is now to convert the SSFM (2.9) into a parameterized "NN-like" model that is then suitable for ML. To that end, we first note that the input signal $s(t)$ in practical implementations has to be time-discretized according to $s[k] \triangleq s(kT)$, where T is the sampling interval. Correspondingly, the CD filters in (2.9) have to be discretized as well. In fact, we are going to replace them with completely general (complex-valued) FIR filters, where the impulses responses are denoted by $h_1[k], h_2[k], \ldots, h_M[k]$. The resulting model is

$$f_\theta(s[k]) \triangleq \sigma(h_M[k] * \cdots * \sigma(h_2[k] * \sigma(h_1[k] * s[k]))), \tag{2.10}$$

where θ collects all tunable parameters (i.e., the filter coefficients of all M FIR filters) and we use $\sigma \triangleq \sigma_\delta$ as a shorthand notation for the fixed Kerr nonlinearity. Eq. (2.10) is essentially a parameterized version of the SSFM. A block diagram can be found in Fig. 2.8a. Applications of this model (e.g., how to train it) will be discussed further below. For now, the two key points can be summarized as follows:

1. Applying spatial discretization methods (e.g., the SSFM) to the NLSE (2.5) results in a multi-layer computation graph that alternates linear steps and element-wise nonlinearities.

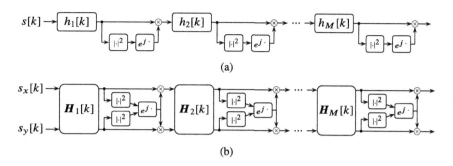

Figure 2.8 Block diagram of the ML models for (a) SP and (b) DP systems.

2. By parameterizing all linear steps and viewing them as general linear functions (similar to the weight matrices in an NN), one obtains the parameterized physics-based ML model (2.10).

In principle, we remark that it is also possible to parameterize the nonlinear steps as well. For example, this can be done by introducing scaling factors in the exponent in (2.8) or, more generally, trainable (real-valued) FIR filters that are applied to the squared magnitude signals prior to computing the nonlinear phase shift similar to filtered DBP or the enhanced SSFM [5,23].

2.3.3 Dual-polarization systems

The standard NLSE (2.5) and the resulting ML model (2.10) apply only to SP systems. In this section, we discuss the extension to DP systems, where we recall that $u_x(t, z)$ and $u_y(t, z)$ are the optical baseband signals in the x and y polarization, respectively. In general, the signal evolution is described by a set of coupled NLSEs that take into account the interactions between the two (degenerate) polarization modes. In birefringent fibers where the state of polarization changes rapidly along the link, an appropriate approximation is given by (2.1), which can be rewritten in vector form according to

$$\frac{\partial u(t, z)}{\partial z} = Du(t, z) + j\frac{8}{9}\gamma \|u(t, z)\|^2 u(t, z), \qquad (2.11)$$

where the signal is now described by a Jones vector $u(t, z) = (u_x(t, z), u_y(t, z))^\top$ and $\|u(t, z)\|^2 \triangleq |u_x(t, z)|^2 + |u_y(t, z)|^2$. The operator D in (2.11) is assumed to be a linear operator that models all linear transmission impairments including, for example, fiber loss and dispersive effects.

Similar to the standard NLSE, (2.11) does not have a general closed-form solution and must be solved numerically. Applying the SSFM in the same manner as before leads again to a feed-forward computation graph that alternates linear and nonlinear steps.

The nonlinear steps are now given by

$$\sigma_z(\boldsymbol{u}(t,z)) = \boldsymbol{u}(t,z)\exp\left(j\frac{8}{9}\gamma z\|\boldsymbol{u}(t,z)\|^2\right), \tag{2.12}$$

where the nonlinear phase shift is computed based on the instantaneous signal power in *both* polarizations. In other words, the nonlinearity in (2.12) does not act element-wise anymore but takes into account both elements of the Jones vector. As for the linear steps, the solution depends on which impairments are assumed to be modeled by the linear operator $\boldsymbol{D}$. In addition to fiber loss and CD, the signal may also experience polarization-dependent impairments such as polarization mode dispersion (PMD) or polarization-dependent loss. In general, the linear steps can always be approximated by (complex-valued) 2×2 MIMO-FIR filters in discrete time, assuming sufficiently long filter lengths.[2] Similar to before, we regard all coefficients of these MIMO-FIR filters as free parameters, where the corresponding impulse responses are denoted by $\boldsymbol{H}_1[k], \boldsymbol{H}_2[k], \ldots, \boldsymbol{H}_M[k]$. The resulting ML model can then be written as

$$f_\theta(\boldsymbol{s}[k]) = \sigma(\boldsymbol{H}_M[k] * \cdots \sigma(\boldsymbol{H}_2[k] * \sigma(\boldsymbol{H}_1[k] * \boldsymbol{s}[k]))), \tag{2.13}$$

where $\boldsymbol{s}[k] = (s_x[k], s_y[k])^\top$ is the time-discretized input signal and the tunable parameters collected in θ are now the filter coefficients of all MIMO-FIR filters. A block diagram of this model is shown in Fig. 2.8b.

In practice, one issue with using MIMO-FIR filters is the potentially high implementation complexity. On the other hand, it has been shown that the complexity of these filters can be significantly reduced by decomposing them into separate filters for each polarization, followed by memoryless rotation matrices. For more details, we refer the interested reader to [24] and [25].

2.3.4 Subband processing via filter banks

The models described in the previous two subsections can be further generalized through the use of subband processing. The general idea is to split the input signal of the model (i.e., either $s[k]$ or $\boldsymbol{s}[k]$) into N parallel signals using a filter bank. Subband processing can provide substantial complexity advantages compared to processing the input signal directly and has been studied extensively in the context of long-haul optical systems, see, e.g., [26,27].

A theoretical foundation for subband processing can be obtained by inserting the split-signal assumption $u(t,z) = \sum_{i=1}^N u_i(t,z)$ into the NLSE (2.1). This leads to a set of coupled equations which can then be solved numerically (the same approach can also be applied to DP systems). In [28], it is shown how an ML model can be obtained

---

[2] Alternatively, 4×4 real-valued MIMO-FIR filters can also be used.

based on the particular SSFM for coupled NLSEs proposed in [29]. This method is essentially equivalent to the standard SSFM for each subband, except that all sampled intensity signals are jointly processed with a (real-valued) MIMO filter prior to the non-linear phase rotation step. In other words, the resulting computation graph effectively alternates between (i) linear steps, where multiple filters are applied in parallel to each subband signal and (ii) modified nonlinear steps, where MIMO filters take into account some of the nonlinear interactions between subband signals. All linear filters including the MIMO filters can again be parameterized [28]. This gives an ML model whose structure resembles that of convolutional NNs commonly used in image processing.

2.3.5 Training and application examples

So far, we have mainly discussed how to obtain ML models from the NLSE (and its variants). In this section, we will provide some more details about how such models can actually be trained and applied in practice. Before we start, we remark that the models could in principle be applied either at the transmitter (as a pre-distortion algorithm) or at the receiver.[3] This is very similar to the NN-based approach discussed in Section 2.2. In the following, we assume that the models will be applied at the receiver side as a nonlinear equalization algorithm, similar to most works on DBP.

As mentioned in Section 2.1, standard DBP exploits the fact that the NLSE is invertible. This means that the transmitted signal can be recovered by solving the NLSE in the reverse propagation direction, using the received signal as a boundary condition. Rather than applying the SSFM for this purpose, the philosophy behind the considered ML models is to regard the standard SSFM merely as a "rough" blueprint for a non-linear equalizer. The inherent model parameterization then allows this blueprint to be "polished" by jointly training all model parameters θ using a large-scale data-driven optimization. This approach is referred to as learned DBP (LDBP) in [19]. To that end, any of the particular models discussed in Secs. 2.3.2, 2.3.3, or 2.3.4 can be integrated into a receiver DSP chain for long-haul fiber transmission, where the model input corresponds to the received and sampled optical waveform. The model output can be further processed by other DSP blocks such as a matched filter.

The training process itself is fairly standard and in fact quite similar to that of training a conventional NN. For example, as a loss function the MSE between the transmitted and equalized data symbols can be used, similar to (2.4). In [19], the Adam optimizer [30] is used for the gradient-based parameter optimization, although other off-the-shelf deep-learning optimizers can also be applied [25]. In general, it is possible to implement all models in standard deep-learning frameworks such as TensorFlow or PyTorch to automatize the computation of gradients with respect to the model parameters θ.[4]

---

[3] Additionally, the models could also serve as the basis for a refined channel model, where the training would be based on experimental data traces.

[4] TensorFlow source code can be found for example at www.github.com/chaeger/LDBP.

At first glance, it might not be obvious how a "parameterized SSFM" could be useful in practice. After all, the standard SSFM can solve the NLSE and perfectly equalize the nonlinear fiber channel. However, there are (at least) two caveats with this reasoning. First, *perfect* channel inversion requires essentially *unlimited* complexity. On the other hand, given a fixed complexity budget, it becomes highly nontrivial to find the best nonlinear equalizer. Second, it is known that in the presence of noise caused by, e.g., optical amplifiers, perfect channel inversion (also known as zero-forcing equalization) may no longer be optimal. In the following, we discuss these issues in some more detail by highlighting three possible use cases of the above ML approach.

1. First, it is worth pointing out that several problems that have been previously studied for the standard SSFM and DBP can be regarded as special cases of the ML approach. For example, it is well known that non-uniform step sizes can improve accuracy [31], which leads to the problem of finding an optimized step-size distribution. Another common problem is to properly adjust the nonlinearity parameter in each step (i.e., using γ' instead of the actual fiber parameter γ) which is sometimes referred to as "placement" optimization of the nonlinear operator [5]. Since the above models implement general linear steps, a step-size optimization can be thought of as being implicitly performed. The same is true also for the nonlinear operator placement because it can be shown that adjusting the nonlinearity parameters is equivalent to rescaling the filter coefficients in the linear steps, thereby adjusting the filter gain [19].

2. Second, the use of ML can lead to significant advantages in terms of complexity compared to traditional hand-designed approaches. Indeed, a major issue with DBP is the large computational burden and reducing the complexity is an important challenge. Approaching this problem from an ML perspective allows for the use of *model compression* techniques. One such technique is called pruning, whereby unimportant parameters are identified and removed from the model [32,33]. This is typically done in an iterative fashion throughout the entire training. Pruning is commonly used to compress a large model (e.g., an image classifier) in order to facilitate deployment (e.g., on mobile devices). In the context of LDBP, it has been shown that the FIR filters in the linear steps can be pruned to remarkably short lengths—as few as 3 symmetric taps/step—without sacrificing performance [19]. In turn, the use of very short FIR filters facilitates hardware-efficient time-domain implementations [34]. As an example, Fig. 2.9 shows the model performance for an SP system as a function of the overall impulse response length which is defined as the length of the filter obtained by convolving all M filters in the model. For each line, points closer to the left correspond to shorter impulses responses, i.e., more aggressive pruning. It can be seen that the impulse response lengths can be close to the theoretically expected minimum lengths according to the memory introduced by CD (indicated by the dashed vertical lines, cf. [19, (15)]). This is a major

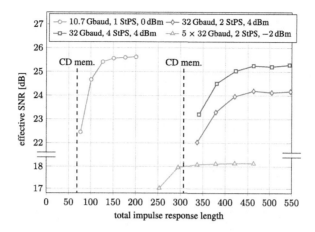

Figure 2.9 Performance in terms of the effective signal-to-noise ratio (SNR) as a function of the total impulse response length of the model (defined as the impulse response length of the filter obtained by convolving all filters in each of the M steps) [19]. The transmission link consists of 25 × 80 km and 10 × 100 km spans for the 10.7 Gbaud and 32 Gbaud systems, respectively.

improvement compared to previous work based on a hand-designed filter-design methodology where the filter lengths in DBP are typically much longer than the CD memory, sometimes by orders of magnitude.

3. Lastly, in the presence of noise, it is not clear if the linear steps in DBP should actually just invert the linear transmission impairments such as CD. Note that the ML models can in principle learn and apply arbitrary filter shapes to the signal in a distributed fashion. In that regard, it is shown in [25] that quite interesting filter shapes can emerge during the training. Importantly, these shapes may deviate significantly from the frequency-flat all-pass response that one might expect if the linear model steps were just inverting CD. A possible explanation for this observation is also provided in [25]. In particular, the authors argue that the learned filter coefficients try to strike a balance between inverting the channel and minimizing additional nonlinear phase-noise distortions due to noise-corrupted signal power levels. This shows that the adoption of ML in the context of DBP can even help to better analyze the nonlinear noise statistics involved in the DBP steps. Note that [25] also conducts an extensive experimental investigation of the discussed approach. Other experimental demonstrations can be found in [35] and [36].

2.4. Outlook of machine learning in long-haul systems

The ML approaches demonstrated in the previous Sections 2.2 and 2.3 are specific examples that illustrate how to apply feed-forward DNNs and other deep-learning tools to compensate for intra-channel fiber nonlinearity. In the literature, there are many previous and on-going efforts to apply other ML models to the NLC field [37–44]. For example, a K-means and clustering algorithm was implemented in a real-time field-programmable gate-array to demonstrate up to 3 dB Q-factor over linear equalization [42]. An NLC algorithm based on a recurrent NN architecture in [43] is capable of reducing the complexity of DNN-based NLC by ~47% without sacrificing performance. A recent paper shows the application of DNN to improve the performance gain of a turbo equalizer to mitigate fiber nonlinearity [44].

Although many success demonstration of various ML models has been applied in the NLC field in long-haul system, the majority of them are still limited to compensate for intra-channel fiber nonlinearity. The challenge still remains to be very difficult when applying these ML models to other channel impairments, in particular inter-channel nonlinearity such as cross-phase modulation. In [24], a more practical ML model is proposed to include PMD when performing ML-based NLC. In addition, another hurdle of applying these ML models in a real-time application-specific integrated circuit (ASIC) chip is its computation complexity considering the limited resources in ASIC chips. Thanks to the technology advancement in semiconductor manufacturing, beyond 3 nm process seems to be achievable in the near future [45,46] so that ML-based algorithm is possible to be implemented in a real-time ASIC chip. The turning point will be that ML algorithm can significantly outperform existing DSP algorithms at an affordable computational complexity.

References

[1] K. Roberts, Q. Zhuge, I. Monga, S. Gareau, C. Laperle, Beyond 100Gb/s: capacity, flexibility, and network optimization, J. Opt. Commun. Netw. 9 (4) (Apr. 2017) C12–C24.
[2] S. Zhang, F. Yaman, Y.-K. Huang, J.D. Downie, X. Sun, A. Zakharian, R. Khrapko, W. Wood, I.B. Djordjevic, E. Mateo, Y. Inada, 50.962Tb/s over 11185 km bi-directional C+L transmission using optimized 32QAM, in: Conference on Lasers and Electro-Optics, Optical Society of America, 2017, p. JTh5A.9.
[3] R.-J. Essiambre, G. Kramer, P.J. Winzer, G.J. Foschini, B. Goebel, Capacity limits of optical fiber networks, J. Lightwave Technol. 28 (4) (Feb. 2010) 662–701.
[4] G.P. Agrawal, Nonlinear Fiber Optics, Academic Press, 2007.
[5] L.B. Du, A.J. Lowery, Improved single channel backpropagation for intra-channel fiber nonlinearity compensation in long-haul optical communication systems, Opt. Express 18 (16) (Aug. 2010) 17075–17088.
[6] Y. Gao, J.H. Ke, K.P. Zhong, J.C. Cartledge, S.S.H. Yam, Assessment of intrachannel nonlinear compensation for 112 Gb/s dual-polarization 16QAM systems, J. Lightwave Technol. 30 (24) (Dec. 2012) 3902–3910.
[7] Z. Tao, L. Dou, W. Yan, L. Li, T. Hoshida, J.C. Rasmussen, Multiplier-free intrachannel nonlinearity compensating algorithm operating at symbol rate, J. Lightwave Technol. 29 (17) (Sept. 2011) 2570–2576.
[8] A. Ghazisaeidi, R. Essiambre, Calculation of coefficients of perturbative nonlinear pre-compensation for Nyquist pulses, in: Proc. European Conf. on Optical Communication (ECOC), Sept. 2014, pp. 1–3.
[9] A. Ghazisaeidi, I.F. de Jauregui Ruiz, L. Schmalen, P. Tran, C. Simonneau, E. Awwad, B. Uscumlic, P. Brindel, G. Charlet, Submarine transmission systems using digital nonlinear compensation and adaptive rate forward error correction, J. Lightwave Technol. 34 (8) (Apr. 2016) 1886–1895.

[10] Y. Gao, J.C. Cartledge, A.S. Karar, S.S.-H. Yam, M. O'Sullivan, C. Laperle, A. Borowiec, K. Roberts, Reducing the complexity of perturbation based nonlinearity pre-compensation using symmetric EDC and pulse shaping, Opt. Express 22 (2) (Jan. 2014) 1209–1219.

[11] J.C. Cartledge, F.P. Guiomar, F.R. Kschischang, G. Liga, M.P. Yankov, Digital signal processing for fiber nonlinearities [invited], Opt. Express 25 (3) (2017) 1916–1936.

[12] A. Mecozzi, C.B. Clausen, M. Shtaif, Analysis of intrachannel nonlinear effects in highly dispersed optical pulse transmission, IEEE Photonics Technol. Lett. 12 (4) (Apr. 2000) 392–394.

[13] S. Zhang, F. Yaman, K. Nakamura, T. Inoue, V. Kamalov, L. Jovanovski, V. Vusirikala, E. Mateo, Y. Inada, T. Wang, Field and lab experimental demonstration of nonlinear impairment compensation using neural networks, Nat. Commun. 10 (1) (2019) 1–8.

[14] H.-T. Cheng, L. Koc, J. Harmsen, T. Shaked, T. Chandra, H. Aradhye, G. Anderson, G. Corrado, W. Chai, M. Ispir, R. Anil, Z. Haque, L. Hong, V. Jain, X. Liu, H. Shah, Wide & deep learning for recommender systems, 2016.

[15] A. Géron, Hands-on Machine Learning with Scikit-Learn, Keras, and TensorFlow: Concepts, Tools, and Techniques to Build Intelligent Systems, O'Reilly Media, 2019.

[16] K. Hornik, M. Stinchcombe, H. White, Multilayer feedforward networks are universal approximators, Neural Netw. 2 (5) (1989) 359–366.

[17] E. Nachmani, E. Marciano, L. Lugosch, W.J. Gross, D. Burshtein, Y. Be'ery, Deep learning methods for improved decoding of linear codes, IEEE J. Sel. Top. Signal Process. 12 (1) (2018) 119–131.

[18] A. Balatsoukas-Stimming, C. Studer, Deep unfolding for communications systems: a survey and some new directions, in: Proc. IEEE Int. Workshop Signal Processing Systems (SiPS), Nanjing, China, 2019.

[19] C. Häger, H.D. Pfister, Physics-based deep learning for fiber-optic communication systems, IEEE J. Sel. Areas Commun. 39 (1) (2021) 280–294.

[20] C. Häger, H.D. Pfister, Nonlinear interference mitigation via deep neural networks, in: Proc. Optical Fiber Communication Conf. (OFC), San Diego, CA, 2018.

[21] H.W. Lin, M. Tegmark, D. Rolnick, Why does deep and cheap learning work so well?, J. Stat. Phys. 168 (6) (2017) 1223–1247.

[22] M. Raissi, P. Perdikaris, G.E. Karniadakis, Physics-informed neural networks: a deep learning framework for solving forward and inverse problems involving nonlinear partial differential equations, J. Comput. Phys. 378 (2019) 686–707.

[23] M. Secondini, S. Rommel, G. Meloni, F. Fresi, E. Forestieri, L. Poti, Single-step digital backpropagation for nonlinearity mitigation, Photonic Netw. Commun. 31 (3) (2016) 493–502.

[24] R.M. Bütler, C. Häger, H.D. Pfister, G. Liga, A. Alvarado, Model-based machine learning for joint digital backpropagation and PMD compensation, J. Lightwave Technol. (2020).

[25] Q. Fan, G. Zhou, T. Gui, C. Lu, A.P.T. Lau, Advancing theoretical understanding and practical performance of signal processing for nonlinear optical communications through machine learning, Nat. Commun. 11 (1) (2020) 3694–3704.

[26] M.G. Taylor, Compact digital dispersion compensation algorithms, in: Proc. Optical Fiber Communication Conf. (OFC), San Diego, CA, 2008.

[27] K.-P. Ho, Subband equaliser for chromatic dispersion of optical fibre, Electron. Lett. 45 (24) (2009) 1224–1226.

[28] C. Häger, H.D. Pfister, Wideband time-domain digital backpropagation via subband processing and deep learning, in: Proc. European Conf. Optical Communication (ECOC), Rome, Italy, 2018.

[29] J. Leibrich, W. Rosenkranz, Efficient numerical simulation of multichannel WDM transmission systems limited by XPM, IEEE Photonics Technol. Lett. 15 (3) (2003) 395–397.

[30] D.P. Kingma, J. Ba, Adam: a method for stochastic optimization, in: Proc. Int. Conf. Learning Representations (ICLR), San Diego, CA, 2015.

[31] G. Bosco, A. Carena, V. Curri, R. Gaudino, P. Poggiolini, S. Benedetto, Suppression of spurious tones induced by the split-step method in fiber systems simulation, IEEE Photonics Technol. Lett. 12 (5) (2000) 489–491.

[32] Y. Lecun, J.S. Denker, S.A. Solla, Optimal brain damage, in: Proc. Advances in Neural Information Processing Systems (NIPS), Denver, CO, 1989.

[33] S. Han, H. Mao, W.J. Dally, Deep compression: compressing deep neural networks with pruning, trained quantization and Huffman coding, in: Proc. Int. Conf. Learning Representations (ICLR), San Juan, Puerto Rico, 2016.

[34] C. Fougstedt, C. Häger, L. Svensson, H.D. Pfister, P. Larsson-Edefors, ASIC implementation of time-domain digital backpropagation with deep-learned chromatic dispersion filters, in: Proc. European Conf. Optical Communication (ECOC), Rome, Italy, 2018.

[35] V. Oliari, S. Goossens, C. Häger, G. Liga, R.M. Bütler, M. van den Hout, S. van der Heide, H.D. Pfister, C. Okonkwo, A. Alvarado, Revisiting efficient multi-step nonlinearity compensation with machine learning: an experimental demonstration, J. Lightwave Technol. 38 (12) (2020) 3114–3124.

[36] B.I. Bitachon, A. Ghazisaeidi, M. Eppenberger, B. Baeurle, M. Ayata, J. Leuthold, Deep learning based digital backpropagation demonstrating SNR gain at low complexity in a 1200 km transmission link, Opt. Express 28 (20) (2020) 29318–29334.

[37] T.S.R. Shen, A.P.T. Lau, Fiber nonlinearity compensation using extreme learning machine for DSP-based coherent communication systems, in: Proc. Optoelectronics and Communications Conf. (OECC), Kaohsiung, Taiwan, 2011.

[38] A.M. Jarajreh, E. Giacoumidis, I. Aldaya, S.T. Le, A. Tsokanos, Z. Ghassemlooy, N.J. Doran, Artificial neural network nonlinear equalizer for coherent optical OFDM, IEEE Photonics Technol. Lett. 27 (4) (2015) 387–390.

[39] D. Wang, M. Zhang, Z. Li, C. Song, M. Fu, J. Li, X. Chen, System impairment compensation in coherent optical communications by using a bio-inspired detector based on artificial neural network and genetic algorithm, Opt. Commun. 399 (2017) 1–12.

[40] O. Sidelnikov, A. Redyuk, S. Sygletos, Equalization performance and complexity analysis of dynamic deep neural networks in long haul transmission systems, Opt. Express 26 (25) (2018) 32765–32776.

[41] S. Deligiannidis, A. Bogris, C. Mesaritakis, Y. Kopsinis, Compensation of fiber nonlinearities in digital coherent systems leveraging long short-term memory neural networks, J. Lightwave Technol. 38 (21) (2020) 5991–5999.

[42] E. Giacoumidis, Y. Lin, M. Blott, L.P. Barry, Real-time machine learning based fiber-induced non-linearity compensation in energy-efficient coherent optical networks, APL Photonics 5 (4) (2020) 041301.

[43] Y. Zhao, X. Chen, T. Yang, L. Wang, D. Wang, Z. Zhang, S. Shi, Low-complexity fiber nonlinearity impairments compensation enabled by simple recurrent neural network with time memory, IEEE Access 8 (2020) 160995–161004.

[44] T. Koike-Akino, Y. Wang, D.S. Millar, K. Kojima, K. Parsons, Neural turbo equalization: deep learning for fiber-optic nonlinearity compensation, J. Lightwave Technol. 38 (11) (2020) 3059–3066.

[45] H. Lee, L.-E. Yu, S.-W. Ryu, J.-W. Han, K. Jeon, D.-Y. Jang, K.-H. Kim, J. Lee, J.-H. Kim, S. Jeon, et al., Sub-5nm all-around gate finfet for ultimate scaling, in: 2006 Symposium on VLSI Technology, 2006. Digest of Technical Papers, IEEE, 2006, pp. 58–59.

[46] Samsung: 3nm process is one year ahead of TSMC in GAA and three years ahead of intel, https://www.elinfor.com/news/samsung-3nm-process-is-one-year-ahead-of-tsmc-in-gaa-and-three-years-ahead-of-intel-p-11201. (Accessed 30 December 2020).

CHAPTER THREE

Machine learning for short reach optical fiber systems

Boris Karanov[a], Polina Bayvel[b], and Laurent Schmalen[c]
[a]Eindhoven University of Technology, Eindhoven, The Netherlands
[b]University College London, London, United Kingdom
[c]Karlsruhe Institute of Technology, Karlsruhe, Germany

3.1. Introduction to optical systems for short reach

Short reach optical fiber communications are integral in modern data center, metro and access networks. Such systems benefit from the simplicity and cost-effectiveness of a signal transmission scheme based on intensity modulation with direct detection (IM/DD) [1–3]. The links typically consist of a single span of fiber, avoiding the associated costs of deploying in-line optical amplification. The IM/DD technology enables the utilization of inexpensive electronic and optical components that have a small footprint and low power consumption. For this reason, it is considered as a prime candidate for *fiber to the home* (FTTH) systems, for instance realized via passive optical networks (PON).

Fig. 3.1 shows a general schematic of an optical IM/DD system, highlighting the parts of the transmission link where digital-to-analog/analog-to-digital, electro-optical/opto-electrical conversions and optical fiber propagation are performed. The optical IM/DD communication is mainly characterized by the presence of fiber chromatic dispersion, which introduces inter-symbol interference (ISI), and non-linear (square-law) opto-electrical conversion by a single photo-diode (PD). Moreover, due to the hardware imperfections and bandwidth limitations in the low-cost components, the signal is corrupted by a significant amount of noise and non-linearity stemming from multiple sources at both transmitter and receiver, i.e. modulators, electrical amplification circuits as well as low resolution digital-to-analog and analog-to-digital converters. Meeting the ever-increasing data rate demands for such systems becomes quite a challenging task because of the limitations imposed by these impairments. For this reason, the application of advanced DSP algorithms for IM/DD has attracted great interest in recent years [4–6].

In particular, after DSP at the transmitter, the information-carrying electrical signal $x(t)$ is obtained using a digital-to-analog converter (DAC). For the electro-optical conversion stage, a Mach-Zehnder modulator (MZM) is often employed to externally modulate the amplitude of an optical carrier wave, generated by a 1550 nm laser source.

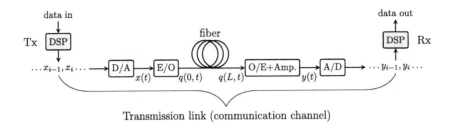

Figure 3.1 Schematic of an IM/DD communication system, indicating the signal transformations across the transmission link. Tx - transmitter, DSP - digital signal processing, D/A - digital-to-analog conversion, E/O - electro-optical conversion, O/E - opto-electrical conversion, Amp. - amplification, A/D - analog-to-digital conversion, Rx - receiver.

The MZM is driven by $x(t)$. This scheme is referred to as *Intensity Modulation* (IM) since varying the amplitude of the laser field, results in modulation of the optical intensity at the fiber input.

The propagation of the optical signal $q(z, t)$ through the fiber medium is characterized by the presence of fiber attenuation, chromatic dispersion and Kerr non-linearity effects. Typically, for IM/DD systems the intensity at the fiber input is small and subsequently attenuated by the medium with no in-line signal amplification. For this reason, the impact of the Kerr non-linearity during propagation along the fiber can be neglected [1]. As a consequence, chromatic dispersion is the dominant distortion effect for IM/DD transmission systems. In particular, it induces a propagation delay between the different components in the frequency content of the propagating pulses, thus causing their temporal broadening. The effect accumulates linearly with distance and quadratically with signal bandwidth. As a result, a pulse will interact and interfere with other *preceding* and *succeeding* pulses during fiber propagation. Such a distortion leads to inter-symbol interference (ISI), introducing *memory* in the communication channel.

The signal $q(z, t)$ at the output of a fiber span of length z is converted from the optical back to the electrical domain by a photo-detector. In IM/DD systems, a simple *p-i-n photo-diode* is used to perform a so-called *square-law* detection of the incoming optical signal, where the photo-current output of the PD is proportional to the intensity of the impinging optical field. Such a non-linear signal reception using a single PD is called *Direct Detection* (DD).

The combination of ISI and non-linear detection imposes severe limitations on the achievable data rates and transmission distances in optical fiber links based on IM/DD. The effects render the short reach IM/DD links a non-linear communication channel with memory – one for which optimal, computationally feasible algorithms are not available. Modern IM/DD systems employ state-of-the-art DSP techniques such as non-linear Volterra equalization [2,5] or maximum likelihood sequence detection [7–9] to address these limitations. However, as the channel memory increases, the implementation of such algorithms quickly becomes infeasible because of the associated computational complexity. Moreover, their performance can be significantly degraded when the complexity is constrained. On the other hand, it was recently shown that artificial neural networks (ANN) and deep learning, approximating complex non-linear functions, are able to provide carefully optimized DSP functions for the short reach fiber links, which have reduced complexity or improved performance over the classical approaches [10–14].

3.2. Deep learning approaches for digital signal processing

State-of-the-art communication systems rely on advanced DSP algorithms to enable the compensation of wide variety of transmission impairments. As a result, DSP has become an effective, ubiquitously employed solution for increasing the data rate and transmission reach [15,16]. Within the scope of communication systems engineering, artificial neural networks and deep learning are considered particularly suitable for DSP applications [17–19].

ANNs are formed by interconnected processing elements called artificial *neurons*, organized into multiple layers with non-linear relation between them [20]. Using a collection of powerful optimization techniques, such computational systems can be tuned to approximate complex non-linear functions — a process which is often referred to as *deep learning*. Deep learning is a *data-driven* optimization approach which relies on the availability of a large set of examples, representative of the function's behavior. It combines well-known algorithms such as back-propagation [21] and gradient descent [22,23] with modern adaptive learning techniques, specifically developed for multi-layer (deep) ANNs [24,25].

Conventionally, the communication transceiver consists of several DSP blocks, each employed to perform an individual task. Fig. 3.2 shows a basic schematic representation of such a system. At the transmitter, signal processing functions such as coding, modulation, pre-compensation and pulse-shaping are applied to the incoming data, shaping the signal before it is launched into the transmission link. At the receiver, the distorted by the communication channel signal is equalized before demodulation and decoding are performed to recover the transmitted data. From an engineering perspective, such a modular transceiver implementation is convenient as it allows separate analysis, optimization and control of the DSP blocks. Deep learning techniques have often been considered for the optimization or the computational complexity reduction of a specific function within the transceiver DSP chain with first examples from the early 1990s for satellite communications [26,27] as well as more recently for other wireless and optical fiber links [28–38]. Within the framework of optical fiber communications for short reach, advanced non-linear equalization schemes using ANN processing of multiple received symbols have been proposed [11,39–44].

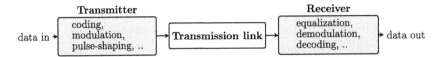

Figure 3.2 Basic schematic of a conventional communication system design, showing some of the main signal processing modules at the transmitter and the receiver.

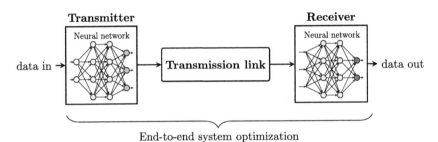

Figure 3.3 Communication system implemented as an end-to-end deep artificial neural network and optimized in a single process.

In all aforementioned examples the machine learning techniques are applied for the optimization of a specific DSP function in the transmitter or receiver. A different approach to deep learning-based DSP, which utilizes the function approximation capabilities of artificial neural networks and deep learning to a greater extent, is to interpret the complete communication chain from the transmitter input to the receiver output as a single deep ANN and optimize its parameters in an end-to-end process. The idea of such fully learnable communication transceivers (known as *auto-encoders*) was introduced for the first time for wireless communications in [17,45,46]. The auto-encoders avoid the modular design of conventional communication systems and have the potential to achieve an end-to-end optimized performance for a specific metric. Fig. 3.3 shows a simple schematic of such an auto-encoder system. The application of end-to-end deep learning is particularly viable in communication scenarios where the optimum pair of transmitter and receiver or optimum DSP modules are not available, for example due to complexity constraints. The approach was quickly utilized for exploiting to a greater extent the potential for data transmission in optical fiber communications [47–50]. In particular, end-to-end deep learning was highlighted as a suitable DSP solution to enhance the system performance at reduced complexity in the IM/DD optical fiber links for short reach applications. The first experimental demonstration of an optical fiber auto-encoder showed that a simple feed-forward neural network (FFNN) design can outperform the conventional IM/DD pulse amplitude modulation (PAM) with linear equalizers [47,48]. Subsequently, an advanced transceiver architecture based on recurrent neural networks (RNN) achieved a substantial performance improvement over the

FFNN design [13,51]. Moreover, it was shown that, when complexity is constrained, the RNN-based auto-encoder can achieve an improved performance compared to PAM transmission with state-of-the-art non-linear receivers based on ANNs [51], classical Volterra equalization [13] or maximum likelihood sequence detection (MLSD) [10].

The rest of this chapter is organized as follows: Section 3.3 introduces the design of ANN-based receivers for the conventional PAM transmission as well as two different auto-encoder designs – the low complexity system based on a simple feed-forward ANN as well as the advanced design using recurrent neural networks for non-linear processing of data sequences. The performance and complexity of these systems is discussed in Section 3.3.3. Moreover, Section 3.3.4 provides a review of an ANN optimization approach, first proposed in [47], which allows transmission over varied distances without reconfiguration of the ANN transceivers – thus enabling efficient and versatile short reach fiber links. Finally, Section 3.4 explains the approaches for optimizing and implementing the ANN-based transmitter and receiver on the actual physical transmission link.

3.3. Optical IM/DD systems based on deep learning

3.3.1 ANN receiver

3.3.1.1 PAM transmission

Conventional optical IM/DD transmission schemes are based on the pulse amplitude modulation technique, which consists in encoding the input data using a finite set of amplitude levels [2]. In particular, two- and four-level PAM (PAM-2/PAM-4) are the predominantly employed formats due to their higher robustness to transmission distortions. At the transmitter, the stream of PAM symbols is pulse-shaped (including up-sampling) typically using a raised cosine filter. The signal is applied to the DAC, obtaining the information-carrying electrical signal $x(t)$ (see Fig. 3.1) which drives the modulator. After propagation through the fiber and direct-detection at the receiver, the signal is sampled before DSP is used for equalization and subsequent recovery of the transmitted data. Due to the effects of the non-linear dispersive channel, IM/DD systems require advanced DSP algorithms which process multiple preceding and succeeding symbols simultaneously. This section describes a general deep learning-based approach for designing such a digital IM/DD receiver.

3.3.1.2 Sliding window FFNN processing

A feed-forward neural network can be conveniently used for the detection of M-level PAM symbols, taking multiple received samples into account [11,14,51]. Fig. 3.4 shows a schematic of such a system. The FFNN is employed in a sliding window scheme, such that at time t it processes the sub-sequence of received PAM symbols $(\mathbf{y}_{t-\frac{W-1}{2}}, \ldots, \mathbf{y}_t, \ldots, \mathbf{y}_{t+\frac{W-1}{2}})$. Note that the processing memory of the algorithm W

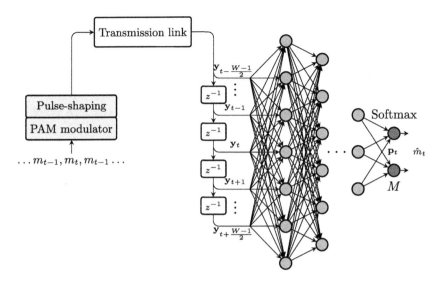

Figure 3.4 Schematic of a sliding window FFNN receiver for PAM symbols.

dictates the size of the input FFNN layer. The signal is further transformed by multiple hidden layers, employing standard activation functions such as ReLU. To achieve a machine learning estimator, the output of the receiver is a probability vector $\mathbf{p}_t \in \mathbb{R}^M$, obtained using a *softmax* layer. The dimensionality M of the output vector is dictated by the number of modulation levels, e.g. $\mathbf{p}_t \in \mathbb{R}^2$ for PAM-2 or $\mathbf{p}_t \in \mathbb{R}^4$ for PAM-4 transmission. The probability vector $\mathbf{p}_t$ is then utilized in two ways: During parameter optimization, it is used for computing the loss of the receiver ANN. The loss at a time instance t is obtained as

$$\mathcal{L}(\theta) = \ell(m_t, f_{\text{RX-FFNN}}((\mathbf{y}_{t-\frac{W-1}{2}}, \dots, \mathbf{y}_t, \dots, \mathbf{y}_{t+\frac{W-1}{2}}))), \tag{3.1}$$

where

$$\theta = \{\mathbf{W}_1, \mathbf{b}_1, \dots, \mathbf{W}_K, \mathbf{b}_K\} \tag{3.2}$$

denotes the set of trainable receiver neural network parameters (weights and biases), and the functions $f_{\text{RX-FFNN}}(\cdot)$ and $\ell(\cdot)$ are used to express the input-to-output mapping of the FFNN and the utilized loss function (e.g. cross entropy), respectively. Note that, without loss of generality each PAM symbol is represented by an integer message m_t from a set $\mathcal{M} = \{1; 2; \dots; M\}$ of $M = |\mathcal{M}|$ messages. Optimization is performed using stochastic gradient descent aimed at minimizing the average loss computed over a mini-batch of received messages. After optimization, the output probability vector $\mathbf{p}_t$ is utilized for making a decision on the received PAM symbol as

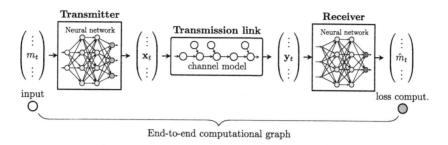

Figure 3.5 Schematic of the optical communication system implemented as an end-to-end computational graph, which represents the process of computing the loss between the transmitter input and the receiver output.

$$\hat{m}_t = \underset{i=1,2,\dots M}{\text{argmax}} (\mathbf{p}_{t,i}). \tag{3.3}$$

The FFNN processor slides by one symbol position ahead to estimate the next symbol $\mathbf{y}_{t+1}$ in the received sequence.

A similar deep learning approach for designing a PAM receiver for optical IM/DD communications was considered in [11,14,41,42]. It has also been investigated for applications in long-haul coherent fiber systems for example in [52].

3.3.2 Auto-encoders

Artificial neural networks and deep learning provide the framework to design the optical communication system by carrying out the optimization of the transmitter and receiver in a single end-to-end process. As explained in Sec. 3.1, such fully learnable communication systems are based on the idea of implementing the complete chain of transmitter, channel and receiver as an end-to-end deep artificial neural network [17,45,46]. Using deep learning, the transmitter and receiver sections are optimized jointly, such that a set of ANN parameters is obtained for which the end-to-end system performance is optimized for a specific metric. The resulting auto-encoder is particularly suitable in scenarios, such as short reach optical fiber communications, where the optimum transmitter-receiver pair is unknown or computationally prohibitive.

Fig. 3.5 shows a schematic of the approach. The transmitter ANN layers encode the sequence of random input messages ($\dots m_{t-1}, m_t, m_{t+1} \dots$) from a finite alphabet $\mathcal{M}$ into a sequence of symbols ($\dots \mathbf{x}_{t-1}, \mathbf{x}_t, \mathbf{x}_{t+1} \dots$), with each symbol $\mathbf{x}$ being a vector (block) of n digital *waveform* samples. The produced digital waveform is applied to the DAC and fed to the transmission link, where each sample acquires noise as well as interference from preceding and succeeding samples, a process that is governed by the adopted optical transmission link model. The output of the channel is distorted waveform samples, accordingly organized into the sequence of received symbols ($\dots \mathbf{y}_{t-1}, \mathbf{y}_t, \mathbf{y}_{t+1} \dots$). The

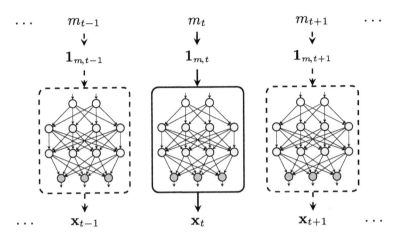

Figure 3.6 Schematic of the FFNN-based transmitter section of the optical fiber auto-encoder. The input messages $(\ldots m_{t-1}, m_t, m_{t+1} \ldots)$ are represented as one-hot vectors $(\ldots, \mathbf{1}_{m,t-1}, \mathbf{1}_{m,t}, \mathbf{1}_{m,t+1} \ldots)$, which are processed independently by the FFNN at each time instance to produce the encoded symbols (blocks of samples) $(\ldots \mathbf{x}_{t-1}, \mathbf{x}_t, \mathbf{x}_{t+1} \ldots)$, transmitted through the communication channel.

symbols are decoded by the receiver layers in order to obtain the sequence of recovered messages $(\ldots \hat{m}_{t-1}, \hat{m}_t, \hat{m}_{t+1} \ldots)$. During optimization, the (average) loss between the transmitted and received messages $\overline{\mathcal{L}} = \overline{\sum_t \ell(m_t, \hat{m}_t)}$ is computed. Using a differentiable channel model (see [53,54]), the complete communication system spanning from the transmitter input to the receiver output can be considered as an end-to-end computational graph. Exploiting the dependencies on the graph, the back-propagation algorithm is then readily applied to obtain a gradient of the loss with respect to every trainable parameter at both transmitter and receiver. The transceiver is optimized via gradient descent aimed at minimizing the loss, e.g. the message (symbol) error rate of the system.

3.3.2.1 Auto-encoder design based on a feed-forward neural network

As proposed in [47], a simple FFNN-based optical auto-encoder can be used to perform the encoding and decoding of messages. Fig. 3.6 shows a schematic of the encoding procedure at the *transmitter*. The transmitter encodes the sequence of input messages $(\ldots m_{t-1}, m_t, m_{t+1} \ldots)$ into a corresponding sequence of robust symbols (blocks of digital waveform samples) $(\ldots \mathbf{x}_{t-1}, \mathbf{x}_t, \mathbf{x}_{t+1} \ldots)$. Each message in the sequence $m \in \{1, 2, \ldots, M\}$ is independently chosen from a set of $|\mathcal{M}| = M$ total messages. It is encoded by the transmitter into a block $\mathbf{x}$ of n transmit samples. Feed-forward neural networks process each input independently and thus an FFNN architecture at the transmitter can be used to encode the input message m_t into a symbol $\mathbf{x}_t$ via a process that does not utilize any information from the encoding of m_{t-k} into $\mathbf{x}_{t-k}$ or the encoding of m_{t+k} into $\mathbf{x}_{t+k}$.

The encoding of messages by this network is performed in the following way: First, m_t is represented as a one-hot vector of size M, denoted as $\mathbf{1}_{m,t} \in \mathbb{R}^M$, where the m-th element equals 1 and the other elements are 0. The one-hot vector $\mathbf{1}_{m,t}$ is then processed by the hidden FFNN layers, whose dimensions are hyper-parameters of the system design. The final FFNN layer prepares the data for transmission and its output is the encoded block of waveform samples $\mathbf{x}_t \in \mathbb{R}^n$. Typically for optical systems based on IM/DD, a unipolar signaling is considered. Thus, the output of the final transmitter layer needs to be positive. Moreover, for practical purposes the output values need to be restricted within the operational range of the subsequent opto-electrical devices, forming the transmission link, for example using a modified clipping activation functions [47].

Every transmitted message m represents the equivalent of $\log_2(M)$ bits of information, since it is one of M equiprobable choices. The message is encoded into n transmit samples, forming a symbol. The system can thus be viewed as an auto–encoder with a coding rate of

$$R_{\text{cod.}} = \frac{\log_2(M)}{n} \quad [\text{bits/Sa}]. \tag{3.4}$$

Denoting the DAC sampling rate in the system as $R_{\text{samp.}}$ [Sa/s], the information rate of the auto–encoder becomes

$$R_{\text{inf.}} = R_{\text{cod.}} \cdot R_{\text{samp.}} \quad [\text{bits/s}]. \tag{3.5}$$

These two expressions show that the coding as well as the information rate of the communication system can be controlled by adjusting the dimensionality of the input and the final layers of the neural network. Increasing the size M of the input layer results in an increased information rate. In contrast, expanding the dimension n of the final layer reduces the number of bits per second transmitted through the link.

After propagation through the communication channel, the sequence of transmit symbols $(\ldots \mathbf{x}_{t-1}, \mathbf{x}_t, \mathbf{x}_{t+1} \ldots)$ is transformed into to the distorted sequence $(\ldots \mathbf{y}_{t-1}, \mathbf{y}_t, \mathbf{y}_{t+1} \ldots)$ which is processed by the receiver neural network. The decoding in the FFNN system design follows the block-by-block procedure employed at the transmitter. The message $\hat{m}_t$ is recovered from the received symbol $\mathbf{y}_t$ without taking into account any of the preceding $\mathbf{y}_{t-k}$ or the succeeding $\mathbf{y}_{t+k}$ symbols. This decoding scheme is depicted in Fig. 3.7. The received block of samples $\mathbf{y}_t$ is transformed to an output probability vector $\mathbf{p}_t$ from which the recovered message is deduced. The input symbols $\mathbf{y}_t \in \mathbb{R}^n$ are transformed by the hidden layers. The activation function at the final receiver layer is typically the *softmax* and thus the output is a probability vector $\mathbf{p}_t \in \mathbb{R}^M$ with the same dimension as the one-hot vector encoding of the message. The vector $\mathbf{p}_t$ is utilized in two ways: during the end-to-end optimization of the transceiver, the loss between transmitted and received message at time t is computed as

$$\mathcal{L}(\theta) = \ell(\mathbf{1}_{m,t}, f_{\text{AE-FFNN}}(\mathbf{1}_{m,t})), \tag{3.6}$$

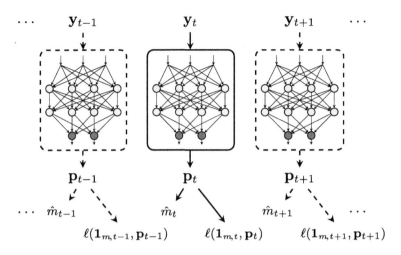

Figure 3.7 Schematic of the FFNN-based receiver section of the optical fiber auto-encoder. The received symbols $(\ldots, \mathbf{y}_{t-1}, \mathbf{y}_t, \mathbf{y}_{t+1} \ldots)$ are processed independently by the FFNN at each time instance to produce the output probability vectors $(\ldots \mathbf{p}_{t-1}, \mathbf{p}_t, \mathbf{p}_{t+1} \ldots)$, which are used to make a decision on the recovered message as well as to compute the loss of the auto-encoder in the optimization stage.

where θ is the set of all trainable neural network parameters in the system (transmitter and receiver) and $\ell(\cdot)$ is the utilized loss function (e.g. cross entropy). The function $f_{\text{AE-FFNN}}(\cdot)$ denotes the complete input–to–output mapping function of the FFNN-based auto–encoder, which can be expressed as

$$\mathbf{p}_t = f_{\text{FFNN-AE}}(\mathbf{1}_{m,t}) \triangleq f_{\text{Dec.-FFNN}}\left(\mathcal{H}\left\{\ldots, f_{\text{Enc.-FFNN}}\left(\mathbf{1}_{m,t}\right), \ldots\right\}\right), \quad (3.7)$$

with the operator $\mathcal{H}\{\cdot\}$ describing the IM/DD channel effects, and $f_{\text{Enc.-FFNN}}(\cdot)$ and $f_{\text{Dec.-FFNN}}(\cdot)$ denoting the encoding and decoding functions, respectively. Optimization can be performed using stochastic gradient descent and computing the average loss over a mini-batch of training messages. The goal is to find a set θ of trainable parameters from the end-to-end computational graph, i.e. the weight matrices and bias vectors at the transmitter and receiver, that effectively minimize the symbol (block) error rate of the system.

After optimization, the output probability vectors $\mathbf{p}_t$ are used for making a decision on the received symbols as

$$\hat{m}_t = \operatorname*{argmax}_{1,2,\ldots M}(\mathbf{p}_{t,i}). \quad (3.8)$$

Similarly to the receiver-only ANN-based system described in Section 3.3.1, the message decision consists in choosing the index $\hat{m}_t$ of the greatest element in $\mathbf{p}_t$ as the recovered message at time t. A symbol error occurs when $m_t \neq \hat{m}_t$.

3.3.2.2 Auto-encoder design based on a recurrent neural network

The dispersion-induced interference is a fiber propagation effect which accumulates linearly with distance and quadratically with the signal bandwidth. It can rapidly extend across multiple preceding and succeeding symbols in high-speed (e.g. multi-gigabit) IM/DD systems, which often transmit data over distances of up to 100 km [2,3]. The presence of inter-symbol interference (ISI) renders the optical IM/DD links a communication channel with memory.

Within the FFNN auto-encoder framework, the symbol memory of the channel is not exploited in the encoding and decoding functions since there is no connection in the processing of neighboring blocks. The FFNN-based auto-encoder encodes/decodes each symbol *independently* and is inherently unable to compensate for interference outside of the symbol block (i.e. ISI). As a consequence, the achievable performance of such systems, in terms of chromatic dispersion that can be compensated and hence transmission distance, is limited by the block size. A possible solution is expanding the FFNN by including more samples within a symbol block. However, in addition to increasing the output FFNN dimension n, for a given coding rate $R_{cod.}$ this approach would require an exponential increase in the input neural network dimension M, which is related to n as $M = 2^{n \cdot R_{cod.}}$. The result is an auto-encoder system that can quickly become computationally infeasible to implement due to the rapidly increased amount of transceiver ANN parameters.

The limitations of the FFNN-based auto-encoder for communication over channels with memory have been addressed by considering a deep learning-based transceiver tailored for the processing of data sequences using a recurrent neural network (RNN) [10, 51]. In particular, a bidirectional RNN (BRNN)-based auto-encoder allowed to utilize information from both pre- and post-cursor symbols in the encoding and decoding processes. As proposed in [51], the optimized system is combined with an efficient sliding window sequence estimation algorithm, which was initially developed for molecular communications [55]. This allows to control the processing memory in the RNN receiver via the estimation window W, external to the end-to-end neural network architecture. As a result, the resilience to ISI can be enhanced, while keeping the number of ANN parameters at the transceiver fixed [13]. Interestingly, the investigation in [10] showed that when complexity is constrained, the BRNN-based optical auto-encoder can outperform IM/DD systems based on the conventional PAM and maximum likelihood sequence detection.

The transceiver design uses a BRNN to adequately handle the inter-symbol interference effects stemming from both preceding and succeeding symbols, induced by the dispersive optical IM/DD channel. The function of the BRNN at the transmitter is to encode the stream of input messages $(\ldots, m_{t-1}, m_t, m_{t+1}, \ldots)$, $m_t \in \{1, 2, \ldots, M\}$, each of which drawn independently from an alphabet $\mathcal{M}$ of $|\mathcal{M}| = M$ total messages, into a sequence of transmit blocks of digital waveform samples (symbols) $(\ldots, \mathbf{x}_{t-1}, \mathbf{x}_t, \mathbf{x}_{t+1}, \ldots)$

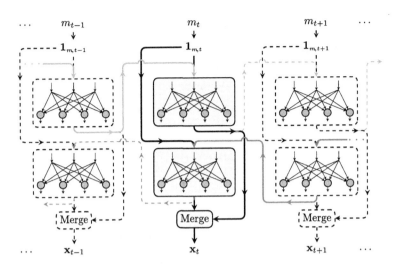

Figure 3.8 Schematic of the BRNN-based transmitter section of the optical fiber auto-encoder. The input messages $(\ldots m_{t-1}, m_t, m_{t+1} \ldots)$, represented as one-hot vectors $(\ldots, \mathbf{1}_{m,t-1}, \mathbf{1}_{m,t}, \mathbf{1}_{m,t+1} \ldots)$, are processed bidirectionally at each time instance by the neural network to produce the sequence of encoded symbols (blocks of samples) $(\ldots \mathbf{x}_{t-1}, \mathbf{x}_t, \mathbf{x}_{t+1} \ldots)$. Thick solid lines are used to highlight the connections to symbols that have an impact on the processing of m_t.

ready for transmission over the communication channel. Unlike the FFNN auto-encoder system, the BRNN transmitter structure utilizes information from both pre- and post-cursor symbols, i.e. $\mathbf{x}_{t-k}$ or $\mathbf{x}_{t+k}$, into the encoding of the message m_t at time t. Thus the BRNN transmitter performs encoding of input *sequences*, which is necessary for communication over communication channels with memory.

Fig. 3.8 shows a schematic of the BRNN encoding procedure. First, the messages m_t are represented as one-hot vectors $\mathbf{1}_{m,t} \in \mathbb{R}^M$, which are fed into the recurrent structure for bidirectional encoding. As a result of fiber dispersion, adjacent symbols impose stronger interference on the current symbol and thus simple recurrent processing based on concatenation between the current input and the previous output achieves good performance for the optical channel [51]. Nevertheless, it should be noted that advanced variants of the recurrent cell such as long short-term memory (LSTM) or gated recurrent units (GRU) have also been implemented within the discussed framework (see [51]).

The recurrent encoder processing in the backward and forward directions can be performed using identical network hyper-parameters and architectures. In the forward direction the input $\mathbf{1}_{m,t}$ at time t is concatenated with the previous output $\overrightarrow{\mathbf{x}}_{t-1}$, obtaining the vector $\left(\mathbf{1}_{m,t}^T \quad \overrightarrow{\mathbf{x}}_{t-1}^T \right)^T$, which is processed by the recurrent cell to produce the current updated output $\overrightarrow{\mathbf{x}}_t \in \mathbb{R}^n$. In the backward direction $\mathbf{1}_{m,t}$ is instead combined with $\overleftarrow{\mathbf{x}}_{t+1}$, yielding $\overleftarrow{\mathbf{x}}_t \in \mathbb{R}^n$. The two unidirectional outputs of the transmitter BRNN

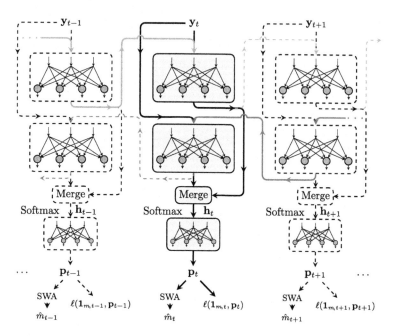

Figure 3.9 Schematic of the BRNN-based receiver section of the proposed optical fiber auto-encoder. The received symbols $(\ldots, \mathbf{y}_{t-1}, \mathbf{y}_t, \mathbf{y}_{t+1} \ldots)$ are processed bidirectionally by the BRNN to produce the output probability vectors $(\ldots \mathbf{p}_{t-1}, \mathbf{p}_t, \mathbf{p}_{t+1} \ldots)$. These are utilized in two ways: to compute the loss of the auto-encoder in the optimization stage as well as to make a decision on the recovered message via the sliding window algorithm (SWA) for sequence estimation.

$\overrightarrow{\mathbf{x}}_t$ and $\overleftarrow{\mathbf{x}}_t$ at time t are merged, e.g. via element-wise averaging, producing the encoded output $\mathbf{x}_t \in \mathbb{R}^n$. Note that, similarly to the previously presented FFNN design, an appropriately chosen activation function needs to be considered at the transmitter such that the digital waveform samples output of the ANN are within the operational range of the opto-electrical modules. It is worth mentioning that the BRNN transmitter encodes the *full* sequence of input messages. Thus, the forward/backward recurrent encoding introduces extra latency of the order of a full data sequence. By termination of the sequences, this latency can be limited to a practically manageable level. In contrast, by operating the optimized auto–encoder in a sliding window sequence estimation scheme, the processing delay at the receiver can be significantly reduced.

The function of the BRNN receiver is to decode the sequence of distorted symbols $(\ldots \mathbf{y}_{t-1}, \mathbf{y}_t, \mathbf{y}_{t+1} \ldots)$ into a sequence of recovered messages $(\ldots \hat{m}_{t-1}, \hat{m}_t, \hat{m}_{t+1} \ldots)$. This operation utilizes information from neighboring symbols. The BRNN decoding scheme is shown schematically in Fig. 3.9. It follows an identical recurrent procedure to the one used at the transmitter. The message $\hat{m}_t$ is recovered from the received symbol $\mathbf{y}_t$ considering both preceding $\mathbf{y}_{t-k}$ and succeeding $\mathbf{y}_{t+k}$ symbols.

In particular, the received block of samples $\mathbf{y}_t$ is transformed to an output probability vector $\mathbf{p}_t$ as follows: $\mathbf{y}_t$ is concatenated with the previous output in the forward direction $\overrightarrow{\mathbf{h}}_{t-1}$ and processed by the cell, producing $\overrightarrow{\mathbf{h}}_t$. Identically, in the backward direction $\mathbf{y}_t$ is concatenated with $\overleftarrow{\mathbf{h}}_{t+1}$ and transformed to $\overleftarrow{\mathbf{h}}_t$. The two outputs $\overrightarrow{\mathbf{h}}_t$ and $\overleftarrow{\mathbf{h}}_t$ are merged (e.g. via element-wise averaging or concatenation), yielding $\mathbf{h}_t$. Subsequently, this output is applied to a softmax layer in order to obtain the output probability vector $\mathbf{p}_t \in \mathbb{R}^M$, having the dimensionality of the input one-hot vector. The vector of probabilities $\mathbf{p}_t$ is utilized for loss computation and message recovery.

The loss computed at the time instance t is given by

$$\mathcal{L}(\boldsymbol{\theta}) = \ell\left(\mathbf{1}_{m,t}, f_{\text{BRNN-AE},t}(\dots \mathbf{1}_{m,t-1}, \mathbf{1}_{m,t}, \mathbf{1}_{m,t+1}\dots)\right), \tag{3.9}$$

where $\ell(\cdot)$ is the loss function (e.g. cross entropy), while $f_{\text{AE-BRNN},t}(\cdot)$ is used to denote the input-to-output mapping function of the BRNN-based auto-encoder at time instance t, which can be expressed as

$$\mathbf{p}_t = f_{\text{BRNN-AE},t}(\dots, \mathbf{1}_{m,t}, \dots) \triangleq f_{\text{Dec.-BRNN}}\left(\mathcal{H}\left\{f_{\text{Enc.-BRNN},t}(\dots, \mathbf{1}_{m,t}, \dots)\right\}\right), \tag{3.10}$$

with the operator $\mathcal{H}$ describing the effects of the optical IM/DD channel, and $f_{\text{Enc.-BRNN}}(\cdot)$ and $f_{\text{Dec.-BRNN}}(\cdot)$ denoting the encoding and decoding functions, respectively. Note that, in contrast to the FFNN auto-encoder, the transmitter and receiver functions take as an input a sequence of data.

The set of transmitter and receiver BRNN parameters (denoted by $\boldsymbol{\theta}$) is optimized using SGD with an objective to minimize the average loss over mini-batches of messages from the training data-set.

After the optimization, the BRNN transceiver is combined with a sliding window algorithm (SWA) for sequence estimation, where the probability vector outputs $\mathbf{p}_t$ are utilized for making a decision on the transmitted messages. The algorithm was first proposed in [55,56] for the detection of sequences in molecular communication systems. Fig. 3.10 shows a basic schematic of the SWA applied on the optimized system. The auto-encoder is represented by the modules *Tx BRNN*, *channel* and *Rx BRNN*. For a given sequence of $T + W - 1$ test messages, the transmitter BRNN encodes the *full* stream of input one-hot vectors $\mathbf{1}_{m,1}, \dots, \mathbf{1}_{m,T+W-1}$. The obtained sequence of symbols is then subject to the channel, yielding the sequence of received symbols $\mathbf{y}_1, \dots, \mathbf{y}_{T+W-1}$. At this stage, the sliding window technique is employed to efficiently obtain probability vectors and recover the transmitted messages. In particular, at a time t, the receiver BRNN processes the window of W symbols $\mathbf{y}_t, \dots, \mathbf{y}_{t+W-1}$, transforming them into W probability vectors $\mathbf{p}_t^{(t)}, \dots, \mathbf{p}_{t+W-1}^{(t)}$ via its final *softmax* layer. Then it slides one time slot ahead to process the symbols $\mathbf{y}_{t+1}, \dots, \mathbf{y}_{t+W}$. Notice that this enables the scheme to provide multiple estimates of the probability vector at time t, for $t \geq 2$.

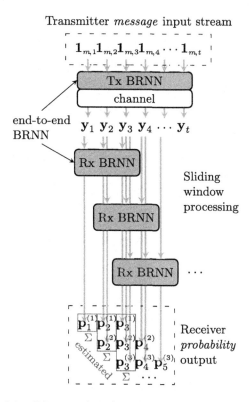

Figure 3.10 Schematic of the sliding window algorithm (SWA) for sequence estimation in which the optimized BRNN transceiver is operated. Note that $W = 3$ is chosen for illustration purposes.

The final output probability vectors for the symbols from the received sequence are estimated as

$$\mathbf{p}_i = \sum_{k=0}^{\min(W,i)-1} [\min(W,i)]^{-1} \cdot \mathbf{p}_i^{(i-k)}. \tag{3.11}$$

The weighting coefficients in the estimation algorithm in this case are assumed equal, as originally proposed in [55]. It was shown in [10] that for the optical IM/DD small performance gains can be obtained via an additional weight optimization, which, however, increases the complexity of the system and is thus not discussed here. After the SWA estimates the final probability vector for a given symbol, a decision on the transmitted message can be performed as

$$\hat{m}_t = \underset{j=1...M}{\mathrm{argmax}}(\mathbf{p}_{i,j}) \tag{3.12}$$

and a symbol error is counted when $m_t \neq \hat{m}_t$.

Table 3.1 Simulations parameters.

LPF bandwidth	32 GHz
DAC/ADC ENOB	6
DAC/ADC sampling rate	84 GSa/s
Information rate	42 Gb/s
MZM operation range	$[0; \pi/4]$
Fiber transmission distance	20-90 km
Fiber dispersion	17 ps/(nm $\cdot$ km)
Fiber attenuation	0.2 dB/km
Receiver noise power	0.245 mW

3.3.3 Performance

This section discusses results from the numerical performance investigation of the described ANN-based optical transmission schemes for short reach. The IM/DD channel model includes low-pass filtering (LPF) at transmitter and receiver to reflect current hardware limitations, digital-to-analog and analog-to-digital converters (DAC/ADC), Mach-Zehnder modulator (MZM), photo-diode (PD) detection, electrical amplifier at the receiver and transmission over a single span of standard single mode fiber (SSMF). Table 3.1 list the parameters for the simulation of the transmission link. A detailed description of the modeling of each channel component can be found in [47,51]. The performance of the systems is evaluated at different transmission distances for the fixed data rate of 42 Gb/s.

For the FFNN auto-encoder, input messages from a set of either $|\mathcal{M}| = M = 64$ (6 bits) or $|\mathcal{M}| = M = 8$ (3 bits) total messages are encoded at the transmitter into symbols of $n = 12$ or $n = 6$ digital waveform samples, respectively. After low-pass filtering, the signal is applied to a DAC with the sampling rate of $R_{\text{samp.}} = 84$ GSa/s. Similarly for the sliding window BRNN auto-encoder, $M = 64$ and $n = 12$ are chosen. For a comparison, PAM-2 transmission with a receiver based on the sliding window FFNN (SFFNN) described in Section 3.3.1.2 is also considered. For this system, the pulse-shaping of PAM symbols is performed at 2 Sa/symbol using a raised cosine filter with 0.25 roll-off. Note that for all systems the resulting coding rate is $R_{\text{cod.}} = 1/2$ b/Sa, resulting in information rates of $R_{\text{inf.}} = 42$ Gb/s because of the $R_{\text{samp.}} = 84$ GSa/s DAC. The hyper-parameter description for the considered systems can be found in [47] (FFNN auto-encoder) and [51] (SBRNN auto-encoder and PAM-2 & SFFNN). It should be noted that, in order to use the bit error rate (BER) as a common performance metric, additional bit-to-symbol mapping is performed for the auto-encoders (see [10]).

Fig. 3.11 shows the BER performance of the deep learning-based IM/DD systems as a function of transmission distance. The $4.5 \cdot 10^{-3}$ hard-decision FEC (HD-FEC) threshold [57] (6.7% overhead) is used as an indicator of the system performance. For the FFNN auto-encoder, it can be seen that the resilience to chromatic dispersion is

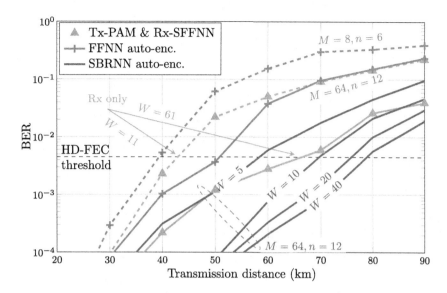

Figure 3.11 BER as a function of transmission distance for 42 Gb/s FFNN and SBRNN auto-encoders with different neural network hyper-parameters M, n (FFNN and SBRNN) and sliding window size W (SBRNN). They are compared to 42 Gb/s PAM-2 systems with sliding window FFNN receivers of different window size W.

enhanced by considering larger processing blocks (increasing n). However, it was already discussed that for a fixed coding/information rate, this approach is associated with a rapid increase in the network dimensions (number of nodes) and thus complexity, dictated by the hyper-parameters n and $M = 2^{n \cdot R_{cod.}}$. On the other hand, the SBRNN design handles sequences of symbols, having a structure that allows mitigation of the inter-symbol interference accumulated during transmission without the necessity to increase the network size. In particular, the processing memory in the SBRNN receiver depends on the sliding window W assumed in the estimation algorithm. The parameter W is external to the BRNN transceiver architecture and thus adjusting it to compensate for more of the interference does not require any increase in the number of transceiver nodes. Indeed, it can be seen that for the SBRNN system, increasing W effectively reduces the BER. As a consequence, the SBRNN system significantly outperforms the FFNN design for fixed network dimensions. In particular, the results indicate that the SBRNN auto-encoder can enable communication below the HD-FEC threshold at distances up to 80 km. It should be noted that as ISI is compensated, noise and non-linearities in the system become dominant and thus the gains from increasing W eventually start to diminish.

The performances of two PAM-2 systems, which use an FFNN at the receiver to estimate a PAM symbol by simultaneously processing a window of $W = 11$ or $W = 61$ received symbols are also shown in Fig. 3.11. The PAM & SFFNN system with

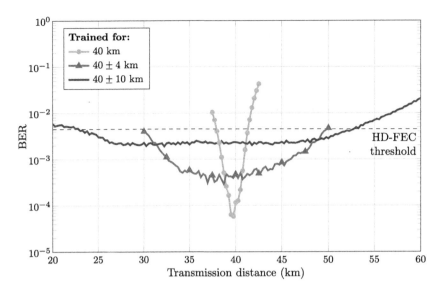

Figure 3.12 Bit error rate as a function of transmission distance for systems where the training is performed at a fixed nominal distance of 40 km or normally distributed distances with mean $\mu = 40$ km and standard deviation σ. The horizontal dashed line indicates the 6.7% HD-FEC threshold.

$W = 61$ can be directly compared to the SBRNN auto-encoder with a sliding window of $W = 10$, since in both scenarios the ANN receivers perform the decoding using relatively the same number of samples. The results show that at shorter distances the SBRNN system achieves a significantly improved performance. As the distance increases the BER becomes comparable, although above the HD-FEC threshold. It is interesting to observe that the FFNN auto-encoder, whose processing blocks use a reduced number of samples, outperformed the PAM-2 system with a window of $W = 11$. This highlights the benefit from end-to-end deep learning-based optimization.

It should be also mentioned that for the PAM & SFFNN system increasing W to capture more of the interference corresponds to an increase in the number of neural network nodes in the receiver. This is in contrast to the architecture based on a recurrent processing. For more details on the complexity of the considered systems, the interested reader is referred to [10,51].

3.3.4 Distance-agnostic transceiver

In the previous section, the deep learning-based DSP for the considered systems is optimized separately for each transmission distance. In particular, the optimization is performed assuming a fixed nominal transmission distance L and thus all training examples within the mini-batches in the SGD algorithm accumulate identical ISI. Since the DSP is optimized exclusively for communication at the fixed nominal distance, operation at distances different from L requires re-initiation of the training stage. Fig. 3.12

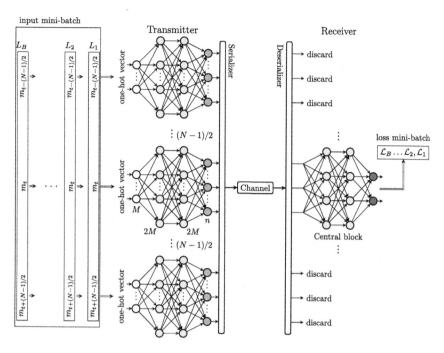

Figure 3.13 Schematic of the training procedure for the FFNN-based auto-encoder, showing how a mini-batch is formed from different transmitted sequences of length N over fiber lengths L_i and the corresponding losses, computed for the central message in every sequence.

shows an example of such behavior for a reference FFNN auto-encoder, optimized only for operation at 40 km. A rapid increase in error rate of the system is observed when the distance changes.

For obtaining a neural network-based transceiver robust to distance variations, a novel multi-distance learning method was proposed and verified in transmission experiments in [47,48]. In particular, it exploits the opportunity during auto-encoder optimization to associate each of the sequences i from the training mini-batch with a specific distance L_i, and thus a specific amount of ISI-induced distortion on the transmitted signal. The distortion has a direct impact on the calculated loss $\mathcal{L}_i$, which would increase with distance. As a result, the aggregated loss over a mini-batch $\overline{\mathcal{L}}(\theta)$ comprises of training examples from multiple transmission scenarios. Performing deep learning aimed at minimizing $\overline{\mathcal{L}}(\theta)$, could thus be viewed as optimizing the set of transmitter and receiver parameters θ for transmission over all different L_i. In other words, the goal is to allow the deep learning algorithm to converge to a set of generalized ANN parameters robust to certain variation of the link dispersion. The method is illustrated in Fig. 3.13. In particular, the simple scheme proposed in [47] involves assigning to the i-th mini-batch sequence a link length L_i which is randomly drawn from a probability distribution with appropriately chosen parameters, e.g. $L_i \sim \mathcal{N}^{(\mu,\sigma)}$, $i = 1, 2, \ldots |S|$, where $\mathcal{N}^{(\mu,\sigma)}$ denotes the Gaussian distribution, whose mean is μ and standard deviation is σ, and $|S|$ is the mini-batch size.

For a direct comparison, the BER performance of systems trained at the mean distance $\mu = 40$ km and different values of the standard deviation is also shown in Fig. 3.12. It can be seen that for the cases of $\sigma = 4$ km and $\sigma = 10$ km the multi-distance training method allows the system to operate at BER values below the HD-FEC threshold in much wider ranges of transmission distances compared to training at 40 km only. It should be noted that in these cases the obtained BERs are higher since training converges to parameters valid for the significantly varied distances. Thus, there exists a trade-off between system robustness and performance. Nevertheless, such a multi-distance learning method has an important practical application as it introduces both robustness and flexibility of the deep learning-based systems to variations in the link parameters, without requiring any re-optimization or reconfiguration of the transceiver.

3.4. Implementation on a transmission link

Both receiver-only [13,41,43] as well as auto-encoder [13,47,48] deep learning-based systems have been successfully demonstrated on IM/DD transmission test-beds in recent years. This section summarizes the different strategies for optimization of the ANN-based transmitter or/and receiver which were considered when implementing the systems on the actual fiber transmission link. Note that, when considering optical fiber transmission experiments, special care should be taken during optimization of the ANNs, such that they do not learn to adapt to the generation/interference patterns of the training sequences (see [40] for detailed analysis).

3.4.1 Conventional PAM transmission with ANN-based receiver

For systems based on the conventional PAM transmission, the ANN receiver is optimized using the digitized photo-current obtained after sampling by the ADC. The optimization can be performed using pilot data, transmitted over the physical link and collected in a training database. Fig. 3.14 shows a schematic of the procedure. The traces are formed by the received blocks of samples (symbols) $\mathbf{y}_t$, which are labeled with the corresponding transmitted message m_t. Optimization can be performed in a supervised manner, as described in 3.3.1.2. In particular, the pairs of received symbols and labels are used to compute the receiver loss, which is then minimized by the optimization algorithm via back-propagation and gradient descent. Note that the optimization of ANN-based receivers is relatively straightforward because the back-propagation algorithm for computing gradients extends only to the link output and thus knowledge of the actual transmission link is not required.

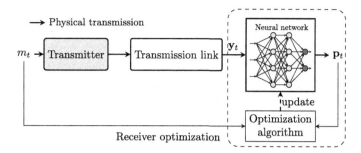

Figure 3.14 Schematic of the procedure for receiver ANN parameters optimization using experimentally collected data.

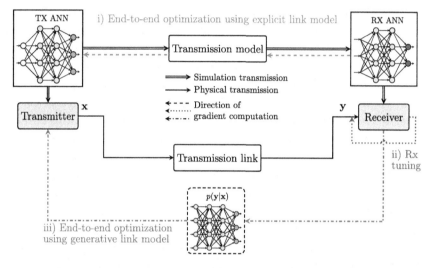

Figure 3.15 Schematic of the experimental optical IM/DD transmission test-bed, showing the methods for system optimization using i) numerical simulation of the link; ii) & iii) experimental traces.

3.4.2 Auto-encoder implementation

The schematic shown in Fig. 3.15 summarizes three different approaches for implementing the auto-encoder systems on a physical transmission link. Within the end-to-end deep learning framework, which was introduced in Sec. 3.3.2, the channel is considered a segment of the end-to-end computational graph representing the complete communication system. Thus, end-to-end learning can be readily applied for communication links accurately described by a differentiable transmission model. In such cases, the ANN-based transceiver is optimized via simulation transmission and applied "as is" to the real system, as shown in Fig. 3.15 i). Such systems present a viable perspective for low-cost optical fiber communications since no optimization process is required after deployment. However, as reported in [47], the performance of transceivers learned on a specific model assumption can be deteriorated when applied to the test-bed. To improve performance, the ANN parameters can be optimized to the specific link properties using pilot experimental data, collected for training.

Similarly to the PAM systems from Section 3.4.1, receiver-side ANN optimization (Fig. 3.15 ii)) can be readily implemented by forming the pairings of input messages m_t and the corresponding received blocks of digital photo-current samples (symbols) $\mathbf{y}_t$ (see e.g. [13,48]). The receiver ANN parameters $\boldsymbol{\theta}_{\mathrm{Rx}}$ are adjusted for minimizing the loss, computed at the receiver output.

However, the transmission link can be considered a *black box* for which only inputs and outputs are observed, precluding gradient back-propagation to the transmitter ANN. As a consequence, optimizing the transmitter parameters without the explicit knowledge of an accurate model for the underlying physical channel remains an open problem. A framework for transmitter optimization using a generative model of the optical IM/DD link was developed and experimentally demonstrated in [58]. The method uses a generative adversarial network (GAN) for approximating the conditional distribution $p(\mathbf{y}|\mathbf{x})$ of the channel. As shown in Fig. 3.15 iii), the obtained ANN-based model is applied *in lieu* of the transmission link during optimization. This enables the use of back-propagation for computation of gradients at the transmitter and thus end-to-end system optimization via gradient descent.

3.5. Outlook

There is a number of future directions for further development of the deep learning and specifically auto-encoder frameworks for short reach optical fiber communications. Distinct lines of research can be related to the design of the deep learning-based system, the optimization procedure or the application of the concepts to different types of short reach systems and models.

This chapter concentrated on the application of deep learning for short reach optical IM/DD communications. Nevertheless, the described methods are general and their application can be extended to other short reach systems, for instance single span coherent links. In such scenarios, deep learning can be used to address the limitations induced by the strong Kerr non-linearity as a result of high input powers. In recent years, a technique known as *probabilistic shaping* (PS), which consists in controlling the occurrence of the modulation symbols, has been demonstrated as an effective method for reach increase and seamless adaptation of the data rates in long-haul coherent optical systems [59]. Nevertheless, finding the optimal symbol distribution for the non-linear fiber channel is an open problem, especially for multi-dimensional modulation formats [60]. Hence, finding a deep learning solution is an interesting perspective, especially for single span coherent links, where the feasibility of such an approach was recently reported in [54].

In Section 3.3.4, a novel method for ensuring tolerance to distance variations was discussed. It consisted in implementing an advanced multi-distance optimization process which allowed the system to achieve error rates below a FEC threshold for a large range of link distances. Importantly, such a distance-agnostic transceiver does not require any reconfiguration or feedback for tuning to the changing link conditions. Nevertheless, it was shown that the method introduces a trade-off between optimal system performance and robustness. Thus, a more detailed investigation on possible enhancements in the learning process is an important direction for future work. For example, in [61] an improved multi-distance learning framework was demonstrated, which includes coarse distance information to the receiver ANN input.

References

[1] G. Agrawal, Fiber-Optic Communication Systems, 4th ed., Wiley, 2010.
[2] K. Zhong, et al., Digital signal processing for short-reach optical communications: a review of current technologies and future trends, Journal of Lightwave Technology 36 (2) (2018) 377–400.
[3] M. Chagnon, Optical communications for short reach, Journal of Lightwave Technology 37 (8) (2019) 1779–1797.
[4] D. Plabst, et al., Wiener filter for short-reach fiber-optic links, IEEE Communications Letters 24 (11) (2020) 2546–2550.
[5] N. Stojanovic, et al., Volterra and Wiener equalizers for short-reach 100G PAM-4 applications, Journal of Lightwave Technology 35 (21) (2017) 4583–4594.
[6] Q. Hu, et al., High data rate Tomlinson-Harashima-precoding-based PAM transmission, in: Proc. of European Conference on Optical Communication (ECOC), 2019, pp. 1–4.
[7] G.D. Forney, The Viterbi algorithm, Proceedings of the IEEE 61 (3) (1973) 268–278.
[8] P. Poggiolini, et al., Branch metrics for effective long-haul MLSE IMDD receivers, in: Proc. of European Conference on Optical Communications (ECOC), 2006, pp. 1–2.
[9] G. Bosco, P. Poggiolini, M. Visintin, Performance analysis of MLSE receivers based on the square-root metric, Journal of Lightwave Technology 26 (14) (2008) 2098–2109.
[10] B. Karanov, et al., Deep learning for communication over dispersive nonlinear channels: performance and comparison with classical digital signal processing, in: Proc. of 57th Annual Allerton Conference on Communication, Control, and Computing, Allerton, 2019, pp. 192–199.
[11] D. van Veen, V. Houtsma, Strategies for economical next-generation 50G and 100G passive optical networks, IEEE/OSA Journal of Optical Communications and Networking 12 (1) (2020) A95–A103.
[12] Z. Xu, et al., Computational complexity comparison of feedforward/radial basis function/recurrent neural network-based equalizer for a 50-Gb/s PAM4 direct-detection optical link, Optics Express 27 (25) (2019) 36953–36964.
[13] B. Karanov, et al., Experimental investigation of deep learning for digital signal processing in short reach optical fiber communications, in: Proc. of IEEE International Workshop on Signal Processing Systems (SiPS), 2020, pp. 1–6.
[14] I. Lyubomirsky, Machine learning equalization techniques for high speed PAM4 fiber optic communication systems, CS229 Final Project Report, Stanford University, 2015 [online], available: http://cs229.stanford.edu/proj2015/232_report.pdf, 2015. (Accessed 21 August 2020).
[15] J. Proakis, D. Manolakis, Digital Signal Processing: Principles Algorithms and Applications, 3rd ed., Prentice Hall, 2001.
[16] E. Agrell, et al., Roadmap of optical communications, Journal of Optics 18 (6) (May 2016) 063002.
[17] T.J. O'Shea, J. Hoydis, An introduction to deep learning for the physical layer, IEEE Transactions on Cognitive Communications and Networking 3 (4) (2017) 563–575.
[18] O. Simeone, A very brief introduction to machine learning with applications to communication systems, IEEE Transactions on Cognitive Communications and Networking 4 (4) (2018) 648–664.

[19] F.N. Khan, C. Lu, A.P.T. Lau, Optical performance monitoring in fiber-optic networks enabled by machine learning techniques, in: Proc. of Optical Fiber Communications Conference (OFC), 2018, pp. 1–3.

[20] I. Goodfellow, Y. Bengio, A. Courville, Deep Learning, MIT Press, 2016.

[21] D. Rumelhart, G. Hinton, R. Williams, Learning representations by back-propagating errors, Nature 323 (1986) 533–536.

[22] A. Cauchy, Méthode générale pour la résolution des syst'emes d'équations simultanées, Comptes Rendus. Academie Des Sciences. Paris 25 (1847) 536–538.

[23] H.B. Curry, The method of steepest descent for non-linear minimization problems, Quarterly of Applied Mathematics 2 (3) (1944) 258–261.

[24] D. Kingma, J. Ba, Adam: a method for stochastic optimization, in: Proc. of International Conference for Learning Representations, 2015, pp. 1–15.

[25] S. Ruder, An overview of gradient descent optimization algorithms, arXiv:1609.04747, 2016.

[26] S. Chen, et al., Adaptive equalization of finite non-linear channels using multilayer perceptrons, Signal Processing 20 (1990) 107–119.

[27] N. Benvenuto, et al., Non linear satellite radio links equalized using blind neural networks, in: [Proceedings] ICASSP 91:1991 International Conference on Acoustics, Speech, and Signal Processing, vol. 3, 1991, pp. 1521–1524.

[28] O. Shental, J. Hoydis, Machine LLRning: learning to softly demodulate, in: IEEE Globecom Workshops (GC Wkshps), 2019, pp. 1–7.

[29] E. Nachmani, Y. Be'ery, D. Burshtein, Learning to decode linear codes using deep learning, in: 201654th Annual Allerton Conference on Communication, Control, and Computing, Allerton, 2016, pp. 341–346.

[30] T. Gruber, et al., On deep learning-based channel decoding, in: 51st Annual Conference on Information Sciences and Systems (CISS), 2017, pp. 1–6.

[31] S. Cammerer, et al., Scaling deep learning-based decoding of polar codes via partitioning, in: GLOBECOM 2017 - 2017 IEEE Global Communications Conference, 2017, pp. 1–6.

[32] D. Tandler, et al., On recurrent neural networks for sequence-based processing in communications, in: 2019 53rd Asilomar Conference on Signals, Systems, and Computers, 2019, pp. 537–543.

[33] A. Buchberger, et al., Pruning neural belief propagation decoders, in: 2018 IEEE International Symposium on Information Theory (ISIT), 2020, pp. 1–5.

[34] C. Häger, H.D. Pfister, Nonlinear interference mitigation via deep neural networks, in: Proc. of Optical Fiber Communications Conference (OFC), 2018, pp. 1–3.

[35] E. Sillekens, et al., Time-domain learned digital back-propagation, in: 2020 IEEE Workshop on Signal Processing Systems (SiPS), 2020, pp. 1–4.

[36] V. Oliari, et al., Revisiting efficient multi-step nonlinearity compensation with machine learning: an experimental demonstration, Journal of Lightwave Technology 38 (12) (2020) 3114–3124.

[37] T. Koike-Akino, et al., Neural turbo equalization: deep learning for fiberoptic nonlinearity compensation, Journal of Lightwave Technology 38 (11) (2020) 3059–3066.

[38] V. Kamalov, et al., Evolution from 8QAM live traffic to PS 64-QAM with neural-network based nonlinearity compensation on 11000 km open subsea cable, in: Proc. of Optical Fiber Communications Conference (OFC), 2018, pp. 1–3.

[39] J. Estaran, et al., Artificial neural networks for linear and non-linear impairment mitigation in high-baudrate IM/DD systems, in: Proc. of European Conference on Optical Communication (ECOC), 2016, pp. 1–3.

[40] T.A. Eriksson, H. Bülow, A. Leven, Applying neural networks in optical communication systems: possible pitfalls, IEEE Photonics Technology Letters 29 (23) (2017) 2091–2094.

[41] V. Houtsma, E. Chou, D. van Veen, 92 and 50 Gbps TDM-PON using neural network enabled receiver equalization specialized for PON, in: Proc. of Optical Fiber Communications Conference (OFC), 2019, pp. 1–3.

[42] L. Zhang, et al., 160-Gb/s Nyquist PAM-4 transmission with GeSi-EAM using artificial neural network based nonlinear equalization, in: Proc. of Optical Fiber Communications Conference (OFC), 2020, pp. 1–3.

[43] Z. Xu, et al., Cascade recurrent neural network enabled 100-Gb/s PAM4 short-reach optical link based on DML, in: Proc. of Optical Fiber Communications Conference (OFC), 2020, pp. 1–3.

[44] Z. Xu, et al., Feedforward and recurrent neural network-based transfer learning for nonlinear equalization in short-reach optical links, Journal of Lightwave Technology 39 (2) (2021) 475–480.

[45] T.J. O'Shea, K. Karra, T.C. Clancy, Learning to communicate: channel auto-encoders, domain specific regularizers, and attention, in: 2016 IEEE International Symposium on Signal Processing and Information Technology (ISSPIT), 2016, pp. 223–228.

[46] S. Dörner, et al., Deep learning based communication over the air, IEEE Journal of Selected Topics in Signal Processing 12 (1) (2018) 132–143.

[47] B. Karanov, et al., End-to-end deep learning of optical fiber communications, Journal of Lightwave Technology 36 (20) (2018) 4843–4855.

[48] M. Chagnon, B. Karanov, L. Schmalen, Experimental demonstration of a dispersion tolerant end-to-end deep learning-based IM-DD transmission system, in: Proc. of European Conference on Optical Communication (ECOC), 2018, pp. 1–3.

[49] R.T. Jones, et al., Deep learning of geometric constellation shaping including fiber nonlinearities, in: Proc. of European Conference on Optical Communication (ECOC), 2018, pp. 1–3.

[50] S. Li, et al., Achievable information rates for nonlinear fiber communication via end-to-end autoencoder learning, in: Proc. of European Conference on Optical Communication (ECOC), 2018, pp. 1–3.

[51] B. Karanov, et al., End-to-end optimized transmission over dispersive intensity-modulated channels using bidirectional recurrent neural networks, Optics Express 27 (14) (2019) 19650–19663.

[52] O. Sidelnikov, A. Redyuk, S. Sygletos, Equalization performance and complexity analysis of dynamic deep neural networks in long haul transmission systems, Optics Express 26 (25) (2018) 32765–32776.

[53] B. Karanov, P. Bayvel, L. Schmalen, End-to-end learning in optical fiber communications: concept and transceiver design, in: Proc. of European Conference on Optical Communication (ECOC), 2020, pp. 1–4.

[54] B. Karanov, et al., End-to-end learning in optical fiber communications: experimental demonstration and future trends, in: Proc. of European Conference on Optical Communication (ECOC), 2020, pp. 1–4.

[55] N. Farsad, A. Goldsmith, Neural network detection of data sequences in communication systems, IEEE Transactions on Signal Processing 66 (21) (2018) 5663–5678.

[56] N. Farsad, A. Goldsmith, Neural network detectors for molecular communication systems, in: 2018 IEEE 19th International Workshop on Signal Processing Advances in Wireless Communications (SPAWC), 2018, pp. 1–5.

[57] Z. Wang, Super-FEC codes for 40/100 Gbps networking, IEEE Communications Letters 16 (12) (2012) 2056–2059.

[58] B. Karanov, et al., Concept and experimental demonstration of optical IM/DD end-to-end system optimization using a generative model, in: Proc. of Optical Fiber Communications Conference (OFC), 2020, pp. 1–3.

[59] F. Buchali, et al., Rate adaptation and reach increase by probabilistically shaped 64-QAM: an experimental demonstration, Journal of Lightwave Technology 34 (7) (2016) 1599–1609.

[60] R. Dar, et al., On shaping gain in the nonlinear fiber-optic channel, in: 2014 IEEE International Symposium on Information Theory, 2014, pp. 2794–2798.

[61] B. Karanov, L. Schmalen, A. Alvarado, Distance-agnostic autoencoders for short reach fiber communications, in: Proc. of Optical Fiber Communications Conference (OFC), 2021, pp. 1–3.

CHAPTER FOUR

Machine learning techniques for passive optical networks

Lilin Yi, Luyao Huang, Zhengxuan Li, Yongxin Xu, and Wanting Xu
Shanghai Jiao Tong University, Shanghai, China

4.1. Background

A passive optical network, or PON [1–3], is a network in which fiber optic cables (instead of copper) bring signals all or most of the way to the end-user. It is sometimes referred to as the "last mile" between an internet service provider and the customer. It is described as passive because no active equipment (electrically powered), like amplifiers or repeating circuits, are required between the central office (or hub) and the customer premises. The use of only passive components such as splitters and combiners makes a PON significantly less expensive than a network built on active components. However, this also results in a shorter overall range of coverage due to limited signal strength—whereas active optical networks (AONs) can reach over 90 km, a PON is typically limited to roughly 20 km.

Depending on where the PON terminates, the system can be described as an FTTx network, for example, fiber-to-the-curb (FTTC), fiber-to-the-building (FTTB), and fiber-to-the-home (FTTH), which typically allows a point-to-multipoint connection from the source to destination. In most PON setups, a point-to-multipoint network is created using a central optical line terminal and multiple optical splitters to provide service for up to 128 users, dramatically reducing network installation, management, and maintenance costs. An optical network unit, or ONU, terminates the PON at a user's location and communicates with an optical line termination (OLT) to connect the PON to a router, telephone, computer, and television.

There are two mainstream PON standardization groups, the ITU-T and IEEE [4]. As a general rule, they endeavor to ensure as much commonality between key aspects of their respective PON standards as possible to minimize divergence and promote convergence. This is especially true for the physical layer to enable the use of common optical components by adopting similar wavelength plans as much as possible. From the timeline of PON standards and the associated maximum aggregated downstream (DS) bandwidth shown in Fig. 4.1 [4], the evolution of PON technologies is presented clearly. Blue (dark gray in print version) dots represent ITU-T PON and orange (mid gray in print version) dots IEEE Ethernet PON. Starting with broadband PON (BPON, 622

Mb/s downstream and 155 Mb/s upstream rate), PON is evolving from Ethernet PON (EPON) and Gigabit PON (GPON) [5] to the following 10G EPON and 10 Gigabit asymmetric/symmetric PON (XG-PON/XGS-PON) standardized by IEEE 802.3 and ITU-T Q2/SG15, respectively. In 2017-2020, both groups release 50G or beyond PON standards. Note that for >10 Gb/s scenario, some standardizations start to turn from a single-wavelength solution to a multi-wavelength solution. The IEEE task force's 50G objective is based on a solution that stacks two fixed wavelengths, which will add more cost to the system and bring more maintenance difficulties. As a counterpart, ITU-T begins to standardize 50G PON, but with a single carrier in early 2018. Compared with the IEEE's wavelength stacked $2 * 25G$ solution [6], the ITU-T's single carrier non-return-to-zero based 50G PON [7] has the advantages of low cost, easy operation, and convenient management [8]. Hence how to increase the line rate per wavelength channel with low-cost devices is the main concern in PON.

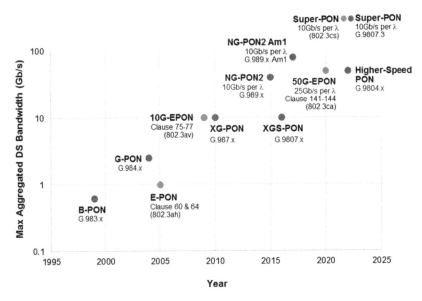

Figure 4.1 Timeline of PON standards and the associated maximum aggregated downstream (DS) bandwidth. For multi-wavelength PONs, data labels also show the maximum DS line rate per wavelength channel. Specific aggregated bandwidth for Super-PON is not shown as it depends on the number of wavelength channels supported [4].

Owing to the preference for low-cost optical devices, the bandwidth of the channel is inevitably limited, resulting in severe inter-symbol interference [9–12]. In a coherent system, chromatic dispersion in fiber can be easily compensated by DSP, but cannot in an IMDD system. Since there is only one PD in the receiver and PD's square law detection discards light's phase information. Moreover, these linear distortions can mix with other effects, such as nonlinearities [13] of the optical devices, which further degrades the signal quality during transmission. In this case, traditional algorithms such as FFE and DFE perform poorly and can not meet the equalization requirements. More powerful tools are needed. So researchers pay their attention to machine learning, which has shown its strength in almost every field.

Machine learning is a huge set of algorithms containing logistic regression, support vector machines (SVM), unsupervised learning, neural network (NN), etc. Because of its powerful ability to express any functions using enough nodes [14] and its versatile structures, NN becomes the most popular machine learning algorithm in PON. In this chapter, we mainly introduce the use of NN in PON.

NN has three main applications in the optical network, pattern recognition, modeling, and equalization. There are many works about NN-enabled modulation format identification [15–17] utilizing its powerful pattern recognition ability. And some networks are used to model fiber channel [18,19] since nonlinear effects in fiber are complexed and classical numerical computation such as split-step Fourier method (SSFM) is in high complexity. But these two applications are mainly deployed in the long-haul coherent optical communication system. In PON, the need for MFI and modeling is not as urgent as in a long-haul system, NN is more suited for equalization. NN has shown its significantly superior performance than the traditional equalizers in many experiments, which makes the researchers believe it will be a promising candidate as an advanced DSP technique in short reach applications.

4.2. The validation of NN effectiveness

Several problems need to be sorted out before using NN for equalization. In-depth studies on the NN-based equalizer find that the performance of NN-based equalization may be overestimated owing to the use of a pseudo-random bit sequence (PRBS) [20]. When data of the same PRBS pattern are used to train and test the NN, the NN will learn and use the generation rules of the PRBS, resulting in abnormally good performance [20–22]. References [20] and [21] reported that an NN with one hidden layer containing at least two nodes can learn PRBS rules. Therefore, the NN misled by the PRBS is thus not suitable for equalization. However, as an NN is often regarded as a black box, no in-depth study has been conducted to reveal the detailed mechanism by which the NN characterizes PRBS rules. So we will explain in detail how an NN learns PRBS rules [23] later. Subsequently, questions naturally transfer to how to validate the effectiveness of training the NN, whether other pseudo-random data generated by more advanced algorithms like the Mersenne twister can also be learned by the NN, and if it is necessary to find true random data for training. Effective methods of validating training and generating random data for training are thus essential and we will clarify these points explicitly. In this section, first, we introduce the generation rule of PRBS. Then we explain how an NN learns PRBS rules. Finally, we prove the effectiveness of NN training using mutual verification.

A PRBS is a binary sequence generated with a deterministic algorithm and has statistical behaviors similar to those of a real random sequence. It is widely used in the fields of telecommunication, encryption, and navigation. Specifically, the PRBS is generated by linear feedback shift registers based on corresponding generating polynomials; e.g., the rules for PRBS15 and PRBS23 are

$$PRBS15 = X^{15} + X^{14} + 1 \tag{4.1}$$

$$PRBS23 = X^{23} + X^{18} + 1 \tag{4.2}$$

This is essentially a simple XOR operation. For instance, the generation rule of PRBS15 can be expressed as

$$x(n) = x(n-15) \oplus x(n-14) \tag{4.3}$$

where $\oplus$ is the XOR operation and x is the bit in the sequence with the index in parentheses. In a basic XOR operation, each of the three variables can be computed using the other two variables by XOR, and we thus deduce properties of PRBS15 as

$$x(n-14) = x(n) \oplus x(n-15), x(n-15) = x(n) \oplus x(n-14) \tag{4.4}$$

$$x(n) = x(n+14) \oplus x(n-1) = x(n+15) \oplus x(n+1) \tag{4.5}$$

A bit can be computed using two other related bits, and we thus continue to expand the above formula into a more complex one, such as

$$\begin{aligned} x(n) &= x(n+15) \oplus x(n+1) \\ &= x(n+15) \oplus \left(x(n+16) \oplus x(n+2)\right) \\ &= x(n+16) \oplus x(n+15) \oplus x(n+2) \end{aligned} \tag{4.6}$$

As all the bits are connected by XOR operations with simple nonlinearity, an NN can easily characterize this feature.

For simplicity, we analyze the mechanism of an NN-based equalizer with a PAM2 signal. For a high-order or more complex modulation format, the principle of the NN in characterizing the PRBS is the same, but the expression is more complicated. We first present the model of the NN expressing an XOR operation. We assume that $a \oplus b = c$, and we use the values 0 and m to represent false and true symbols respectively. According to Eq. (4.7), we can simply express the XOR operator using the ReLU function as

$$c = ReLU(a - b) + ReLU(b - a) \tag{4.7}$$

or

$$c = m - (\text{ReLU} (a + b - m) + \text{ReLU}(-a - b + m)) \tag{4.8}$$

The two equations express the XOR by subtraction and addition of the related bits respectively. Table 4.1 presents the related logical operation and results. According to the results, we claim that one hidden layer with two nodes activated by the ReLU function can achieve this operation.

Table 4.1 Value table of the ReLU-based XOR operation.

a	b	$c = a \oplus b$	**Eq. (4.7)**	**Eq. (4.8)**
0	0	0	0	0
0	m	m	m	m
m	0	m	m	m
m	m	0	0	0

Eq. (4.7): $c = \text{ReLU} (a - b) + \text{ReLU} (b - a)$.
Eq. (4.8): $c = m - (\text{ReLU} (a + b - m) + \text{ReLU}(-a - b + m))$.

In the case of an equalizer, the network can also receive a sample of the target itself. The sample of c can also be involved in the operation. Letting $d_1 = a - b$ and $d_2 = b - a$, we then have

$$c = \text{ReLU} \left(\frac{d_1 + c}{2} \right) + \text{ReLU} \left(\frac{d_2 + c}{2} \right) \tag{4.9}$$

As the two parts can have different weights, a more general expression is

$$c \approx \text{ReLU} \left(\alpha_1 d_1 + \beta_1 c \right) + \text{ReLU} \left(\alpha_2 d_2 + \beta_2 c \right) \tag{4.10}$$

where $\alpha_1 + \beta_1 = 1$ and $\alpha_2 + \beta_2 = 1$. When $c = m$, one of the ReLU operators in (4.9) will be m while the other may not be zero. The expression is thus only approximation. We modify the bias to obtain strict equality as

$$c \approx \text{ReLU} \left(\alpha_1 d_1 + \beta_1 c - \gamma_1 m \right) + \text{ReLU} \left(\alpha_2 d_2 + \beta_2 c - \gamma_2 m \right) \tag{4.11}$$

where $\alpha_i + \beta_i - \gamma_i = 1$ and $\gamma_i > \beta_i - \alpha_i$, with $i = 1, 2$. In this case, when $c = m$, the two ReLU operations obtain the results m and 0.

This equation can be easily leveraged to equalize the NRZ signal. In the case of regression, Eq. (4.11) can be the output of the equalizer. In the case of classification, we let $\hat{c}$ denote the result of the right side of the equation, and the values before output activation of the two classes are then $y_1 = b_1 - \hat{c}$ and $y_2 = \hat{c} - b_2$, corresponding to symbols 0 and 1 respectively, where b_1 and b_2 are two adjustable biases. For instance, letting $b_i = \frac{m}{2}$, it follows that (y_1, y_2) is $(\frac{m}{2}, -\frac{m}{2})$ when $\hat{c} = 0$. After activation, the first value will be appreciably greater than the second one, and the decision result is the symbol "0".

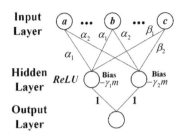

Figure 4.2 Configuration of the NN-based equalizer for regression.

Eq. (4.11) is still a special solution for the NN-based equalizer to detect the rules of the PRBS. The configuration of the NN for regression is shown in Fig. 4.2. We also consider that any solution of the NN in characterizing the PRBS rules will follow a similar mechanism. To validate this claim, we conduct a simulation using the similar NN structure shown in Fig. 4.2, which has only one hidden layer containing two nodes. The parameters of the NN are trained using data from the AWGN channel, where the NRZ format and PRBS15 are adopted. We set the input size as 16, while the target is the second bit, whose related bits are the first and sixteenth bits as expressed by Eq. (4.5). We refer to a related bit as a dependent bit, and the two bits constitute a dependent pair. No other bit can have a complete dependent pair in the input vector, and in this way, we entirely cover the only dependent pair and prevent a knock-on effect. The length of the training data set is 40,000. The loss function is the mean square error (MSE). Critical parameters of the NN after convergence are given in Table 4.2.

Table 4.2 Critical parameters of the NN in equalizating the PRBS15-based NRZ signal.

hidden node	1-st weight	2-nd weight	16-th weight	bias	weight of output
1	-0.1478	-0.3889	-0.1386	0.6735	-1.2305
2	0.4910	-0.1176	0.4805	-0.4561	-0.6935

Parameters that are not shown are less than 0.02.
The bias of the output is 1.0636.

According to the results, we construct an approximate expression of the NN as

$$y_n = -1.23 ReLU \left(-0.14 \left(x_{n+1} + x_{n+15}\right) - 0.39 x_{n+2} + 0.67\right)$$
$$-0.69 ReLU \left(0.48 \left(x_{n+1} + x_{n+15}\right) - 0.12 x_{n+2} - 0.46\right) + 1.06 \tag{4.12}$$

The weights of the first and sixteenth bits are approximately equal, indicating that the addition operation is used to conduct the XOR operation, which exhibits the same form as we analyze.

The effectiveness of NN-based training is controversial as many works have revealed that a PRBS can mislead NN training. This raises the question of whether an NN can still detect rules when using different data for training, such as a random sequence with a more complex generation rule. PRBS rules can also be learned by a Volterra nonlinear equalizer if the equalizer has enough second-order items. It seems that an advanced equalization algorithm based on statistics and recognition probably learns such internal features of data. Therefore, the first important point is to establish an effective method that verifies the training effectiveness of the NN or some other advanced algorithm.

We here propose a mutual verification strategy to address the above problem. The strategy requires one more data set with different generation rules to conduct the experiment, which will be used for testing the performance of the algorithm. If the NN does not learn the data rules, then the BER performance on these secondary data should exactly equal the performance on the primary test data.

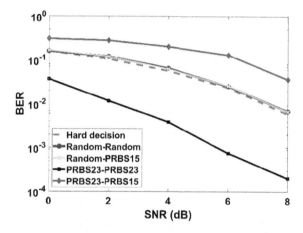

Figure 4.3 BER performance of the NN on different training and test data combinations for different SNRs in basic cases. The notation A-B means training on A and test on B.

We first verify this mutual verification on NN training using common data. PRBS data are used to exhibit misleading training, and random data generated using Matlab® and the generator Mersenne twister are applied to show a positive case. We conduct a simulation of different training and test data combinations. The results are shown in Fig. 4.3, where notation of measurement data A–B means that the model is trained on A and tested on B. All the training and test data for the same case belong to different data patterns, though some have the same generation rule. The hard-decision curve representing the theoretically optimal performance of such an AWGN system is provided as a benchmark.

The results show that, when the training data are Mersenne twister random data, all test BER performance curves are close to the hard-decision curve, which confirms that such Mersenne twister random data will not be characterized by the evaluated NN model. However, when the training data are PRBS23, the test results present different performances on different test data. In the case PRBS23-PRBS23 where the test data have the same generation rule as the training data, the test BER performance is much better than the hard-decision performance. This abnormally high performance does not exist in the case of PRBS23-PRBS15, because the test data are generated by a different rule. The different performances on the same training data show that the NN learns many generation rules, which misleads the equalization. Since NN cannot learn the generation rule of Mersenne twister random data, it is an effective equalization algorithm if random data is transmitted.

4.3. NN for nonlinear equalization

After the validation of its effectiveness, NN can be regarded as an equalizer in PON. There are both linear and nonlinear distortions in PONs. For linear distortions, many compensation methods have been proposed [24–26]. Nonlinear distortions [27], however, are much more complex problems and require more powerful equalization algorithms. The nonlinearity of PONs is mainly induced by the interplay between optics and optical fiber. For example, a directly modulated laser (DML) can lead to frequency chirp, and semiconductor optical amplifier (SOA) [28] induces the nonlinear pattern effect. And these nonlinear effects will further interact with the chromatic dispersion and nonlinearity of square-law PD.

The most commonly used NN structure for equalization is the fully-connected neural network (FCNN, called NN for short). Convolutional neural network (CNN) and recurrent neural network (RNN) are also used in the short-reach system, but none of them are used in PON, so we will mainly introduce ANN in this section. ANN is the simplest and most widely-used neural network, which is employed in many works as an equalizer and shows a good performance [22,29]. It can be used for post-equalization at the receiver or for pre-equalization at the transmitter. In this section, we will first introduce NN as post-equalizer, then decision feedback neural network (DFNN)-based equalizer, finally NN as pre-equalizer.

The typical structure of the post-equalizer based on NN is shown in Fig. 4.4. It is a 3-layer network, containing two hidden layers. Utilizing a NN as an equalizer needs to convert the time series signal into a 1-D input array. Considering the inter-symbol interference (ISI) in the system, the input array comprises the samples of the past and post $(N-1)/2$ symbols. The output layer has 8 nodes, corresponding to 8 kinds of symbols of 8-level pulse amplitude modulation (PAM8) signal.

Though an NN can achieve regression as FFE or Volterra nonlinear equalizer (VNE), in this NN-based equalization, classification is applied to replace the equalization and decision. The activation function of hidden layers is ReLU, while the output activation is Softmax. The label of symbols is one-hot vectors. And the loss function of the network is the cross-entropy loss.

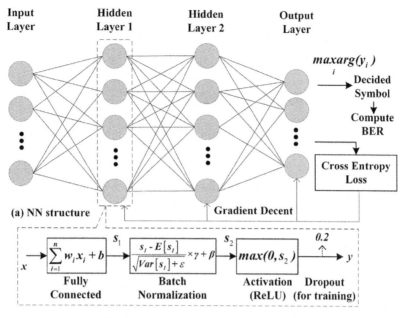

Figure 4.4 Structure of NN-based equalizer.

The 100 Gb/s/λ IMDD PON using NN-based equalizer is shown in Fig. 4.5. An AWG generates a 33 GBaud PAM-8 random sequence, then is directly modulated by a Mach–Zehnder modulator (MZM). The laser from a 1550 nm DML is injected to the MZM biased at its quadrature point, while a 100 Mb/s PRBS signal from a pulse pattern generator (PPG) is modulated on the laser to broaden the optical spectrum for suppressing the stimulated Brillouin scattering (SBS) under the condition of high launch power. An EDFA follows the MZM to control the launch power. After 20-km standard single-mode fiber (SSMF) transmission, a variable optical attenuator (VOA) is used to modify the power of the received signal to measure the receiver sensitivity. With another EDFA as a preamplifier in ONU, the input optical power of the PD can be adjusted to achieve the best receiver sensitivity. An optical filter (OF) is applied to suppress the out of band amplification spontaneous emission (ASE) noise induced by the preamplifier. The received electrical signal is firstly amplified by An electrical amplifier (EA), and then sampled by a digital sampling oscilloscope (DSO). The sampled signal will be processed in Matlab and Python. The offline DSP first resamples the signal to one sample per symbol. The original random sequence help to extract the sample to keep synchronous. The extracted transmitted sequence is sent into the FCNN-based equalizer together with the original sequence. After training and test, the equalizer computes and outputs the BER.

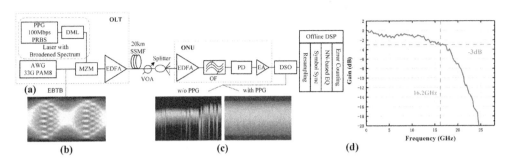

Figure 4.5 (a) Experimental setup. (b) EBTB eye diagram. (c) The waveforms w/o and with broadened spectrum laser. (d) The frequency response of the system.

To evaluate the efficiency of NN in our system, an NN-based equalizer is compared with FFE and VNE under the same conditions. The BER performance versus the injected power to PD, launch power, and received power injected to pre amplifier are presented in Fig. 4.6. The sensitivity of PD in the back-to-back (BtB) case is firstly investigated. The signal adjusted by VOA is directly injected to PD without using EDFA. The BER under different injected power is presented in Fig. 4.6(a). The BER rises rapidly as the injected power decreases. Since the amplitudes of adjacent levels of PAM8 are very close, the device noise including thermal noise and shot noise that cannot be eliminated by any algorithm tends to create a huge error for the symbol equalization. The BER performance versus the launch power for 20-km SSMF transmissions with −5 dBm received optical power after the VOA is shown in Fig. 4.6(b). The EDFA at the receiver adjusts the power injected to PD to 3 dBm. All the BER curves of the three algorithms first go down and then go back up, while NN gets the best performance and VNE overwhelms FFE, which are consistent with the simulation results. When the launch power is small, the nonlinearity of the transmission system is weak so that the performances of three equalizers are similar. As NN is more powerful to equalize nonlinearity, it will get significantly better performance when the nonlinearity of the system improves to a limited extent. However, as the launch power continues to grow, the nonlinearities increase beyond the capability of the algorithms, so that the BER of all equalizers rises.

As NN-based equalizer can perform better than FFE and VNE in the transmission system with proper nonlinearities, it becomes possible to increase the system loss budget by increasing the launch power to an appropriate level. To measure the maximal loss budget, we set 18 dBm as the launch power to test the BER performance with different received power. Besides, we still keep the PD injected power 3 dBm. As shown in Fig. 4.6(c), an NN-based equalizer can achieve a sensitivity of about −12 dBm at the 7% FEC limit. With the 18-dBm launch power, the total link loss budget can reach 30 dB.

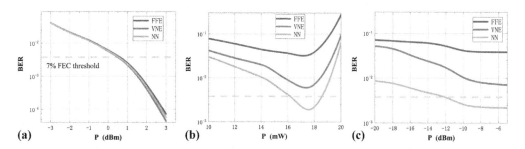

Figure 4.6 (a) BER performance comparison with different injected power to PD. (b) BER performance comparison with different launch power for 20-km SSMF transmission. (c) BER performance comparison of different received power for 20-km SSMF transmission.

30-dB loss budget for 100 Gb/s/λ PON is achieved, meeting IEEE802.3av PR30 requirement. Even though the value is much lower than the coherent 100 Gb/s/λ PON with 38.9 dB [30], we prove that assisted by an NN-based equalizer, IMDD technology is also feasible to achieve 100 Gb/s/λ PON with an acceptable loss budget. In this experiment, EDFA is used in the ONU, which is unacceptable for practical applications. If using a 20G-class APD+TIA with high receiver sensitivity, EDFA may be replaced by SOA. The feasibility of the system will be substantially improved. Besides, an AWG with higher bandwidth an d enhanced signal quality will also improve the system performance.

We have also evaluated some novel models of NN such as CNN for equalization. Generally, these models cannot demonstrate better performance than a basic FCNN. As the PON system is simple, a basic NN model is enough for characterizing the signal features. Since the behavior of NN depends on the features exhibited by the data, feature extraction may be a more effective way to improve equalization performance. This way exposes some hidden critical features of data to NN so that it can be trained well and model the channel more accurately.

Based on the observations that the feedback filter is used to remove that part of the ISI from the presently estimated symbol caused by previously detected symbols, we propose a decision feedback neural network (DFNN)-based equalizer [29], which introduces decision feedback function to extract more data features without taking any additional computational overhead. DFNN originates from NN with a little difference in the input data structure. The input vector of the NN-based equalizer is a sequence of samples, while the input vector of the DFNN-based equalizer is composed of a sample sequence and a sequence of decision feedback symbols.

The diagram of the proposed DFNN-based equalizer is illustrated in Fig. 4.7. Its basic structure is an FCNN with two hidden layers. The input size is 51. Each hidden layer contains 64 nodes with the activation function of ReLU. For the output layer, 4 nodes denote 4 kinds of symbols of the PAM4 signal respectively. Activated by

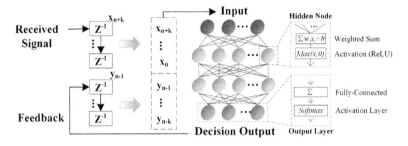

Figure 4.7 Structure of DFNN-based equalizer.

Softmax, the symbol corresponding to the node with the greatest value is the decision result. The difference between FCNN and DFNN is in the organization of input data. For the basic NN, the input data only consist of the consecutive sampled symbols resampled from the received signal, like $X_n = (x_{n-k}, \ldots, x_n, \ldots, x_{n+k})$ where $X_n X_n$ is the input vector, in which each element $x_i x_i$ is a received sample of a symbol at time i. Differently, the input vector of DFNN contains both the sampled symbols and the decision symbols, as $X_n = (y_{n-k}, \ldots, y_{n-1}, x_n, \ldots, x_{n+k})$ where the first k elements are replaced with $y_{n-k}, y_{n-k+1}, \ldots, y_{n-1} y_{n-k}, y_{n-k+1} \ldots, y_{n-1}$, which are the decision symbols that have been processed at previous k times.

For the training process, we replace the decision symbol y_i with its actual target $\hat{y}_i$, where $\hat{y}_i$ is in the set $\{0,1,2,3\}$ corresponding to the symbols for PAM4. For real test or practical applications, DFNN with well-trained parameters can accept the decision feedback value to reconstruct the input vector and equalize the subsequent symbols. Note that DFNN cannot run in parallel when decision feedback is used. This modification will not increase the computational overhead, as practical equalization is completely a sequential process.

We conduct the experiments on 50 Gb/s/λ IMDD PON to measure the performance of the proposed scheme. As shown in Fig. 4.8(a), an arbitrary waveform generator (AWG) with a sampling rate of 65 GSa/s is deployed to generate a 25 GBaud PAM4 random sequence. The PAM4 signal is modulated on a 10G-class O-band directly modulated laser (DML). After 20-km SSMF transmission, a variable optical attenuator (VOA) is applied for receiver sensitivity measurement. The optical signal is detected by a 20G-class avalanche photodiode (APD). The detected signal is finally sampled by a digital sample oscilloscope (DSO) with 45-GHz bandwidth and 120-GSa/s sample rate.

To entirely evaluate the equalization performance of the DFNN-based equalizer, we experimentally compare it with FFE, VNE, and a basic NN-based equalizer under the same transmission circumstance. With parameters optimizing, the number of FFE taps is set to 51. The lengths of the three order l1, l2, and l3 of VNE are 51, 21, and 9 respectively. As for NN and DFNN, we set the same configurations for them as explained above. The numbers of nodes for the input layer, two hidden layers, and the

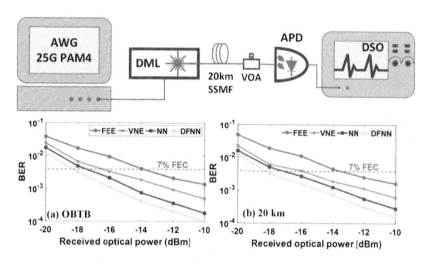

Figure 4.8 (a) experimental setup of 25 Gbaud PAM-4 IMDD system based on 10G-class optical devices. BER performance comparison with different equalization algorithms for (b) OBTB and (c) 20-km SSMF.

output layer are 51, 64, 64, and 4 respectively. We compare the BER performance of different equalization algorithms for 20-km SSMF and optical back-to-back (OBTB) transmission. The launching power into the fiber is 8 dBm. The results are shown in Fig. 4.8(b) and (c). In both cases, DFNN outperforms all other algorithms. Since the signal from AWG has a high peak–to–peak RF output voltage (Vpp), resulting in the nonlinear modulation of DML. Therefore, FFE as a linear equalizer shows its poor capability. NN-based equalizer has a lower BER than VNE, proving that NN has its advantage in equalizing nonlinear distortions, attributed to its potential to characterize nonlinearities. As expected, DFNN introduces more features for the network and effectively guides the training. As a result, it increases about 1–dB sensitivity compared to NN for transmission over 20-km SSMF.

In this work, the feature extraction is conducted at the receiver. We consider that feature engineering at the transmitter may be more effective. For example, linear pre-equalization can be regarded as a kind of feature engineering at the transmitter. It helps to expose more signal features after transmission and can simplify the PON system complexity because if NN is deployed at the OLT, all connected ONUs can share the deployment cost.

We propose to use a well-trained NN-based pre-equalizer to mitigate the SOA pattern effect in a 50G PON system. Thanks to the capability of nonlinearity compensation, the pattern effect induced dynamic range limitation can be significantly improved by the NN. By introducing training data from multiple sampling, the NN model shows better performance resisting system dynamics. Experiment demonstration is conducted to verify the effectiveness of the NN in SOA nonlinearity mitigation.

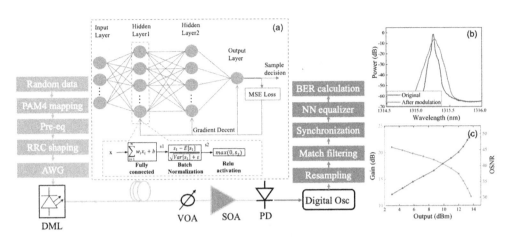

Figure 4.9 NN-based equalizer for SOA nonlinearity mitigation for 50 Gb/s/λ PAM4 PON; (a) Flow chart of NN architecture. (b) Optical spectra of the signal. (c) SOA gain and OSNR versus output power.

The experimental setup is shown in Fig. 4.9. At the transmitter side, the 25 Gbaud PAM4 signal is generated by an AWG working at 65 GSa/s with a 3-dB analog bandwidth of 20 GHz. The output signal is first amplified to 1.8 Vpp by an electrical amplifier (EA) before driving the commercial 20 Gbps DML with a 3-dB bandwidth of 12 GHz and a center wavelength of 1315 nm, and the spectra before and after modulation are shown in Fig. 4.9(b). Then, the 25 Gbaud PAM4 optical signals are transmitted over 20 km SSMF with an average loss of 0.33 dB/km at 1310 nm. An O-band SOA pre-amplifier with a noise figure of 7 dB is used at the receiver side before direct detection. The optimal SOA current is 250 mA and the corresponding operating state is shown in Fig. 4.9(c). A variable optical attenuator (VOA) is applied to emulate the splitter loss and also test the SOA performance. Note that the optical filter in our lab with 6-dB insertion loss cannot further improve the receiver sensitivity, so it is not used after the SOA for filtering amplified ASE noise. A 40G PD is used to convert the optical signal to the electrical signal. Following the PD, the output signal is captured by a 40 GSa/s oscilloscope with 20 GHz bandwidth and processed by offline DSP.

The proposed NN-based pre-equalization architecture is presented in Fig. 4.10. In the first step, the sampled signal $R_t(n)$ is used to train the NN by minimizing the mean square error (MSE) between the output $R_t(n)$ and original data $T_t(n)$. When the MSE is small enough, we consider NN is well trained until it successfully fits the inverse function $h(n)^{-1}$ of the system response. Note that $R_t(n)$ must be normalized before training to ensure the pre-NN and post-NN can provide unit gain. Once the network is converged, it can be directly put at the transmitter side without additional manipulation to deal with the nonlinear distortions induced by the SOA. In the second step, the transmitted signal is processed with the well-trained NN to generate $T_t(n)$'

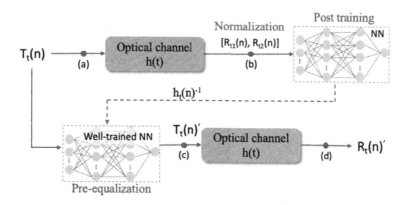

Figure 4.10 The architecture of NN-based pre-equalizer.

before launching into the transmission system. Finally, the received data $R_t(n)'$ can be recovered without using the NN for post-equalization.

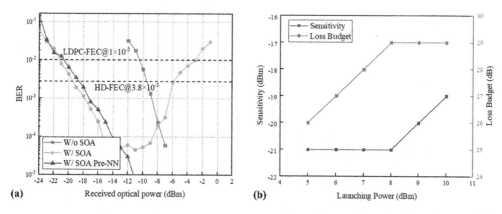

Figure 4.11 (a) BER of 25 Gbaud PAM4 signals over 20 km SMF at O-band with and without SOA. (b) Receiver sensitivity and system power budget versus DML launching power into the fiber.

To evaluate the system performance of using the SOA as a pre-amplifier, we first measure and compare the receiver sensitivity before and after using the SOA. The results are shown in Fig. 4.11(a). Without the SOA, the optical signal is directly injected into the PD for detection. The sampled signal is processed offline by a 37-tap FFE, and the receiver sensitivity of -10 dBm is achieved at the low-density parity-check forward error correction (LDPC) FEC limit of 1×10^{-2} [31]. After employing the SOA as a pre-amplifier, 10 dB sensitivity improvement can be obtained at the LDPC FEC limit. However, the BER curve changes with the input power. In the beginning, the BER drops rapidly with the increase of the received optical power thanks to the improved optical signal-to-noise ratio (OSNR). When the input power is higher than -14 dBm, the SOA is saturated and the BER stops decreasing due to the nonlinear pattern effect. When the input power surpasses -8 dBm, degraded BER performance can be observed, which can be attributed to the more serious nonlinear effect. The dynamic range of the receiver is an important performance indicator for the receivers deployed in PON. Considering the hard decision (HD) BER limit at 3.8×10^{-3}, a dynamic range is limited to 14 dB when the SOA is employed. To enlarge the dynamic range of the receiver, the nonlinear effect needs to be compensated. Following the previous analysis, the NN is first trained by 100,000 sampled at the receiver side and then applied to the transmitter side for pre-equalization. As shown in Fig. 4.11(a), the BER decreases continually with the increase of the received optical power. When the power is higher than -10 dBm, the BER is lower than 1×10^{-5} which cannot be displayed in the curve, indicating a significantly improved dynamic range of the receiver. Note that there is no optical filter used between the SOA and PD in the experiment, therefore the SOA+PD solution will be easy for integration as a high-sensitivity receiver with low cost.

We evaluate the system power budget and receiver sensitivity at different launching power for 25 Gbaud PAM4 PON (i.e., 50G PON). As shown in Fig. 4.11(b), the optimal launching power of the DML is 8 dBm and the corresponding loss budget can reach 29 dB which can meet the PR-30 requirement [32]. Higher launch power makes the DML bias at nonlinear region causing nonlinear distortion, and the receiver sensitivity and corresponding power budget will be degraded accordingly.

In conclusion, the NN-based pre-equalizer is used at the OLT to compensate for the nonlinear pattern effect in the high-speed PON system where the SOA+PD is used as a low-cost optical receiver at the ONU. Applying the equalizer at the transmitter side can relax the complexity of DSP needed at each ONU and allows for sharing the cost with all ONUs while maintaining system performance. The potential to further extend the PON system with a higher baud rate is also investigated. The experiment validation reveals that the NN-based pre-equalizer effectively compensate the SOA nonlinear pattern effect, improving the dynamic range of the receiver and offering a power budget as high as 29 dB for the 50G PON system.

In this section, we introduce the NN application at the single end in PON. NN can be deployed as pre and post equalizer and both outperform traditional equalization filters, which proving NN's powerful ability and promising future in PON.

4.4. End to end deep learning for optimal equalization

In the last section, NN has been deployed to mitigate impairments in the fiber-optic system, which itself consists of several signal processing blocks at both transmitter and receiver, carrying out equalization tasks. In principle, such a modular implementation allows the system components to be analyzed and the channel model can be partly characterized by NN. However, this approach can be sub-optimal, especially for communication systems where channel models and loss functions for which the optimal solutions are unknown. As a consequence, in some systems, a block-based receiver with one or several sub-optimum modules does not necessarily achieve the optimal end-to-end system performance. Especially if the optimum joint receiver is not known or too complex to implement, carefully chosen approximations are required.

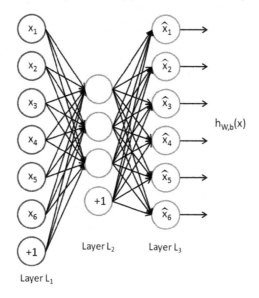

Figure 4.12 The structure of a basic autoencoder.

An autoencoder [33] shown in Fig. 4.12 is a type of artificial neural network in deep learning to learn a representation (encoding) for a set of data, typically for dimensionality reduction. It tries to learn a function $h_{w,b}(x) = x$. In other words, it is trying to learn an approximation to the identity function, to output $\hat{x}$ that is similar to x. The aim of the autoencoder is consistent with the equalization goal of the communication system, which is to equalize received signal to estimate transmitted signal $\hat{x}$ close to transmitted signal x. In this case, $h_{w,b}(\cdot)$ can be regarded as an unknown channel model. By representing transmitter, channel, and receiver as one deep neural network, an autoencoder can be trained and it is possible to learn full transmitter and receiver implementations for a given channel model which are optimized for a chosen loss function (e.g., minimizing block error rate (BLER)).

Such a novel design based on full system learning avoids the conventional modular structure, because the system is implemented as a single deep neural network, and has the potential to achieve optimal end-to-end performance. The objective of this approach is to acquire a robust representation of the input message at every layer of the network. Importantly, this enables a communication system to be adapted for the optimum receivers or optimum blocks are not known or not available due to complex reasons. The viability of such an approach has been introduced for wireless communications [34,35] and also demonstrated experimentally with a wireless link [36,37].

In Chapter 5 end-to-end deep learning will be introduced in detail. In this section, we focus on one end-to-end learning structure used in PON. As introduced in the last section, DFNN outperforms NN as an equalizer at the receiver, DFNN can also replace traditional FCNN in an autoencoder.

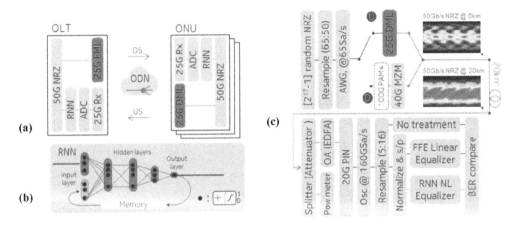

Figure 4.13 (a) 50G-NRZ (100G-PAM4) PON system architecture, (b) Insight of DFNN structure, and (c) Experimental setup for both uplink and downlink transmission.

The end-to-end learning algorithm scheme based on DFNN structure [38] is depicted in Fig. 4.13. The NN structure in the autoencoder is not FCNN. Though the author calls it RNN as shown in Fig. 4.13(b), it is the same with DFNN in the last section. So we will call this structure DFNN in this chapter. The PON experimental setup is demonstrated in Fig. 4.13(c). Signal sampled by PD is equalized by FFE and DFNN based end-to-end learning algorithms. The number of FFE taps is set to 8. The neuron number of each layers in DFNN is 15, 8, 4, 2 and 1, while for the input layer 8 are current symbols and 7 are previous time delay outputs. Data generation for experiment transmission is the same as traditional end-to-end learning and only the decoder can be retrained for new incoming datasets.

The results of upstream tests with 25G DML are shown in Fig. 4.14. We can see, only end-to-end learning can make the BER lower than the 3.8×10^{-3} threshold at the speed of 50 Gb/s and 60 Gb/s. DFNN based autoencoder outperforms FFE and hard decision under the case of BtB and 20 km fiber transmission. For larger data flow downstream tests, 40G MZM is used. The sensitivity of -22 dbm and -17.5 dbm is achieved respectively for 50 Gb/s NRZ and 100 Gb/s PAM4 at 3.8×10^{-3} threshold.

In this section, an autoencoder structure implemented in PON is introduced. It uses an improved DFNN structure in autoencoder and outperforms FFE algorithm proving its potentiality to extend to more complex models.

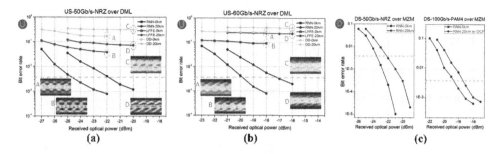

Figure 4.14 BER curves of DML based upstream (a)50 Gb/s-NRZ and (b) 60 Gb/s-NRZ detections using 3 different methods, (c) BER curves of MZM based 50 Gb/s-NRZ and 100 Gb/sPAM4 detections using RNN in downstream.

4.5. FPGA implementation of NN equalizer

In previous sections, NN's good performance as an equalizer has been proved. However, in these deep learning processors including NN equalizer and end-to-end learning equalizer, data is acquired by oscilloscope and algorithms are running offline in Matlab or Python on powerful hardware such as CPU clusters and GPUs with high power consumption. Offline algorithms can only be used to verify their effectiveness. For feasible communication applications, NN algorithms must be running in real-time platforms. The limited precision, the speed of actual DSP hardware, and the actual equipment cost also need to be considered.

Application-specific integrated circuits (ASIC) and field programmable gate arrays (FPGA) are two of the most common hardware to meet the real-time requirement. While ASIC can be customized and have a small size, but it has high initial investments and a long design cycle, and most importantly, a lack of reconfigurability, which is necessary given the NN's variable architectures and parameters [39]. Compared with ASIC, FPGAs are reconfigurable which allows quick search of vast NN configurations space and also benefit for fine-tuning of NN parameters. Hence, FPGAs present as promising platforms for real-time NN in PON. In this section, we will introduce the FPGA implementation of NN for PON.

Nokia Bell lab demonstrates an INT5 quantized 9-layer NN as the receiver DSP with 12.2 Gbit/s running in real-time on an FPGA [40]. NN structure is presented in Fig. 4.15. It's 9 layers fully connected NN, and the most different part compared to traditional NN is its activation function. The activation function of the hidden layer is clipping realized by bit-shifts which quantized the input of the next layer to INT5. The activation function of the output layer performs hard decision. Eight DNNs at a clock rate of 512 MHz are paralleled to process the inputs.

The real-time communication system experiment setup is shown in Fig. 4.16. 4.096-GBaud signed integers $1170 \times [-7, -5, -3, -1, 1, 3, 5, 7]$ PAM8 signal generated by

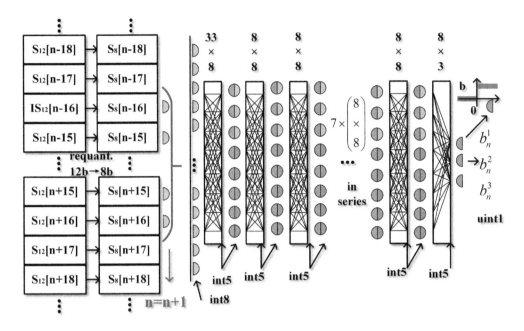

Figure 4.15 DSP directly after ADC, running in real-time on the ZCU111.

DAC is transmitted by a 1550 nm and 10 GHz bandwidth DML with -1.5-dBm launch power. After 10 km SMF, the signal is detected by PIN-TIA and sampled by an ADC at 1sps. Then the signal is quantized from 12-bit to 8-bit signed integers (from -128 to +127) and inputted into INT5 quantized 9-layer NN in which parameters are 5-bit signed integers from -16 to +15. The system can push 3 bits per channel use and deliver 12.28 Gb/s at a long-accumulated counted BER of 4×10^{-5}. This is the first demonstration of quantized NN running in real-time on FPGA for optical communication equalization.

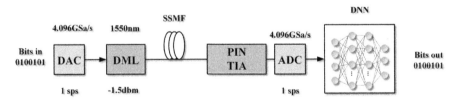

Figure 4.16 Live demonstration with NN FPGA receiver.

The practicability has been proved by the real-time demonstration of NN. However, the rate of 12.28 Gbps is much lower than nowadays 50G PON standardization. Hence, Nokia Bell lab proposes an improved NN architecture applicable for 50 Gb/s PON link [41].

The NN deployed in FPGA is a 4-layer fully connected NN with 2 symbol outputs, which means it can process 2 symbols at the same time. The experimental setup of 50 Gb/s PON is shown in Fig. 4.17. 50G NRZ signal is transmitted by 88 GSa/s DAC. A 35 GHz MZM is used to modulate the signal. The wavelength of the tunable laser is set to 1342 nm. After 30 km SMF transmission, 50.2857 Gb/s signals are received by 25 GHz APD+TIA and captured by sampling scope. After offline resample, signal, as well as the weights and bias values, are stored first in SD card memory, then read into the DDR4 memory and finally loaded into FPGA for DNN online test. The FPGA clock is 325 MHz and 4 NNs are parallelized. The total data rate is 2.6 Gb/s. To meet the 50 Gb/s requirements, the same 20 NNs are needed.

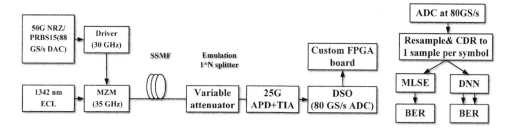

Figure 4.17 Left: Experimental setup of 50 Gb/s downstream PON at 1342 nm. Right: DSP block diagram for the MLSE and DNN.

NN performance is compared with 6 taps MLSE. When the BER threshold is set to 1×10^{-2}, the NN outperforms MLSE by 0.7 dB in sensitivity with 30 km fiber transmission and ~0.2 dB with BtB transmission.

In this section, two NNs implemented in FPGA are achieved to equalize signal in PON. A 9 layers INT5 quantized NN is used to process real-time 12.28 Gbps PAM8 signal and BER of 4×10^{-5} is reached. Further, a fixed-point NN embedded in FPGA for 50 Gbps PON downstream link outperforms MLSE in receiver sensitivity.

4.6. Conclusions and perspectives

In this chapter, firstly we introduce the definition of PON and the evolution of PON standardization. Two general features of PON are a passive network and a point-to-multipoint topology. Typically, it takes 6–7 years for one generation of PON's wide deployment. After knowing the application scenario, the effectiveness of NN is verified before regarding as an equalizer in PON. In Section 4.2, the mechanism that NN characterizes the PRBS rule is clarified and an effective method to verify training effectiveness is provided. Then, NN is utilized as pre and post equalizer to mitigate non-linearity in fiber and optic devices in Section 4.3, which presents excellent performance compared with traditional FFE and VNE algorithms. Though NN can compensate part of impairments in the system, it has the sub-optimal problem. An improved structure based on autoencoder attempts to design transceiver as a single NN is proposed and proved to be suitable for communication over channels where the optimum transceiver is unknown. For real-time implementation, the successful FPGA implementation of NN as an equalizer in the communication system is presented in Section 4.5, which paves the road for the practical application of NN in future PON.

NN has been proved to be an effective and powerful tool in PON, its potentialities will be brought into full play if advanced structures, various loss functions, and diverse optimization algorithms are carefully exploited. Novel structures are encouraged to be investigated in the advanced image processing and speech recognition field, but these structures working together with existing mature algorithms deserve more consideration. On the other hand, NN is a universal tool that can model any complex function but it works like a black box. Our research is far from figuring out the detailed working principle of NN. Therefore, to fully comprehend the deep learning theory and construct explainable NN is also a key research point. Moreover, NN coefficients training always take too much time and hardware resources compared with traditional algorithms when dealing with specific problems. How to simplify the computation complexity of NN achieving real-time FPGA implementation is also a potential research direction, this research will help to accelerate the real application of NN algorithms in the business scenario and make NN a useful tool in everyone's life.

References

[1] D. Hood, E. Trojer, Gigabit-Capable Passive Optical Networks, Wiley, 2012.
[2] J.S. Zou, S.A. Sasu, M. Lawin, et al., Advanced optical access technologies for next-generation (5G) mobile networks, Journal of Optical Communications and Networking 12 (10) (2020) D86–D98.
[3] C. DeSanti, L. Du, J. Guarin, et al., Super-PON: an evolution for access networks, Journal of Optical Communications and Networking 12 (10) (2020) D66–D77.
[4] J.S. Wey, The outlook for PON standardization: a tutorial, Journal of Lightwave Technology 38 (1) (2020) 31–42.
[5] D. Nesset, PON roadmap, IEEE/OSA Journal of Optical Communications and Networking 9 (1) (2017) A71–A76.
[6] Draft standard for Ethernet amendment: physical layer specifications and management parameters for 25 Gb/s and 50 Gb/s passive optical networks, in: IEEE P802.3ca D3.1, 2020.
[7] E. Harstead, R. Bonk, S. Walklin, et al., From 25 Gb/s to 50 Gb/s TDM PON: transceiver architectures, their performance, standardization aspects, and cost modeling, Journal of Optical Communications and Networking 12 (9) (2020) D17–D26.
[8] B. Li, K. Zhang, D. Zhang, et al., DSP enabled next generation 50G TDM-PON, Journal of Optical Communications and Networking 12 (9) (2020) D1–D8.
[9] Z. Wan, J. Li, L. Shu, M. Luo, X. Li, S. Fu, K. Xu, Nonlinear equalization based on pruned artificial neural networks for 112-Gb/s SSB-PAM4 transmission over 80-km SSMF, Optics Express 26 (8) (2018) 10631–10642.
[10] L. Yi, P. Li, T. Liao, W. Hu, 100Gb/s/λ IM-DD PON using 20G-class optical devices by machine learning based equalization, in: European Conference on Optical Communication, 2018, paper Mo4B.5.
[11] J. Estaran, R. Rios-Müller, M.A. Mestre, F. Jorge, H. Mardoyan, A. Konczykowska, J.Y. Dupuy, S. Bigo, Artificial neural networks for linear and non-linear impairment mitigation in high-baudrate IM/DD systems, in: European Conference on Optical Communication, 2016, paper M.2.B.2.

[12] T. Jones, S. Gaiarin, M.P. Yankov, D. Zibar, Noise robust receiver for eigenvalue communication systems, in: Optical Fiber Communication Conference, 2018, paper W2A.59.

[13] C. Hager, H.D. Pfister, Nonlinear interference mitigation via deep neural networks, in: Optical Fiber Communications Conference, 2018, paper W3A.4.

[14] M.A. Nielsen, Neural Networks and Deep Learning, Determination Press, 2015 [online], available: http://neuralnetworksanddeeplearning.com/.

[15] F.N. Khan, K. Zhong, X. Zhou, W.H. Al-Arashi, C. Yu, C. Lu, A.P.T. Lau, Joint OSNR monitoring and modulation format identification in digital coherent receivers using deep neural networks, Optics Express 25 (15) (2017) 17767–17776.

[16] F.N. Khan, Y. Yu, M.C. Tan, W.H. Al-Arashi, C. Yu, A.P.T. Lau, C. Lu, Experimental demonstration of joint OSNR monitoring and modulation format identification using asynchronous single channel sampling, Optics Express 23 (23) (2015) 30337–30346.

[17] L. Huang, L. Xue, Q. Zhuge, et al., Modulation format identification under stringent bandwidth limitation based on an artificial neural network, OSA Continuum 4 (1) (2021) 96–104.

[18] D. Wang, Y. Song, J. Li, et al., Data-driven optical fiber channel modeling: a deep learning approach, Journal of Lightwave Technology (2020).

[19] H. Yang, Z. Niu, S. Xiao, et al., Fast and accurate optical fiber channel modeling using generative adversarial network, Journal of Lightwave Technology (2020).

[20] A. Eriksson, H. Bülow, A. Leven, Applying neural networks in optical communication systems: possible pitfalls, IEEE Photonics Technology Letters 29 (23) (2017) 2091–2094.

[21] L. Shu, J. Li, Z. Wan, W. Zhang, S. Fu, K. Xu, Overestimation trap of artificial neural network: learning the rule of PRBS, in: European Conference on Optical Communication, 2018, paper Tu4F.

[22] L. Yi, T. Liao, L. Huang, W. Hu, Machine learning for 100Gb/s/λ passive optical network, Journal of Lightwave Technology 37 (6) (2018) 1621–1630.

[23] T. Liao, L. Xue, L. Huang, et al., Training data generation and validation for a neural network-based equalizer, Optics Letters 45 (18) (2020) 5113–5116.

[24] H. Xin, et al., 120 GBaud PAM-4/PAM-6 generation and detection by photonic aided digital-to-analog converter and linear equalization, Journal of Lightwave Technology 38 (8) (2020) 2226–2230.

[25] P. Torres-Ferrera, H. Wang, V. Ferrero, M. Valvo, R. Gaudino, Optimization of band-limited DSP-aided 25 and 50 Gb/s PON using 10G-class DML and APD, Journal of Lightwave Technology 38 (3) (2020) 608–618.

[26] A. Masuda, S. Yamamoto, H. Taniguchi, M. Nakamura, Y. Kisaka, 255-Gbps PAM-8 transmission under 20-GHz bandwidth limitation using NL-MLSE based on Volterra filter, in: Optical Fiber Communication Conference, 2019, paper W4I.6.

[27] C. Hager, H.D. Pfister, R.M. Butler, G. Liga, A. Alvarado, Revisiting multi-step nonlinearity compensation with machine learning, in: European Conference on Optical Communication, 2019, paper W.3.B.

[28] K. Wang, et al., Mitigation of pattern-dependent effect in SOA at O-band by using DSP, Journal of Lightwave Technology 38 (3) (2020) 590–597.

[29] L. Yi, T. Liao, L. Xue, et al., Neural network-based equalization in high-speed PONs, in: 2020 Optical Fiber Communications Conference and Exhibition (OFC), IEEE, 2020, pp. 1–3.

[30] L. Xue, L. Yi, P. Li, W. Hu, 50-Gb/s TDM-PON based on 10G-class devices by optics-simplified DSP, in: Proc. Opt. Fiber Commun. Conf., 2018, paper M2B. 4.

[31] IEEE P802.3ca 50G-EPON Task Force, Physical layer specifications and management parameters for 25 Gb/s and 50 Gb/s passive optical networks, http://www.ieee802.org/3/ca/index.shtml.

[32] V. Houtsma, D. Veen, E. Harstead, Strategies for economical next-generation 50G and 100G passive optical networks, Journal of Lightwave Technology 35 (2017) 1228.

[33] A. Ng, Sparse autoencoder, CS294A Lecture Notes 72 (2011) 1–19.

[34] T.J. O'Shea, K. Karra, T.C. Clancy, Learning to communicate: channel auto-encoders, domain specific regularizers, and attention, in: 2016 IEEE International Symposium on Signal Processing and Information Technology (ISSPIT), 2016, pp. 223–228.

[35] T.J. O'Shea, J. Hoydis, An introduction to deep learning for the physical layer, IEEE Transactions on Cognitive Communications and Networking 3 (4) (2017) 563–575.

[36] S. Dörner, S. Cammerer, J. Hoydis, S. ten Brink, Deep learning based communication over the air, IEEE Journal of Selected Topics in Signal Processing 12 (1) (2018) 132–143.

[37] B. Zhu, J. Wang, L. He, J. Song, et al., Joint transceiver optimization for wireless communication PHY using neural network, IEEE Journal on Selected Areas in Communications 37 (6) (2019) 1364–1373.

[38] C. Ye, D. Zhang, X. Hu, et al., Recurrent neural network (RNN) based end-to-end nonlinear management for symmetrical 50Gbps NRZ PON with 29dB+ loss budget, in: 2018 European Conference on Optical Communication (ECOC), IEEE, 2018, pp. 1–3.

[39] S. Himavathi, D. Anitha, A. Muthuramalingam, Feedforward neural network implementation in FPGA using layer multiplexing for effective resource utilization, IEEE Transactions on Neural Networks 18 (3) (2007) 880–888.

[40] M. Chagnon, J. Siirtola, T. Rissa, et al., Quantized deep neural network empowering an IM-DD link running in realtime on a field programmable gate array, in: 2019 European Conference on Optical Communication (ECOC), IEEE, 2019, pp. 1–3.

[41] N. Kaneda, Z. Zhu, C.Y. Chuang, et al., FPGA implementation of deep neural network based equalizers for high-speed PON, in: 2020 Optical Fiber Communications Conference and Exhibition (OFC), IEEE, 2020, pp. 1–3.

## CHAPTER FIVE

End-to-end learning for fiber-optic communication systems

Ognjen Jovanovic, Francesco Da Ros, Metodi Yankov, and Darko Zibar
Department of Photonic Engineering, Technical University of Denmark, Kgs. Lyngby, Denmark

5.1. Introduction

The modern digital communication system is designed in a modular fashion, as shown on Fig. 5.1. Both transmitter and receiver consist of a chain of multiple independent signal processing blocks, such as channel coding, modulation, pulse shaping, etc. In principle, such approach presents a beneficial way of implementing a communication system because each of the signal processing blocks can be optimized and controlled individually. Even though this approach has proven to be efficient, it is unclear if the end-to-end performance of such a system is optimal since the signal processing blocks are not optimized jointly. Furthermore, most of these signal processing algorithms have been developed under the assumption that component and channel models are stationary, linear and Gaussian. In practice this is often not the case. Most of the component and channel models are nonlinear and have imperfections, which can only be approximated using mathematical models. If telecommunication systems are going to sustain the ever-growing data rate demand of consumers, joint optimization of the signal processing blocks, as well as accounting for nonlinearities and system imperfections need to be addressed.

In the last decade, the research in machine learning has exploded and its application has found its way into optimization of communication systems [1,2]. Artificial neural networks (ANNs) are known as universal function approximators [3] which stand out in their ability to model nonlinearities. Therefore, ANNs have found their application in fiber optical channel equalization in the presence of chromatic dispersion and nonlinear Kerr effects, which are considered the main limitation to the achievable information rate (AIR). These ANNs have been applied to optimize individual functions of the receiver with respect to nonlinear channel models, but not to their joint optimization with the transmitter and other signal processing blocks.

This ANN shortcoming can be addressed by another type of deep neural network known as an autoencoder (AE). It consists of an encoder and decoder with the goal of copying its input to the output. The encoder processes the input data and encodes it into a new feature space, depending on the desired purpose. The decoder reproduces

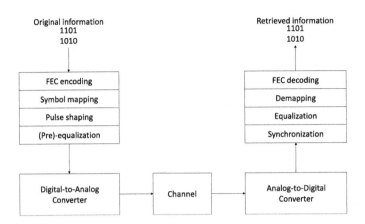

Figure 5.1 Key blocks of a communication system, including digital signal processing (DPS) chain.

the input data of the encoder using this new space. This process is identical to a typical communication system, where the transmitter processes the data in such a way that it can be efficiently and reliably transmitted over a channel and then the receiver reproduces what was sent. Therefore, in [4] it was suggested to replace the above-mentioned block chain structure of the transceiver by an AE, which can be used to jointly optimize the desired subset of blocks for the channel of interest. For instance, the constellation shaping and modulation block used in optical communication, where the channel model is nonlinear, could be optimized for this explicit channel. The gains of using this approach for a nonlinear channel model are shown in [5–9]. The AE approach can be expanded to optimize the pulse shaping alongside the constellation shaping, i.e. to produce optimized waveforms. Such an AE has been applied to short range optical communication channel which mainly suffers from chromatic dispersion and relies on the intensity modulation/direct detection (IM/DD) transmission [10–12]. The AE setup was also combined with the split-step Fourier method (SSFM) [13] in order to capture the behavior of the optical fiber. It has also been shown that end-to-end learning AE approach was highly beneficial for the optimization soliton-based transmission [14,15].

All of the aforementioned examples use a differentiable channel model because the optimization of an AE usually relies on backpropagation of error gradients. The requirement of a differentiable channel model is a limiting factor because not all channel models are differentiable. Furthermore, it can be challenging to apply the AE approach in an experimental environment where gradients cannot be computed. The experimental environment can be modeled with a simple differentiable model which can result into model discrepancy or an accurate differentiable model which can be too complicated for reliable optimization. The issue of modeling the experimental environment can be overcome by using generative adversarial network (GAN) which can accurately represent a wide range of stochastic channel effects. This approach provides a differentiable channel model and allows AE training to be performed on real world measurement data as shown in [16–19]. However, instead of replacing the channel model, the training of the AE setup could be done by deploying gradient-free optimization algorithms instead of the standard gradient descent optimization approach as in [20]. This method is described later in this Chapter.

This Chapter is organized as follows: In Section 5.2, we discuss the idea of end-to-end learning/optimization, where we provide a more detailed description of this principle. Section 5.3 provides the state-of-the-art in end-to-end learning for fiber-optic communication based on several typically used channel model. Section 5.4 details how an AE setup could be used in practice to perform online training on non-differentiable channel models. Finally, the Chapter is concluded in Section 5.5 which outlines its main takeaways and key points.

5.2. End-to-end learning

The concept behind a communication system is such that the data fed to the transmitter should be reproduced with high fidelity at the output of the receiver. This is achieved by employing different digital signal processing (DSP) and error-correction methods, where a typical transmitter-channel-receiver chain is presented on Fig. 5.1. Even though this has proven to be an efficient and controllable system, it is unclear if its end-to-end performance is optimal. The overall system and performance for different channel models are sub-optimal, since each of the DSP blocks is optimized individually and mainly assuming Gaussian statistics.

Autoencoders have the same principle as a communication system, where the input to the encoder is reproduced at the output of the decoder. Therefore, by constructing the encoder and decoder in such a way that the output of the encoder and input of the decoder are attached to the channel model, they can replace the transmitter and receiver, respectively. In such a structure, a channel model is embedded in between the encoder and decoder. This provides a model resembling a transmitter-channel-receiver chain as shown in Fig. 5.2. The chain can be mathematically described as follows

$$s \xrightarrow{f_{w_e}(\cdot)} x \xrightarrow{channel} y \xrightarrow{g_{w_d}(\cdot)} r, \tag{5.1}$$

where s is the information to transmit which the encoder $f_{w_e}(\cdot)$ maps to the latent variable x, and then from its impaired version y the decoder $g_{w_d}(\cdot)$ typically produces posterior probabilities of the transmitted information, denoted r. From the posteriors, an estimate $\hat{s}$ of s can be obtained. The trainable weights of the encoder and decoder neural networks are described by $\mathbf{w_e}$ and $\mathbf{w_d}$, respectively. The set that holds all the trainable parameters will be represented as $\mathbf{w} = \{\mathbf{w_e}, \mathbf{w_d}\}$. From a deep learning perspective, structuring the transmitter-channel-receiver chain in such a way allows it to be

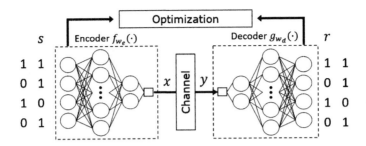

Figure 5.2 Example of End-to-end autoencoder setup.

interpreted as a computational graph. Making it a requirement for the channel model to be differentiable in order to perform joint optimization. The interpretation of the transmitter-channel-receiver chain as a computation graph is important for the optimization which is commonly done using gradient based algorithm.

The computational graph can be used to jointly optimize all of the functionalities of the transmitter and receiver, and is thus referred to as *end-to-end learning*. In principle, the AE does not necessarily have to replace the transmitter and receiver entirely, but they can also replace just certain DSP block. Replacing only specific DSP blocks with AE allows for those DSP blocks to be optimized for different configurations of the rest of the system. Examples are different channel models, different FEC decoders, and/or other transceiver functionalities, such as equalization. The overall goal of the optimization is to produce an encoder which generates a signal robust to channel impairment, and a decoder which reconstructs the original signal from the received data with high fidelity. Depending on the application, the system can have different structures and can be optimized with respect to different cost functions $L(\mathbf{w})$. The following discussion includes the topology and application of AE. First, when the optimization is applied on a symbol level by maximizing the mutual information (MI) cost function. Second, when generalized mutual information (GMI) is maximized, i.e. optimizing on a bit level. Discussion about optimization methods and machine learning frameworks follows.

The AE must be constructed in such a way that the input and output fit the dimension of the desired application. The input and output of the AE can be one-hot encoded vectors, $s \in S = \{e_i | i = 1, \dots, M\}$, where M is the size of the constellation alphabet and S represents the set of all possible symbols. In this case, the last layer of the decoder is a softmax function, which produces a posterior probability of each received symbol. The AE can be optimized by minimizing the cross-entropy cost function [21] Eq. (5.2)

$$L(\mathbf{w}) = \frac{1}{K} \sum_{k=1}^{K} \left[-\sum_{i=1}^{M} s_i^{(k)} \log \left(r_i^{(k)} \right) \right], \tag{5.2}$$

where K is the batch size and the index (k) indicates the symbol from the batch. For uniformly distributed data, minimizing the cross-entropy is equivalent to maximizing the MI between the input and output.

A more general approach is to use block of bits, $s \in S = \{0, 1\}^m$ where $m = \log_2 M$, as input and output of the AE. In this case the last decoder layer is a logistic sigmoid function, which allows the decoder to provide the posterior probability of the bits being 0 or 1, $p(s|y)$, provided the input signal to the decoder. The goal of this setup is to maximize the GMI, in which case the cross-entropy cost function is simplified to the log-likelihood cost function [22]

$$L(\mathbf{w}) = \frac{1}{K} \sum_{k=1}^{K} \left[-\frac{1}{m} \sum_{i=1}^{m} s_i^{(k)} \log (r_i^{(k)}) + (1 - s_i^{(k)}) \log (1 - r_i^{(k)}) \right]. \tag{5.3}$$

These two approaches, MI and GMI maximization, are most commonly found in research papers covering this topic since they are used for classification. There is also the possibility of using the mean square error (MSE) as the cost function [22]

$$L(\mathbf{w}) = \frac{1}{K} \sum_{k=1}^{K} \left[\frac{1}{2} \frac{1}{N} \sum_{i=1}^{N} (s_i^{(k)} - r_i^{(k)})^2 \right] \tag{5.4}$$

where N is the number of input/outputs of the AE. The MSE is most commonly used for regression, and in this case the input and output of the AE are the complex-valued constellation points or the time-domain waveforms (typically represented by 2D real-valued vectors). Using MSE as cost function is more general since it is not constrained to a given constellation format, but it typically suffers from performance penalty with regards to the cross-entropy cost function. The penalty comes from the requirement of a subsequent decision function that produces the posterior, which potentially does not match the channel.

The input and the output space of the AE is important because it dictates if the processing is done on symbols, bits, or samples level, simultaneously determining the cost function in use. Depending on how the latent variable x (channel input) is chosen, the optimization of a constellation or waveform can be performed. It should be emphasized that the commonly used ANNs are real-valued, therefore each of the complex-valued samples are commonly represented with a 2D real-valued vectors.

Apart from the topology of the AE setup, it is crucial to determine which optimization algorithm should be used. The optimization of the system is usually done by using backpropagation and the steepest gradient decent method that uses momentum, such as the Adam optimizer [23]. To apply such an optimizer, machine learning framework such as TensorFlow [24] and PyTorch [25] have been developed. These frameworks backpropagate the gradients using automatic differentiation [26,27] on the graph representation of the cost function. Essentially, the chain rule for differentiation is applied repeatedly to a sequence of functions with respect to the set of parameters. Thus, the gradient of arbitrarily deep differentiable mathematical structures can be calculated and used for optimization.

5.3. End-to-end learning for fiber-optic communication systems

The description of end-to-end learning was kept general because the method can be used in both wireless and optical communication. The main focus of this Chapter is on the application of end-to-end learning in optical communication. End-to-end learning has been applied in optical communication for different combinations of modulations and physical channels and in this section, we will go through the state-of-the-art in this topic. This section is organized mainly based on the modulation type and afterwards by the channel models in use.

The waveform propagation through the optical fiber is described using the Nonlinear Schrödinger equation (NLSE) as given in [28] Eq. (5.5),

$$\frac{\partial A}{\partial z} = \underbrace{-\frac{\alpha}{2}A}_{Fiber\ loss} \overbrace{-\frac{\beta_2}{2}\frac{\partial^2 A}{\partial t^2}}^{Chromatic\ dispersion} \underbrace{+i\gamma|A|^2 A}_{Nonlinear\ Kerr\ effect}, \tag{5.5}$$

where A is the amplitude of the signal, α is the attenuation coefficient, β_2 the group-velocity dispersion (GVD) parameter and γ is nonlinearity parameter. The NSLE consists of three main components fiber loss, chromatic dispersion and the nonlinear Kerr effect.

When it comes to short reach communication such as data centers, access and metro links, IM/DD systems are the preferred choice [29]. In such a system intersymbol interference (ISI) occurs due to the chromatic dispersion and it is the main limiting factor. Nonlinear distortion also exists in the IM/DD channel model in the form of square-law detection, as a result of photodiodes which are used to detect the intensity of the received optical signal and perform opto-electrical conversion.

Further, moving onto long haul communication coherent transmission is the preferred choice, for which many different channel models have been developed. These models approximate the NLSE or focus on one of its components, such as the nonlinear Kerr effect. The fiber-optic channel can be modeled using approximate perturbation models, which represent the distortion which the channel introduces as additive Gaussian noise. From these channel models two are singled out, Gaussian noise (GN) model [30,31] and the nonlinear interference noise (NLIN) model [32–34]. The GN model takes into account the dependency of the nonlinear distortion with respect to launch power and the channel memory is modeled as additional noise. Since it is represented as noise, it comes at a low computational cost. However, the NLIN model takes all of this into account, but also models the dependency of the nonlinear distortion to the high-order moments of the modulation with almost no extra computation expense.

Other models that will be considered are based on analytical expressions of parts of the NLSE or its numerical solution. A memoryless channel, which takes only the nonlinear effect for fiber-optics communication, can be used as defined in [35], so called nonlinear phase noise channel (NLPN). The NLSE does not have an analytical solution, but can be numerically solved using Split-Step Fourier method (SSFM) [28,36,37] that most accurately models the fiber channel. The SSFM models the complete wavelength division multiplexing (WDM) waveform and includes the channel memory and the nonlinear effect dependency on the modulation, but at a high computational cost.

As it was mentioned before end-to-end learning typically uses gradient based back-propagation algorithm, therefore it is limited by the fact that the channel models have to be known and differentiable. The drawbacks of requiring a known and differentiable channel model will be discussed with more detail in the following section. A potential solution to these drawbacks is to model channels with generative adversarial network (GAN) [38], which was demonstrated in [16–18]. The generative model of the channel will be considered as a separate channel model.

5.3.1 Direct detection

When it comes to IM/DD systems for short range communication, end-to-end learning was introduced in [10]. End-to-end learning here is used to optimize the transceiver so it mitigates the nonlinear effect of the photodiode detection and fiber dispersion while also having the presence of noise produced by the amplifier and the analogue-to-digital converter (ADC) and digital-to-analogue converter (DAC). The AE is a fully-connected feed-forward neural network (FFNN), where a one-hot encoded vector of size M is the input to the encoder which directly maps it to a set of robust waveforms instead of modulation symbols. The encoder consists of two hidden layers with $2M$ nodes each. The two hidden layers have ReLU as the activation function [39]. The output layer has n nodes, where n is the number of samples that each of the waveforms has. In order to properly simulate the fiber dispersion introduced memory N consecutive symbols are transmitted. The encoder is followed by a low pass filter (LPF) of 32 GHz bandwidth and a DAC with an effective number of bits (ENOB) that is 6. The DAC introduces quantization noise, which is modeled by a additive, uniformly distributed noise with a ENOB depending variance [40]

$$\sigma_q^2 = 3P \cdot 10^{-(6.02 \cdot \text{ENOB} + 1.76)/10} \tag{5.6}$$

where P is the average power of the input signal. The signal is modulated by an Mach-Zehneder modulator (MZM) modeled by a sine function [41]. The analytical solution in the frequency domain to the chromatic dispersion part of Eq. (5.5) is

$$D(z, w) = e^{i\frac{\beta_2}{2} w^2 z} \tag{5.7}$$

where w is the angular frequency. The distorted signal is received by square-law detection and amplified. The amplification adds additive white Gaussian noise (AWGN) with a variance dictated by the signal-to-noise ratio (SNR). The signal is filtered with a LPF and afterwards fed to an ADC, where noise is added as with the DAC. The central symbol with n samples is extracted and processed by the decoder. The decoder consists of two hidden layers with $2M$ nodes each with ReLU activation function. The output layer has M nodes and a softmax activation function allowing it to make a decision which symbol has been transmitted. The system, i.e. autoencoder, has an effective information rate $R = \log_2(M)$ bits/symbol and a sample rate of $F_s = 336$ GSa/s.

The autoencoder was initially trained for particular distances and the achieved bit error rate (BER) was observed in comparison to the hard decision forward error correction (HD-FEC) threshold. It was shown that a consequence of this training approach is rapid increase of BER when the transmission distance changes, meaning lack of robustness to distance variation. Afterwards, it was shown that if training on varying distance, the model would be more robust over a certain distance span at the cost of having poorer performance at the mean distance. The system was trained for different information rates by changing the input size and the size of the waveform. Performance was observed by examining BER with respect to transmission distance, showing that the BER decreases when the waveform size increases or the input sizes decreases. The results achieved in simulation have been experimentally tested and compared to standard transceiver setup showing potential improvement of around two orders of magnitude in respect to BER at a given distance. It should be emphasized that the decoder had to be retrained with experimental data because it could not fully compensate the distortion occurring in the experimental setup.

The limitation of using feed-forward neural networks is that they cannot address the memory that is introduced in the channel. Therefore, as an improvement, bidirectional recurrent neural networks (BRNN) were adopted in [11] since they use internal states to process sequential data. In [11], a sliding window method at the receiver was applied, where they have processed W blocks of samples at a time, meaning that for each input to the encoder the decoder provides W probability estimates. These probabilities are averaged in order to obtain the final estimate on which the decision would be made. Using this approach provided around 20 km distance increase at the HD-FEC threshold compared to end-to-end learning with FFNN. Further improvement was shown in [12], where a weighted average of the W probability estimates was performed for the sliding window at the receiver, as well as optimizing bit-to-symbol mapping in order to reduce the BER of the system. Here, they have also shown that the sliding window BRNN (SBRNN) can outperform a setup where PAM modulation is transmitted and maximum likelihood (ML) sequence detection used at the receiver. Furthermore, another advantage was shown in computational complexity, where the detector has exponential dependence on the processed memory, while the AE has linear dependence. This work was also discussed in [42,43].

In order to perform training on experimental data, the channel model in [10] was replaced by GAN in [19]. The FFNNs were initialized with the weights obtained from [10], and the encoder generates the waveform sequence, which is fed to the experimental setup. Afterwards, the received data is processed using the decoder to estimate the BER. The transmitted and received waveform sequences were used to train the generative model, which allowed the calculation of the gradient backpropagation and the optimization of the transceiver parameters. It was required to re-train the GAN after each training iteration of the transceiver, which was time consuming. This proved to be the time-limiting factor in the setup and only 10 iterations were performed. It was shown that the BER decreases with each iteration, providing proof-of-concept.

5.3.2 Coherent systems

Amplified and unamplified links including binary mapping were combined with end-to-end learning in [44] to improve the performance of an optical coherent system. The latent variables of the AE represent the constellation symbols and the optimization was performed with respect to GMI. The constellations obtained by training on the amplified link, including bit mapping with and without label extension, achieved gain of 1 and 0.5 dB of optical signal-to-noise ratio (OSNR) at the Turbo Product Codes (TPC) FEC limit of $2 \cdot 10^{-2}$ for 32QAM in 800 Gb/s, respectively. While gain of 1.2 dB was achieved at the same limit by using the constellation optimized on amplified link with bit mapping for 128QAM in 1 Tb/s transmission. Considering the constellation acquired by training on unamplified systems the gain was 0.25 and 0.55 dB in peak OSNR [45] for coherent 800 Gb/s and 1 Tb/s transmission, respectively.

5.3.2.1 Nonlinear phase noise channel

In [7], end-to-end learning was performed on a nonlinear phase noise channel, which is a per-sample recursion as defined in Eq. (5.8),

$$x_{k+1} = x_k e^{iL\gamma|x_k|^2/K} + n_{k+1}, \quad 0 \le k \le K, \tag{5.8}$$

where x_0 is the output of the encoder, x_k is the input to the decoder, $n_{(k+1)}$ is Gaussian noise, L is the total length of the link and K is the number of iterations to simulate the fiber model. This model assumes ideal distributed amplification. This channel was chosen due to the fact that its channel probability density function is known analytically [35,46,47]. The encoder maps one-hot encoded vectors of size M onto a single constellation point, performing geometrical constellation shaping. The encoder consists of six hidden layers with M nodes each, applying tanh activation function, whereas the output layer is linear. The decoder has seven hidden layers with M nodes each. The nodes of the hidden layer use tanh activation function, whereas the output layer uses a sigmoid function. Since the standard softmax function was not used, the output of the decoder is normalized in order to provide the probability of the symbols. It has been shown that by applying end-to-end learning the nonlinear effect of the channel could be mitigated significantly, approaching the same performance as a ML detector [46,47]. The used channel is not a practical one, but the benefit of this work is that it shows that with the autoencoder ML detection can be achieved. This shows potential that ML detection could be achieved for the channels without a known ML detector. The implementation can be found online [48].

5.3.2.2 Perturbation models (NLIN and GN)

Perturbation models were used as the embedded channel in an AE in [5,6] to perform geometric constellation shaping for $M = \{64,256\}$. The performance of the AE optimized constellations was compared to standard modulation formats, such as quadrature amplitude modulation (QAM) and iterative polar modulation (IPM) [49–51] in terms of MI. Both the encoder and decoder had the same architecture. In the case of M=64, it was a single hidden layer NN with 16 nodes using ReLU activation function. In the case of M=256, it was a two hidden layer NN with 32 nodes each using ReLU activation function. The discussion also included the comparison of acquired constellation when training using GN and NLIN models. These channels are quite suitable for training of an AE because they do not add extra computational expenses to the simulation. In order to validate the performance of the learned constellations, they were tested over the fiber channel modeled with the SSFM. The channel model uses following fiber parameters $\alpha = 0.2 \frac{\text{dB}}{\text{km}}$, $D = 16.48 \frac{\text{ps}}{(\text{nm}\cdot\text{km})}$, $\gamma = 1.3 \frac{1}{(\text{W}\cdot\text{km})}$ with erbium-doped fiber amplifier (EDFA) with noise figure of 5 dB, where one span is 100 km long. The channel model includes chromatic dispersion and nonlinear compensation at the receiver.

Observing the learned constellations for different launch powers for these two models we can notice that the AE trained with GN model relies only on maximizing the MI at a given effective signal-to-noise ratio (SNR). The effective SNR is governed by the launch power in this model. Thus, none of the nonlinear effects are mitigated. While this is not the case when the AE was trained using the NLIN model since it takes into account the high-order moments of the constellation. This can be seen on Fig. 5.3, where the two model learn similar constellations at per channel launch power of -6.5 and 0.5 dBm, which achieve similar MI. But at a high presence of nonlinear effect the AE trained with GN fails in finding a suitable constellation to overcome this, which can be best seen by the difference of 0.54 bits/4D-symbols in MI between the two models.

Fig. 5.4 illustrates the effective SNR and MI for M = 64, 256 optimized with the AE and compares it to QAM and IPM when used with SSFM. It can be noticed that

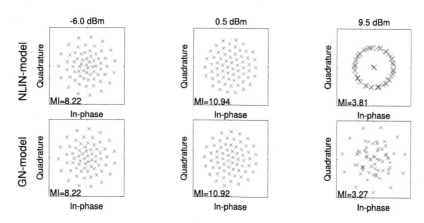

Figure 5.3 Constellation learned with **(top)** the NLIN-model and **(bottom)** the GN-model for M=64 and 1000 km (10 spans) transmission length at per channel launch power **(left to right)** -6.5 dBm, 0.5 dBm and 9.5 dBm. The MI is represented in bits/symbol. Adapted from [5].

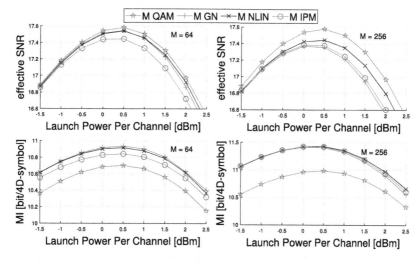

Figure 5.4 Performance in effective SNR **(top)** and MI **(bottom)** in respect to launch power after 1000 km transmission (10 spans) for M =64 **(left)** and M = 256 **(right)**. Adapted from [6].

even though in both cases QAM achieves the best effective SNR value, its performance in terms of MI is the lowest. The effective SNR of IPM is one of the lowest, but it has an upper edge compared to QAM when it comes to MI. Finally, the best performing constellation was the one acquired by training the AE with the NLIN channel model, reaching up to 0.13 bit/4D-symbol of gain in MI compared to IPM.

Fig. 5.5 illustrates the gain of M-NLIN, M-GN, M-IPM compared to the standard M-QAM at optimal launch power with respect to the number of spans. It can be observed that with the increase of distance the gain of the constellation acquired with

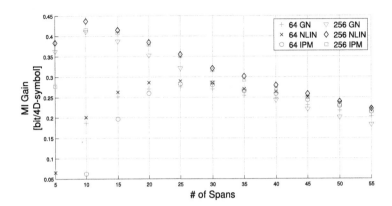

Figure 5.5 Gain compared to the standard M-QAM constellation at the respective optimal launch power in respect to the number of spans estimated. Adapted from [6].

AE overlaps with the gain of the IPM. Furthermore, the gain does not depend on the modulation order at 50 and 55 spans. Compared to the IPM, the constellations learned by the AE achieve the highest gain at shorter transmission distances.

This could have been further improved by including bit mapping as shown in [8], where the goal was to maximize the GMI using Eq. (5.3). The embedded channel used for training was the NLIN channel model, while the testing has been performed on the fiber channel modeled with the SSFM. The channel model uses standard single mode fiber parameters $\alpha = 0.2 \ \frac{\mathrm{dB}}{\mathrm{km}}$, $D = 17 \ \frac{\mathrm{ps}}{(\mathrm{nm \cdot km})}$, $\gamma = 1.3 \ \frac{1}{(\mathrm{W \cdot km})}$, where one span is of 80 km length and includes chromatic dispersion and nonlinear compensation at the receiver. This model optimizes simultaneously the bit mapping and the position of the constellation point, achieving a non-rectangular geometric constellation with a Gray-like code. The setup was adopted from [52] and is a QPSK-hybrid quasi-pilot-based DSP chain which takes advantage of the modulation format. The transmitter includes FEC-encoding, interleaving of the modulation symbols with 10% QPSK symbol. Frame synchronization is done by inserting a Zaddoff-Chu preamble sequence before the first transmitted block. The waveform that is sent to the channel is obtained by applying a square-root cosine pulse shape. The pre-amble is used to detect the start of transmission at the receiver. The Viterbi&Viterbi method is applied to the QPSK symbols for carrier phase estimation and detection. The AE decoder is used for optimization and it is replaced by Gaussian mismatched receiver for MI and GMI estimation as in [52,53]. The testing setup is presented on Fig. 5.6, where the MI and GMI estimation is performed after the standard ML receiver and BER estimation after the FEC decoder.

The learned constellation of order M=256 is shown on Fig. 5.7 left, and on the right is the zoomed in version that includes a subset of the bit mapping to highlight that it is a Gray-like mapping. The learned constellations have been evaluated in a realistic scenario which includes the following impairments at the transmitter and receiver: 1)

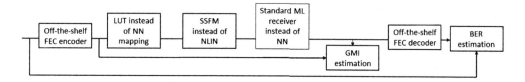

Figure 5.6 Setup of pilot based system including transceiver impairments.

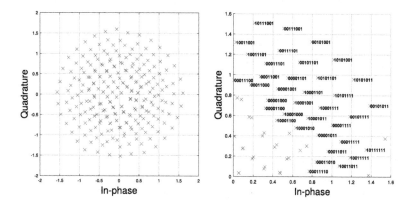

Figure 5.7 GMI optimized constellation **(left)**, zoomed-in version with labeling **(right)**. Reprinted with permission from [8].

laser linewidth of 10 kHz, modeled with a Gaussian random walk; 2) frequency offset of 50 MHz between the transmitter laser and local oscillator; 3) ADC with sampling frequency of 80 GSa/s and resolution of 6 bits. A standard and a probabilistic shaped Gray-coded 256QAM were used as references. The probabilistic shaping is achieved with Maxwell–Boltzamann probabilistic mass function [54], optimized at each distance and each launch power.

The MI of the mentioned formats at the optimal launch power is presented on Fig. 5.8 with regards to transmission distance. The dashed lines show the results of the scenario with impairments, whereas the solid lines show the results without impairments. The geometric constellation shaping achieves up to 90% of the probabilistic shaping gain. If the proper DSP is employed geometric shapes do not suffer extra penalty under the influence of the transceiver impairments compared to rectangular constellation. At short distances the penalty of the transceiver impairments is more noticeable, reaching up to 0.2 bits/symbol for all constellations. The learned constellation achieves up to 0.3 bits/symbol compared to the standard Gray-coded 256QAM.

On Fig. 5.9, the data rate with respect to distance is shown, where the solid line represents the performance in terms of the GMI, while the dashed line shows the maximum distance where the BER is lower than 10^{-5} for the given data rate. Observing the GMI it can be noticed that the constellation acquired with the AE reaches up to

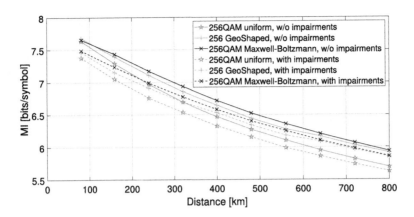

Figure 5.8 Results for MI vs. distance for the studied modulation formats with and without impairments. Reprinted with permission from [8].

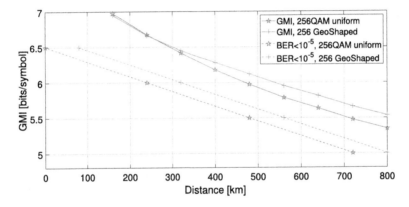

Figure 5.9 Results for GMI and achieved error-rate data rates as a function of the transmission distance. Reprinted with permission from [8].

0.2 bits/symbol of gain, which is translated into gain of one span (80 km) of error-free distance. This is achieved with no additional implementation cost or complexity to the system and can be used with conventional bit-interleaved coded modulation (BICM). The work presented in [5,6,8] has implementations of the AE setup available online [55].

In [9], GMI optimization was also considered, but by mapping bits to one-hot encoded vectors first, which were then used as input to the encoder. They have incorporated the binary switching algorithm (BSA), performed it on each 200 iterations, to find the best position for different labels and overcome potential barriers in the optimization landscape [56]. This was applied on the GN channel model. One of the main achievements of this work was that it was shown that a linear mapping between the input and the output of the transmitter with no biases achieves state-of-the-art GMI-optimized constellation. This way the number of free encoder parameters was decreased. Further they have shown that the initialization of the AE is an important design parameter. The implementation presented in this work is available online [57].

5.3.2.3 Split-step Fourier method (SSFM)

Split-step Fourier method is the numerical solution to the NLSE and embedding it into the AE setup can prove to be challenging since its implementation has to ensure numerical stability, that each operation is differentiable and keeps the memory requirements feasible. Memory requirements issue could be tackled by applying memory saving gradient techniques [58,59]. Nonetheless, the training complexity becomes the practical limitation of this learning-based system [60]. Here, two different applications will be considered. First is the conventional optical fiber communication system, where standard geometric constellation and pulse shaping is used. Whereas the second approach relies on transmission using the theoretically optimal solutions to nonlinear dispersive channel that does not include attenuation. These solutions are based on nonlinear Fourier transform (NFT) [61,62], also known as inverse scattering transform (IST). It is used to construct the nonlinear frequency division multiplexing (NFDM) system.

The standard communication system AE setup with the fiber channel modeled with the SSFM for training was presented in [13]. It was pointed out that it is hard to interpret what an encoder network does. Therefore, instead of the encoder neural network, two of the conventional DSP blocks have been transformed into trainable parameters. Those DSP blocks are symbol mapping, where the IQ-symbols are trainable parameters, and the other is the pulse shaping filter, where its coefficients are the trainable parameters. The decoder is realized as a DNN, where the input is a block of received symbols regulated by a sliding window and the output is the probability of a symbol given the input. The results acquired by the mapping matched the results obtained in [8], which was obtained by training on the NLIN model. It was shown that the pulse-shaping filter matches the analytical solution for a chromatic dispersion compensating filter almost perfectly. Finally, gain of up to 4.46 bits/s/Hz in terms of spectral efficiency compared to a classic DSP implementation with 256QAM mapping and Nyquist-pulse shaping was achieved.

Moving onto the NFT based transmitter and NN based receiver for NFDM system, end-to-end learning was used in [14,15]. The application of end-to-end learning allowed this method to achieve significant gain in a scenario that includes fiber loss, i.e. a realistic scenario where the NFT theory is not strictly defined. Fig. 5.10 illustrates the setup of the optical communication system when using NFDM, where the receiver is either the NN or the actual NFDM receiver, based on the NFT theory. The transmitter is a single polarization NFDM transmitter operating at a symbol rate of 1 GBd. The transmitter input is a stream of symbols from an alphabet of size 16. These symbols are

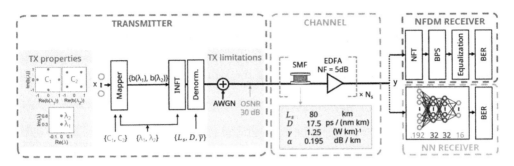

Figure 5.10 Setup of the optical communication system used for both the end-to-end optimization of the system parameters and the BER performance evaluation. Reprinted with permission from [15].

mapped to two complex constellations QPSK, where each of them corresponds to an NFT eigenvalue, which can be interpreted as a nonlinear carrier. The two complex constellations and the corresponding eigenvalues form the nonlinear spectrum and they are the trainable parameters of the transmitters, together with the launch power into the channel. An inverse NFT (INFT) operation then maps a 16-valued symbol (two QPSK points) and two eigenvalues into a normalized time–domain waveform [15]. The launch power into the channel is optimized considering an effective nonlinear coefficient $\bar{\gamma}_{LPA}$ = 0.34 for the denormalization step of the waveform generated by the INFT. The final output of the transmitter are waveforms with 96 samples-per-symbol representing soliton-like pulse-shapes.

At the receiver, a NN receives in input a full time-domain waveform of one symbol, and returns the probability of each of the symbols. The input layer of the receiver NN consists of 192 nodes because the waveform of each symbol has 96 complex-valued samples. The input layer is followed by two hidden layers of 32 nodes each using scaled exponential linear unit (SELU) [63] as the activation function. The output layer provides the posterior probability of the symbols using softmax activation function [21]. Glorot uniform initialization [64] is used to initialize the weights of the NN. The system has been trained on 5520 km ($69 \cdot 80$ km span) and it preserves the improvement at least up to ±10 spans. The same NN architecture has been used for the optimization and the performance evaluation of the system. At the output of the NN the most likely transmitted symbol has been chosen on which the BER is finally estimated through error counting.

The end–to–end learning in this work aims at approaching the final goal of a full auto–encoder, i.e. an encoder defining transmitter waveforms robust to channel impairments and a receiver decoder able to reconstruct the information from the noisy and distorted waveforms after the channel.

As applying directly NNs as both encoder and decoders provide a vast space of degrees of freedom to define the waveforms, in this work, the encoder was guided by the

NFT theory, thus restricting the transmission to soliton–like pulse shapes. Furthermore, the choice of transmitting solitons is critical in enabling the use of a low-complexity memoryless (1-symbol in input) feed-forward NN at the receiver side. Solitons are not significantly affected by pulse broadening during transmission, thus the need for more complex and challenging to train RNNs [11,12] is avoided.

The transmitter and receiver were then trained over a fiber channel consisting of multiple spans of standard single-mode fiber (SSMF) and erbium–doped fiber amplifiers (EDFA) to provide lumped loss compensation. The noise figure of the EDFAs has been sent to 5 dB and the propagation through the SSMF (0.195 dB/km of loss, 17.5 ps/nm km of dispersion and $1.25(\text{W km})^{-1}$ of nonlinear coefficient) is simulated by solving the NLSE of (5.5) through the SSFM. A fixed step–size, logarithmically increasing through a span [65], was used in order for Tensorflow to be able to generate a computational graph for automatic differentiation.

The system is trained by using the Adam optimizer with the Nesterov gradient [66]. The training batch size was limited to 64 symbols by the maximum memory of the available GPUs. The entire computational graph for the whole batch needed to fit within the GPU's memory, and even using memory-saving gradient technique [58,59], the batch size could not be increased. Such a limited batch size caused noisy estimation of the gradient. This limitation was partially overcome with a retraining step of the receiver alone after convergence of the AE training. Finally, the AE training used an adaptive learning rate which was decreased in steps of 1600 iterations to allow for a faster initial convergence followed by a finer-tuning of the system parameters when close to the optimum. Remark that, regardless of the limited batch size, which may result in similar symbols between different batches, the realization of the noise added by the channel is randomly generated at each training iteration (online learning) [67], thus reducing overfitting challenges without the need for a separate cross-validation loss analysis.

The BER results are shown in Fig. 5.12, and compared with an optimized NFT-based system, i.e. with standard NFT transmitter and receiver [15,61]. Four different optimization configurations are considered: conf0 optimizing receiver NN and launch power; conf1 optimizing receiver NN, launch power and constellations; conf2 optimizing receiver NN, launch power and eigenvalues; conf3 optimizing receiver NN, launch power, eigenvalues and constellations. These configurations are shown in Fig. 5.11. As can be seen in Fig. 5.12, the use of a NN receiver already improves significantly the reach compared to a standard NFT system. The performance is further improved by the joint transmitter-receiver optimization, with a BER which is decreased by (in order) optimizing constellations, eigenvalues and constellations, and eigenvalues alone. The slightly degradation in performance when both eigenvalues and constellations are optimized compared to optimizing only eigenvalues is believed to be caused by the presence of local optima within the loss function where the optimization could not escape even

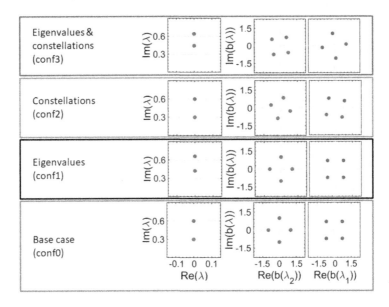

Figure 5.11 Trained transmitter parameters - Eigenvalues and constellations for the four configurations. Adapted from [15].

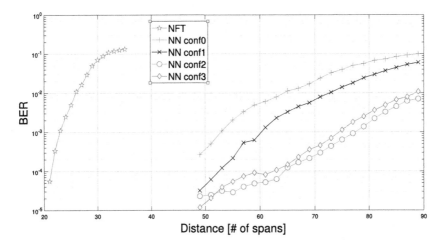

Figure 5.12 Test performance of the NN receiver and NFT receiver - BER as a function of number of spans (transmission distance). Adapted from [15].

with the receiver re-training. This result further highlights the challenges in increasing the degrees of freedom in the overall end-to-end optimization and justifies our choice to fix the nonlinear mapping from symbols to waveforms through the NFT theory, instead of the encoder also taking care of pulse shaping.

5.4. Gradient-free end-to-end learning

All of the aforementioned work used differentiable channel models so that the AE can be observed as a single computational graph and can be optimized using back-propagation. Backpropagation relies on calculating the gradient of the loss function with respect to the trainable parameters of the AE. Therefore, if the channel was non-differentiable the encoder would be cut off from the rest of the computational graph and the gradient of the loss function with respect to the encoder weights could not be calculated. The requirement of a differentiable channel model can be seen as a limitation to the end-to-end AE approach because of few reasons:

1. Not all channel models are differentiable;
2. Developing an accurate differentiable channel model that captures the effects of physical channels can be challenging and lead to complex models;
3. Approximating a physical channel with a simple differentiable model can lead to model discrepancy.

Non-numerical black box channels (e.g. experimental test-beds) and channel models which include non-differentiable DSP algorithms (e.g. blind phase search) are some of the examples of 1). A good example of 2) can be the highly complex SSFM model used in [14,15] where even though memory-saving techniques were applied the computational graph consumed the GPU's memory, limiting the batch size. Furthermore, [10] can be used as an example of 3) since the decoder had to be retrained for the experimental validation.

It was already mentioned that this issue could be potentially overcome by modeling the channel with a GAN, since it would provide a differentiable model which can approximate the channel response. However, using GANs can be time consuming as it was already discussed previously in Subsection 5.3.1. Alternatively, the issue can be tackled by applying derivative free algorithms for optimization, allowing the AE to be trained in a setup without an analytical mathematical model. A distinct advantage of the derivative-free methods occurs when the architecture is such that differentiation with respect to the parameters is not easily derived or not possible to compute. Different algorithms can be used and for now there is no straight forward way on how to pick one. Some of the options are:

- Cubature Kalman Filter (CKF) [68,69],
- Covariance matrix adaptation evolution strategy (CMA-ES) [70],
- Particle Swarm Optimization (PSO) [71],
- Differential Evolution (DE) [49,72],
- Policy Gradients with Parameter-based Exploration (PGPE) [73],
- Simultaneous perturbation stochastic approximation (SPSA) [74].

The SPSA algorithm was applied in wireless communication for optimization without channel knowledge [75]. In [20], the optimization was performed using the CKF algorithm. The concept has been validated by comparing the optimization results of AE using CKF with the optimization results of AE using backpropagation on a simple AWGN channel.

The CKF algorithm is a Bayesian filtering technique which can be used to optimize the trainable parameters $\mathbf{w}$. It uses a state-space modeling framework to describe the system under consideration, in this case the AE structure. The state-space model is described by a pair of equations, referred to as the *process* equation and *measurement* equation. The process equation describes the progression of non-observable variables (states) and the measurement equation relates them with the observable variables. These two equations are used to estimate the states, which in this case are the trainable parameters $\mathbf{w}$. For the AE, the process equation assumes a Gaussian random walk:

$$\mathbf{w}_k = \mathbf{w}_{k-1} + q_{k-1}, \tag{5.9}$$

where k is the k-th iteration and q_{k-1} represents the process-noise term modeled as a zero-mean Gaussian distribution with covariance $\mathbf{Q}_{k-1}$. The measurement equation takes into account the propagation of the input $\mathbf{s}$ throughout the whole AE, denoted as $\mathbf{h}(\cdot)$:

$$\mathbf{r}_k = \mathbf{h}(\mathbf{w}_k, \mathbf{s}_k) + r_k, \tag{5.10}$$

where $\mathbf{r}_k$ is the measurement-noise term modeled as a zero-mean Gaussian distribution with covariance $\mathbf{R}_k$. The trainable parameters are optimized by an iterative estimation of their true mean and covariance using a fixed number of deterministically chosen sample points, so-called sigma points. Using the sigma points, multiple AE realizations are formed and the same input is propagated through all of them. The outputs of these AEs are the posterior sigma points, which are used to calculate the predicted mean and covariance of the AE model and the cross-covariance of the parameters w and the AE output. These results are used to calculate the Kalman gain and the parameter mean and covariance of the CKF algorithm.

Fig. 5.13 shows the MI with respect to the SNR when training on an AWGN channel for a constellation of order $M = 64$. It illustrates the results obtained when using backpropagation with Adam optimizer and CKF for training of the AE and compares them to IPM and QAM constellations. It can be noticed that the results obtained by the two algorithms are similar and overlap with IPM. Even though it has been used on such a simple channel as AWGN, there is potential in gradient-free optimization and it could possibly one day be used to train an AE in an experimental setup.

5.5. Conclusion

This Chapter introduced the concept of end-to-end learning using autoencoders (AEs) for the fiber optic communication system. The basic idea and topology of AEs

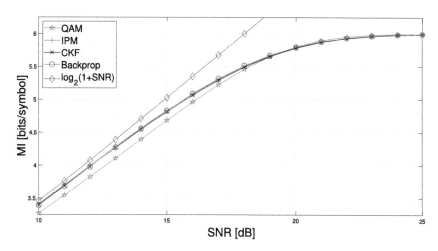

Figure 5.13 Performance in MI with respect to SNR when training on AWGN channel. Adapted from [20].

were discussed together with how AEs can be used to replace different functionalities of interest in the typical communication system chain. The standard ways of optimizing AEs was then introduced from a deep learning perspective, i.e. treating the AE as a deep neural network and applying gradient descent to optimize its weights.

The state of the art in the literature was discussed focusing on the application of AEs to the fiber optic channel. The examples included:

- The application of an AE to combat dispersion in IM/DD links, including the introduction of recurrent neural networks as encoders and decoders.
- The application of an AE for geometric constellation shaping under approximate fiber channel models, with coherent transmission. Examples here include perturbation-based models and nonlinear phase noise channels.
- The application of an AE for geometric constellation and pulse shaping under the split-step Fourier method channel and its challenges.
- The application of an AE to the optimization of soliton-based transmission and nonlinear frequency division multiplexing systems.

It was also demonstrated how the AE can be specified for the optimization of both symbol-level performance through the mutual information cost function, as well as bit-level performance through the generalized mutual information cost function.

Finally, the main limitation of gradient descent-based optimization of AEs was discussed, stemming from the requirement of tractable and differentiable channel model. Two potential approaches to this problem were presented. First, in the form of generative adversarial network (GAN) which can accurately reflect the probability distribution functions of the stochastic channel behaviors. Using GAN allows for training on experimental data. Some proof-of-concept experiments were discussed demonstrating successful training and bit-error rate improvement with successive iterations. Second, the approach of replacing the classical gradient descent optimization approach with a derivative-free optimization algorithms was presented. An example was presented using the cubature Kalman filter for parameter optimization of the AE, which in some cases was demonstrated to obtain solutions with similar performance to backpropagation on the additive white Gaussian noise (AWGN) channel model.

Even-though AEs and end-to-end learning are still in their early days of application to optical communications, the above approaches have shown impressive benefits in a variety of scenarios. While some challenges remain with training and applying AEs efficiently in real-time optical communications, especially for non-differentiable and cumbersome channel models, AEs show promise for the optimization and design of the next generation of optical transceivers.

Acknowledgments

This work was financially supported by the European Research Council through the ERC-CoG FRECOM project (grant agreement no. 771878), the Villum Fonden through the Villum Young Investigator OPTIC-AI project (grant no. 29334), and DNRF SPOC, DNRF123.

References

[1] C. Jiang, H. Zhang, Y. Ren, Z. Han, K. Chen, L. Hanzo, Machine learning paradigms for next-generation wireless networks, IEEE Wireless Communications 24 (2) (2017) 98–105.

[2] F.N. Khan, C. Lu, A.P.T. Lau, Machine learning methods for optical communication systems, in: Advanced Photonics 2017 (IPR, NOMA, Sensors, Networks, SPPCom, PS), Optical Society of America, 2017, http://www.osapublishing.org/abstract.cfm?URI=SPPCom-2017-SpW2F.3, p. SpW2F.3.

[3] K. Hornik, M. Stinchcombe, H. White, et al., Multilayer feedforward networks are universal approximators, Neural Networks 2 (5) (1989) 359–366.

[4] T. O'Shea, J. Hoydis, An introduction to deep learning for the physical layer, IEEE Transactions on Cognitive Communications and Networking 3 (4) (2017) 563–575.

[5] R.T. Jones, T.A. Eriksson, M.P. Yankov, D. Zibar, Deep learning of geometric constellation shaping including fiber nonlinearities, in: 2018 European Conference on Optical Communication (ECOC), IEEE, 2018, pp. 1–3.

[6] R.T. Jones, T.A. Eriksson, M.P. Yankov, B.J. Puttnam, G. Rademacher, R.S. Luis, D. Zibar, Geometric constellation shaping for fiber optic communication systems via end-to-end learning, arXiv preprint, arXiv:1810.00774, 2018.

[7] S. Li, C. Häger, N. Garcia, H. Wymeersch, Achievable information rates for nonlinear fiber communication via end-to-end autoencoder learning, in: 2018 European Conference on Optical Communication (ECOC), IEEE, 2018, pp. 1–3.

[8] R.T. Jones, M.P. Yankov, D. Zibar, End-to-end learning for GMI optimized geometric constellation shape, arXiv preprint, arXiv:1907.08535, 2019.

[9] K. Gümüş, A. Alvarado, B. Chen, C. Häger, E. Agrell, End-to-end learning of geometrical shaping maximizing generalized mutual information, in: 2020 Optical Fiber Communications Conference and Exhibition (OFC), IEEE, 2020, pp. 1–3.

[10] B. Karanov, M. Chagnon, F. Thouin, T.A. Eriksson, H. Bülow, D. Lavery, P. Bayvel, L. Schmalen, End-to-end deep learning of optical fiber communications, Journal of Lightwave Technology 36 (20) (2018) 4843–4855.

[11] B. Karanov, D. Lavery, P. Bayvel, L. Schmalen, End-to-end optimized transmission over dispersive intensity-modulated channels using bidirectional recurrent neural networks, Optics Express 27 (14) (2019) 19650–19663.

[12] B. Karanov, G. Liga, V. Aref, D. Lavery, P. Bayvel, L. Schmalen, Deep learning for communication over dispersive nonlinear channels: performance and comparison with classical digital signal processing, in: 2019 57th Annual Allerton Conference on Communication, Control, and Computing, Allerton, IEEE, 2019, pp. 192–199.

[13] T. Uhlemann, S. Cammerer, A. Span, S. Dörner, S.t. Brink, Deep-learning autoencoder for coherent and nonlinear optical communication, arXiv preprint, arXiv:2006.15027, 2020.

[14] S. Gaiarin, R.T. Jones, F. Da Ros, D. Zibar, End-to-end optimized nonlinear Fourier transform-based coherent communications, in: CLEO: Science and Innovations, Optical Society of America, 2020, p. SF2L–4.

[15] S. Gaiarin, F. Da Ros, R.T. Jones, D. Zibar, End-to-end optimization of coherent optical communications over the split-step Fourier method guided by the nonlinear Fourier transform theory, Journal of Lightwave Technology 39 (2) (2020) 418–428.

[16] T.J. O'Shea, T. Roy, N. West, Approximating the void: learning stochastic channel models from observation with variational generative adversarial networks, in: 2019 International Conference on Computing, Networking and Communications (ICNC), IEEE, 2019, pp. 681–686.

[17] H. Ye, G.Y. Li, B.-H.F. Juang, K. Sivanesan, Channel agnostic end-to-end learning based communication systems with conditional GAN, in: 2018 IEEE Globecom Workshops (GC Wkshps), IEEE, 2018, pp. 1–5.

[18] A. Smith, J. Downey, A communication channel density estimating generative adversarial network, in: 2019 IEEE Cognitive Communications for Aerospace Applications Workshop (CCAAW), IEEE, 2019, pp. 1–7.

[19] B. Karanov, M. Chagnon, V. Aref, D. Lavery, P. Bayvel, L. Schmalen, Concept and experimental demonstration of optical IM/DD end-to-end system optimization using a generative model, in: 2020 Optical Fiber Communications Conference and Exhibition (OFC), IEEE, 2020, p. Th2A.48.

[20] O. Jovanovic, M.P. Yankov, F. Da Ros, D. Zibar, Gradient-free training of autoencoders for non-differentiable communication channels, Journal of Lightwave Technology 39 (20) (2021) 6381–6391.

[21] I. Goodfellow, Y. Bengio, A. Courville, Y. Bengio, Deep Learning, vol. 1, MIT Press, Cambridge, 2016.

[22] C.M. Bishop, Pattern Recognition and Machine Learning, Springer, 2006.

[23] D.P. Kingma, J. Ba Adam, A method for stochastic optimization, arXiv preprint, arXiv:1412.6980, 2014.

[24] M. Abadi, P. Barham, J. Chen, Z. Chen, A. Davis, J. Dean, M. Devin, S. Ghemawat, G. Irving, M. Isard, et al., Tensorflow: a system for large-scale machine learning, in: 12th Symposium on Operating Systems Design and Implementation, 2016, pp. 265–283.

[25] A. Paszke, S. Gross, S. Chintala, G. Chanan, E. Yang, Z. DeVito, Z. Lin, A. Desmaison, L. Antiga, A. Lerer, Automatic differentiation in PyTorch, 2017.

[26] A.G. Baydin, B.A. Pearlmutter, A.A. Radul, J.M. Siskind, Automatic differentiation in machine learning: a survey, The Journal of Machine Learning Research 18 (1) (2017) 5595–5637.

[27] M. Bartholomew-Biggs, S. Brown, B. Christianson, L. Dixon, Automatic differentiation of algorithms, Journal of Computational and Applied Mathematics 124 (1–2) (2000) 171–190.

[28] G.P. Agrawal, Fiber-Optic Communication Systems, vol. 222, John Wiley & Sons, 2012.

[29] M.H. Eiselt, N. Eiselt, A. Dochhan, Direct detection solutions for 100g and beyond, in: Optical Fiber Communication Conference, Optical Society of America, 2017, p. Tu3I–3.

[30] P. Poggiolini, The GN model of non-linear propagation in uncompensated coherent optical systems, Journal of Lightwave Technology 30 (24) (2012) 3857–3879.

[31] P. Poggiolini, G. Bosco, A. Carena, V. Curri, Y. Jiang, F. Forghieri, The GN-model of fiber non-linear propagation and its applications, Journal of Lightwave Technology 32 (4) (2013) 694–721.

[32] R. Dar, M. Feder, A. Mecozzi, M. Shtaif, Properties of nonlinear noise in long, dispersion-uncompensated fiber links, Optics Express 21 (22) (2013) 25685–25699.

[33] R. Dar, M. Feder, A. Mecozzi, M. Shtaif, Accumulation of nonlinear interference noise in fiber-optic systems, Optics Express 22 (12) (2014) 14199–14211.

[34] R. Dar, M. Feder, A. Mecozzi, M. Shtaif, Inter-channel nonlinear interference noise in WDM systems: modeling and mitigation, Journal of Lightwave Technology 33 (5) (2014) 1044–1053.

[35] K.-P. Ho, Phase-Modulated Optical Communication Systems, Springer Science & Business Media, 2005.

[36] G.P. Agrawal, Nonlinear fiber optics, in: Nonlinear Science at the Dawn of the 21st Century, Springer, 2000, pp. 195–211.

[37] O.V. Sinkin, R. Holzlohner, J. Zweck, C.R. Menyuk, Optimization of the split-step Fourier method in modeling optical-fiber communications systems, Journal of Lightwave Technology 21 (1) (2003) 61–68.

[38] I. Goodfellow, J. Pouget-Abadie, M. Mirza, B. Xu, D. Warde-Farley, S. Ozair, A. Courville, Y. Bengio, Generative adversarial nets, in: Advances in Neural Information Processing Systems, 2014, pp. 2672–2680.

[39] V. Nair, G.E. Hinton, Rectified linear units improve restricted Boltzmann machines, in: ICML, 2010.

[40] C. Pearson, High-speed, analog-to-digital converter basics, Texas Instruments Application Report, SLAA510, January, 2011.

[41] A. Napoli, M.M. Mezghanni, S. Calabro, R. Palmer, G. Saathoff, B. Spinnler, Digital predistortion techniques for finite extinction ratio IQ Mach–Zehnder modulators, Journal of Lightwave Technology 35 (19) (2017) 4289–4296.

[42] B. Karanov, M. Chagnon, V. Aref, D. Lavery, P. Bayvel, L. Schmalen, Optical fiber communication systems based on end-to-end deep learning, arXiv preprint, arXiv:2005.08785, 2020.

[43] B. Karanov, M. Chagnon, V. Aref, F. Ferreira, D. Lavery, P. Bayvel, L. Schmalen, Experimental investigation of deep learning for digital signal processing in short reach optical fiber communications, arXiv preprint, arXiv:2005.08790, 2020.

[44] M. Schaedler, S. Calabrò, F. Pittalà, G. Böcherer, M. Kuschnerov, C. Bluemm, S. Pachnicke, Neural network assisted geometric shaping for 800Gbit/s and 1Tbit/s optical transmission, in: Optics InfoBase Conference Papers Part F174- (DM), 2020, pp. 3–5.

[45] G. Böcherer, Labeling non-square QAM constellations for one-dimensional bit-metric decoding, IEEE Communications Letters 18 (9) (2014) 1515–1518.

[46] K. Turitsyn, S. Derevyanko, I. Yurkevich, S. Turitsyn, Information capacity of optical fiber channels with zero average dispersion, Physical Review Letters 91 (20) (2003) 203901.

[47] M.I. Yousefi, F.R. Kschischang, On the per-sample capacity of nondispersive optical fibers, IEEE Transactions on Information Theory 57 (11) (2011) 7522–7541.

[48] End-to-end learning of optical communication systems: concept and transceiver design, https://github.com/kit-cel/ecoc_20_learning.

[49] Z.H. Peric, I.B. Djordjevic, S.M. Bogosavljevic, M.C. Stefanovic, Design of signal constellations for Gaussian channel by using iterative polar quantization, in: MELECON'98. 9th Mediterranean Electrotechnical Conference. Proceedings (Cat. No. 98CH36056), vol. 2, IEEE, 1998, pp. 866–869.

[50] I.B. Djordjevic, H.G. Batshon, L. Xu, T. Wang, Coded polarization-multiplexed iterative polar modulation (PM-IPM) for beyond 400 Gb/s serial optical transmission, in: Optical Fiber Communication Conference, Optical Society of America, 2010, p. OMK2.

[51] H.G. Batshon, I.B. Djordjevic, L. Xu, T. Wang, Iterative polar quantization-based modulation to achieve channel capacity in ultrahigh-speed optical communication systems, IEEE Photonics Journal 2 (4) (2010) 593–599.

[52] H. Hu, M.P. Yankov, F.D. Ros, Y. Amma, Y. Sasaki, T. Mizuno, Y. Miyamoto, M. Galili, S. Forchhammer, L.K. Oxenløwe, T. Morioka, Ultrahigh-spectral-efficiency WDM/SDM transmission using PDM-1024-QAM probabilistic shaping with adaptive rate, Journal of Lightwave Technology 36 (6) (2018) 1304–1308, http://jlt.osa.org/abstract.cfm?URI=jlt-36-6-1304.

[53] M.P. Yankov, F. Da Ros, E.P. da Silva, S. Forchhammer, K.J. Larsen, L.K. Oxenløwe, M. Galili, D. Zibar, Constellation shaping for WDM systems using 256QAM/1024QAM with probabilistic optimization, Journal of Lightwave Technology 34 (22) (2016) 5146–5156.

[54] F. Buchali, F. Steiner, G. Böcherer, L. Schmalen, P. Schulte, W. Idler, Rate adaptation and reach increase by probabilistically shaped 64-QAM: an experimental demonstration, Journal of Lightwave Technology 34 (7) (2016) 1599–1609.

[55] Claude - end-to-end learning of optical communication systems, https://github.com/Rassibassi/claude.

[56] F. Schreckenbach, N. Gortz, J. Hagenauer, G. Bauch, Optimization of symbol mappings for bit-interleaved coded modulation with iterative decoding, IEEE Communications Letters 7 (12) (2003) 593–595.

[57] Geometric constellation shaping, https://github.com/kadirgumus/Geometric-Constellation-Shaping.

[58] T. Chen, B. Xu, C. Zhang, C. Guestrin, Training deep nets with sublinear memory cost, arXiv preprint, arXiv:1604.06174, 2016.

[59] Gradient checkpoint, https://github.com/cybertronai/gradient-checkpointing.

[60] T. Gruber, S. Cammerer, J. Hoydis, S. ten Brink, On deep learning-based channel decoding, in: 2017 51st Annual Conference on Information Sciences and Systems (CISS), IEEE, 2017, pp. 1–6.

[61] S.K. Turitsyn, J.E. Prilepsky, S.T. Le, S. Wahls, L.L. Frumin, M. Kamalian, S.A. Derevyanko, Nonlinear Fourier transform for optical data processing and transmission: advances and perspectives, Optica 4 (3) (2017) 307–322.

[62] Z. Dong, S. Hari, T. Gui, K. Zhong, M.I. Yousefi, C. Lu, P.-K.A. Wai, F.R. Kschischang, A.P.T. Lau, Nonlinear frequency division multiplexed transmissions based on NFT, IEEE Photonics Technology Letters 27 (15) (2015) 1621–1623.

[63] G. Klambauer, T. Unterthiner, A. Mayr, S. Hochreiter, Self-normalizing neural networks, Advances in Neural Information Processing Systems 30 (2017) 971–980.

[64] X. Glorot, Y. Bengio, Understanding the difficulty of training deep feedforward neural networks, in: Proceedings of the Thirteenth International Conference on Artificial Intelligence and Statistics, 2010, pp. 249–256.

[65] G. Bosco, A. Carena, V. Curri, R. Gaudino, P. Poggiolini, S. Benedetto, Suppression of spurious tones induced by the split-step method in fiber systems simulation, IEEE Photonics Technology Letters 12 (5) (2000) 489–491.

[66] T. Dozat, Incorporating Nesterov momentum into Adam, 2016.

[67] S. Ruder, An overview of gradient descent optimization algorithms, arXiv preprint, arXiv:1609.04747, 2016.

[68] S. Haykin, I. Arasaratnam, Cubature Kalman filters, IEEE Transactions on Automatic Control 54 (6) (2009) 1254–1269.

[69] I. Arasaratnam, S. Haykin, Nonlinear Bayesian filters for training recurrent neural networks, in: Mexican International Conference on Artificial Intelligence, Springer, 2008, pp. 12–33.

[70] A. Auger, N. Hansen, A restart CMA evolution strategy with increasing population size, in: 2005 IEEE Congress on Evolutionary Computation, vol. 2, IEEE, 2005, pp. 1769–1776.

[71] J. Kennedy, R. Eberhart, Particle swarm optimization, in: Proceedings of ICNN'95-International Conference on Neural Networks, vol. 4, IEEE, 1995, pp. 1942–1948.

[72] K.V. Price, Differential evolution vs. the functions of the 2/sup nd/ICEO, in: Proceedings of 1997 IEEE International Conference on Evolutionary Computation (ICEC'97), IEEE, 1997, pp. 153–157.

[73] F. Sehnke, C. Osendorfer, T. Rückstieß, A. Graves, J. Peters, J. Schmidhuber, Parameter-exploring policy gradients, Neural Networks 23 (4) (2010) 551–559.

[74] J.C. Spall, A stochastic approximation technique for generating maximum likelihood parameter estimates, in: 1987 American Control Conference, IEEE, 1987, pp. 1161–1167.

[75] V. Raj, S. Kalyani, Backpropagating through the air: deep learning at physical layer without channel models, IEEE Communications Letters 22 (11) (2018) 2278–2281.

CHAPTER SIX

Deep learning techniques for optical monitoring

Takahito Tanimura[a] and Takeshi Hoshida
Fujitsu Limited, Kawasaki, Japan

6.1. Introduction

Traffic demand in optical networks has been growing continuously for decades. New technologies and deployment approaches for optical communication have been introduced to accommodate such growing demand. Advanced modulation formats, digital equalizers, and coding schemes, including forward error correction and probabilistic constellation shaping, are some methods that have been developed rapidly to approach the theoretical limit of fiber-optic communication [1,2].

With the introduction of such new technologies, new deployment approaches have also been developed with openness and additional hardware flexibility. The new architecture is based on a disaggregation of an optical transport network system where network elements from different vendors can interoperate. However, the new architecture gives rise to new network operating issues, such as how to take the best advantage by combining disaggregated components into one network. A natural consequence of moving from an integrated architecture to a disaggregated system is reduced transmission performance, e.g., reduced transmission reach for the same bitrate, due to the accumulated performance margin of each network element. Generally, each network element requires its own design margin so that it can pass performance tests on its own under the worst-case operating condition over its lifetime. The resulting end-to-end system margin would become much more conservative than it would be in a conventional single-vender system where the system vendor can provide holistically optimized engineering at the system level, e.g., by considering statistical aspects and management of worst-case conditions in the integrated design. Thus, the open architecture may suffer from inferior system performance without sophisticated and collaborative optimization among its elements [3].

In other words, new technologies and deployment approaches would be more attractive when network operators would have a means to handle increasing complexity, in particular, the control of increasing degrees of freedom, e.g., routing and spectral

[a] Currently with Hitachi Ltd., Tokyo, Japan.

Machine Learning for Future Fiber-Optic Communication Systems
https://doi.org/10.1016/B978-0-32-385227-2.00013-9

141

slot assignment for light paths, settings of modulation format, symbol rate, and digital signal processing algorithms [2], including settings for forward error correction and coding. However, the optimization over such multidimensional problems may increase the operational expenditure (OPEX) for such systems.

Automation of network operation is a promising solution to this issue [4–6]. Fig. 6.1 shows a schematic of network automation. The idea behind this automation is to repeatedly optimize the system according to the estimated and/or predicted status of the physical layer. This concept differs from the conventional approach that integrates many subsystems according to their guaranteed worst-case specifications. This automation is realized to utilize a closed-loop control comprising an analysis component, a network control component, and a programmable optical network. The remaining part of the introduction describes each part in the closed loop.

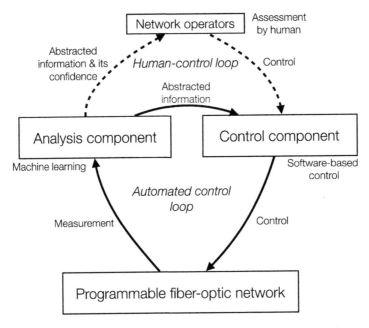

Figure 6.1 Concept of optical network automation with closed-loop control optimization.

First, the programmable fiber-optic network in Fig. 6.1 consists of an optical network equipment that can be controlled remotely using a common interface, e.g., a flexible optical transceiver that supports multiple modulation formats, symbol rates, and coding schemes [7–10], and a flexible optical node, such as flex-grid colorless, directionless, contentionless, and gridless (CDCG) reconfigurable optical add/drop multiplexer (ROADM) [11], that reconfigures light paths and allocated wavelengths of the networks. All equipments need to be controlled through a common application programming interface (API) architecture.

Second, the analysis component in Fig. 6.1 is vital for automating the operation and management of optical networks; this part is also known as a perception part or optical monitoring and analysis (OMA) part. The analysis component can measure the physical parameters of the components and analyze the measured data to extract useful information (or even insight) for the network operator. For example, the analysis component provides a set of summarized indicators, e.g., modulation format, chromatic dispersion, polarization mode dispersion, and optical signal-to-noise ratio (OSNR) of each signal, by measuring the signal in the network. The indicators can be used for making network control decisions in the upper control layer to enable more efficient resource allocation and light path establishment, thus ensuring highly efficient optical network operation.

Practical design and implementation of the analysis component have been an open issue. Although several monitors have been investigated to realize a dedicated monitoring function for specific parameters [12], e.g., optical channel monitor in ROADM, how to deploy and manage these monitors for various parameters across numerous monitoring points distributed over optical networks in a cost-effective manner remains an issue.

Recently, an alternative and more flexible approach has been proposed and investigated to address this issue. This approach uses digital signal processing (DSP)-based monitoring rather than relying on optical hardware dedicated to specific parameter to be monitored. The approach can leverage digital coherent transceivers for monitoring, that are deployed in networks anyway for data transmission. One of the most straightforward implementations to enable this approach is to take advantage of the internal parameters in the demodulation DSP at a digital coherent receiver, e.g., tap coefficients of an adaptive equalizer [13]. Most recently, machine learning (ML) [14] and deep learning (DL) [15] techniques have been applied in this approach, which has advanced this approach in terms of the availability of various unstructured input and automatic learning without domain knowledge [16–28]. In this work, the digital coherent-based monitoring assisted by ML, DL, and/or DSP is referred to as data analytic-based optical monitoring, while the digital coherent-based monitoring assisted by DL is referred to as DL-based optical monitoring. Data analytic-based monitoring can be used to estimate the physical status of optical signals and/or links in optical networks, such as OSNR, chromatic dispersion (CD), polarization mode dispersion (PMD), and fiber nonlinearity, without additional and dedicated optical components for additional parameters to be monitored.

An ML/DL model for the data analytic-based optical monitor that is trained using the dataset captured from an optical network can provide insights on the current status of the network and thus facilitate efficient network operation or even automate some part of it. Network automation using ML/DL models might be even more beneficial in cross-layer analysis and optimization in situations where changes in the physical layer trigger changes in the network layer because such cross-layer phenomena require insights based on expertise across multiple domains for the operation, which are hardly available in conventional human-based operation. The data analytic-based optical monitor sometimes provides the extracted insight (e.g., monitoring results) to the control component that automatically commands the programmable optical network as well as to human network operators to assist decision-making for easier network operation.

Finally, the network control component in Fig. 6.1, also known as the software-defined network controller, sets control parameters on the network equipment. This control can be performed automatically but can be canceled if any concerns arise in the automated decision-making. In this case, the system can immediately fall back to "human operation mode" and assist the decision-making by human network operators by providing relevant data and analysis results because at least as a preliminary step toward fully automated operation, mission-critical systems, such as optical communication systems, require accountability and responsibility for the actions to be taken, as in self-driving automobiles. A model uncertainty estimator, which is one of the essential techniques to assist in this aspect, will be discussed in Section 6.3.4.2.

The remainder of this chapter describes the concept and recent progress in DL-based optical monitoring. In Section 6.2, the building blocks of DL-based optical monitoring, namely, digital coherent reception and deep neural network (DNN), are reviewed. In Section 6.3, the DL-based optical monitor's implementation in both training and inference modes together with advanced options, including data augmentation, transfer learning, federated learning, and model uncertainty estimation, is described. In Section 6.4, the guidelines for designing DNNs for DL-based optical monitoring are discussed. Experimental verifications as a proof of the DL-based optical monitor concept are presented in Section 6.5. The possible direction of the future evolution of DL-based optical monitoring by introducing latest studies on other data analytic-based monitoring, such as DSP-based fiber-longitudinal monitoring, is discussed in Section 6.6, followed by the conclusion in Section 6.7.

6.2. Building blocks of deep learning-based optical monitors

In this section, the principal building blocks of the DL-based optical monitors—digital coherent reception and deep learning—are briefly reviewed. Coincidentally, these two technologies were revolutionizing optical communication and artificial intelligence in around 2006. After briefly describing these two technologies from the perspective of data analytic-based optical monitoring, we describe how a fusion of these technologies opened the new research field of deep learning in optical communications.

6.2.1 Digital coherent reception as a data-acquisition method

In the fiber-optic communication research community, the few years around 2006 can be remembered as the advent of digital coherent reception technologies, which is a revival of coherent optical communication enabled by DSP [29,30]. The rapid progress of semiconductor technologies brought about the introduction of large-scale DSP into optical coherent reception to revive existing but unused analog coherent techniques investigated in the 1980s. DSP enables the use of phase and polarization components of light wave through adaptive equalization and digital phase synchronization between signal and local laser. This development also enabled the introduction of advanced multilevel modulation formats and DSP-based equalization of waveform distortion [2] to optical fiber communication.

DSP has not only improved the capacity of fiber-optic communication systems but also opened up their programmability via the development of technologies that enable finer manipulation of light wave on the basis of DSP. Such powerful function achieved by applying DSP relies on a fundamental function of digital coherent receivers to convert real optical signal waveforms into equivalent digital information that is computable by DSP. Recently, a large number of digital coherent transceivers have been deployed in optical networks throughout the world. Such a massive deployment has introduced the functionality to convert the optical waveform in the network into a computable dataset and can open new viewpoints in which the optical network can be viewed as a massive collection of embedded sensors.

Here, we outline the mechanism applied by a digital coherent receiver for converting the optical waveform over fiber-optic networks into a computable digital dataset. Fig. 6.2 shows an overview of a standard digital coherent receiver from this viewpoint [30].

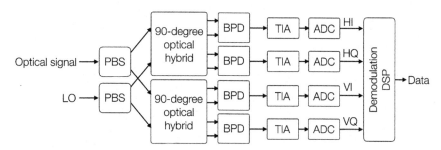

Figure 6.2 Overview of a digital coherent receiver. LO: local oscillator, PBS: polarizing beam splitter, BPD: balanced photodetector, TIA: transimpedance amplifier, ADC: analog-to-digital converter, DSP: digital signal processing, HI: in-phase component of horizontal polarization, HQ: quadrature component of horizontal polarization, VI: in-phase component of vertical polarization, VQ: quadrature component of vertical polarization.

In Fig. 6.2, the digital coherent receiver first converts an incoming optical signal into an analog electrical signal by mixing the signal light with the local oscillator (LO) light in the 90-degree optical hybrid. The detailed steps are as follows. The incoming signal and local laser light are split into each orthogonal polarization using a polarization beam splitter (PBS) to obtain polarization diversity. Next, each polarization's local and signal light is combined in a 90-degree optical hybrid for each polarization. Finally, the eight output beams from the two 90-degree hybrids are converted from optical signals to analog electric signals using four balanced photodetectors (BPDs). After the common noise is removed by the differential output on the BPDs, the resulting four analog electric outputs of the BPDs are written as

$$HI = Re(E_H E_{LO}^*), \qquad (6.1)$$

$$HQ = Im(E_H E_{LO}^*), \qquad (6.2)$$

$$VI = Re(E_V E_{LO}^*), \qquad (6.3)$$

$$VQ = Im(E_V E_{LO}^*), \qquad (6.4)$$

where $Re(x)$ and $Im(x)$ are the mathematical functions that output the real and imaginary parts of complex value x, respectively. E_H and E_V are the horizontal and vertical polarization components of the complex electric field of the input optical signal, respectively. E_{LO} represents the complex electric field of local laser.

Most of the carrier frequencies of the signals are canceled out by taking the product between the signals' electric fields and the complex-conjugated LO because the carrier frequencies of the signals and LO are roughly the same. If the phase and carrier frequency of the signal and LO are exactly the same, then the phase offset and frequency offset between the signal and LO are canceled out. Thus, we can directly find the in-phase and quadrature components of the incoming signal from the output of the BPDs.

The signal and local lasers are not frequency-locked in typical digital coherent receivers. Thus, the resulting outputs from the BPDs have the residual frequency offset $\delta f = f_{sig} - f_{LO}$ and a phase offset between the signal and LO laser (f_{sig} and f_{LO} are the laser-emitting frequencies of the signal and LO, respectively).

The residual frequency offset can be equalized through the DSP performed after the ADCs. However, the stage for monitoring point is likely to be located before this digital equalization part because some monitoring results, e.g., modulation format, are required before digital equalizations, such as adaptive equalizer, frequency offset compensation, and carrier phase recovery, are initiated. Thus, the monitoring algorithm needs to be tolerant against the frequency and phase offsets with respect to the measured waveform dataset. Section 6.3.2.1 describes this matter in detail.

Digital coherent receiver can transform an input optical signal into computable dataset that can be straightforwardly treated as an optical electric field that comprises phase and amplitude of both horizontal and vertical polarization, while there are residual phase and frequency offsets between signal and LO lasers.

6.2.2 Deep learning and representation learning

DNNs, also known as DLs, are neural networks with multiple hidden layers and are one of the most successful ML frameworks in the previous decade [15,31]. A theoretical work explaining the success of DLs is ongoing, which is one of the most exciting research areas in this field. In contrast, key enablers that make DLs useful from the viewpoint of applications are relatively tangible. The enablers are the availability of large amounts of data and powerful computing resources facilitated by specifically designed accelerators, e.g., general-purpose graphics processing units (GPU).[1]

This section briefly describes DNNs from the viewpoint of representation learning, which is one of the most potent properties of DNNs [15]. Representation learning also plays an essential role in combining DNNs with the digital coherent receivers to realize DL–based optical monitors.

First, we start from the definition of DNN in the ML technologies. ML is a set of techniques that enable a computer to perform a task (e.g., classify images and recognize speech) without specific programming by human engineers. DNN is a subset of ML techniques. Fig. 6.3 shows the relationship among ML, neural network, and DNN.

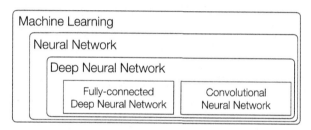

Figure 6.3 Machine learning, neural networks, and deep neural networks.

ML techniques, including DNNs, aim to obtain a function f that is the approximation of function f^* that solves a given task by optimizing a parameter set θ in the relation $y = f(x; \theta)$, where x, y, and θ are the input, output, and the set of the parameters of f, respectively. In a neural network framework, the function f is implemented using the neural network scheme. The function f, i.e., the corresponding neural network, is fitted to the function f^* by tuning the parameter set θ on the training dataset. The training data in the supervised learning are given by a set of input x and label $\hat{y}$ that is an output of f^* at the input point x, i.e., $\hat{y} = f^*(x)$. The function f generates an output y that is close to label $\hat{y}$, and the parameter set θ is optimized to minimize the difference between y and label $\hat{y}$ in the training phase.

Such training is usually not easy in practical situations because no clues are available about which parts of the input data x must be focused on. Thus, a two–stage processing

---

[1] Advanced semiconductor technology has played a significant role in training DNN in the revival of the existing neural network as in the case of the revival of coherent receiver.

pipeline is traditionally employed to mitigate this issue. Fig. 6.4(a) shows the traditional two-stage pipeline with a conventional ML algorithm. The first stage of the pipeline involves the extraction of the features of the input data x. Human engineers who have domain knowledge of the given task design this stage to extract the portions of the data that are needed to perform the task. The features extracted in the first stage are then fed to the ML model in the second stage. The ML model is trained to obtain a corresponding function that executes the given task from the input features, not raw input data x. When ML is applied using this pipeline to real-world problems, specific features need to be designed for each task.

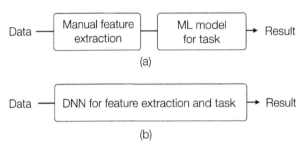

Figure 6.4 Processing pipeline of (a) conventional machine learning and (b) deep learning.

On the contrary, a DNN integrates these two stages into a unified stage, as shown in Fig. 6.4(b), with feature extraction and task processing being executed simultaneously. In other words, the input dataset x is used for two purposes: feature extraction and actual task execution. This process is enabled by the advantage of the nonlinear hierarchical structure of DNNs to automatically learn the representation of the features in a layer-by-layer manner. Feature extraction in DNNs is performed by capturing the input data's more specific features in the DNN's early layers and then combining them hierarchically to extract more complex features in the latter layers. This feature extraction is also known as representation learning.

Representation learning would make DNN useful for many tasks where designing the features manually is challenging for human engineers. A DNN could learn better feature representations than those heuristically designed in many applications.

6.2.3 Combination of digital coherent reception and deep learning

In this section, a scheme that combines digital coherent reception with DL to realize DL-based optical monitors is discussed. The novel scheme utilizes representation learning by DL using large amounts of data acquired by a digital coherent receiver.

First, we review existing ML-harnessed optical monitors that use conventional ML to estimate the optical status to be monitored. The ML-harnessed monitors can learn their monitoring functionality without explicit prior formulation. However, the conventional ML involves a preprocessing step in which features are extracted by human engineers (see Fig. 6.4). The features that describe the desired output well are then fed to the ML models in the ML-harnessed optical monitors. Among the features that are used in [24] include Q-factor, eye closure, root-mean-square jitter, and crossing amplitude obtained from eye diagrams. In [25,26], a distribution of the asynchronous constellation diagram is proposed for features and empirical moments used for features in [27]. In [28], the eye diagram variance is selected as a feature for OSNR estimation. The manual process that is used for extracting the features before applying the ML model leads to challenges in scaling up the optical monitoring for various types of parameters in automating network operation.

To address the issue, representation learning by DNNs, which is explained in Section 6.2.2, can be introduced to data analytic-based monitoring. However, the DNN would require a greater amount of training data than that required for training typical conventional ML algorithms. A data-acquisition method that has limited ability to inexpensively collect a sufficient amount of data for training DNN poses some challenges. In actual practice, collecting a large amount of raw waveform data on a conventional framework of fiber-optic networks, such as 10-Gbit/s intensity modulation-direct detection (IM-DD) systems, is difficult. In modern optical networks where digital coherent receivers are already deployed, digital coherent receivers can be used to collect a large amount of raw waveform data for DL [16].

To highlight this aspect, we discuss the volume aspect of the data measured using digital coherent receivers. Through a simple calculation, we estimate an upper limit of the available data volume collected using a single digital coherent receiver. Let F_s [sample/s] be the ADC's sampling rate in the digital coherent receiver, and $b_{physical}$ [bit] be ADC's physical number of bits. With four simultaneous outputs of the optical front end, i.e., HI, HQ, VI, and VQ, a single digital coherent receiver can potentially generate a waveform dataset with a maximum data rate $D_{max} = 4F_s b_{physical}$ as its upper limit of data volume. Assuming $F_s = 150$ Gsample/s and $b_{physical} = 10$ bits, one digital coherent receiver converts optical signals into computable data at a rate of 6 Tbits/s, which is probably one of the fastest analog-to-digital conversion rates today. The effective data rate might be lower than the upper limit due to a limitation of SNR on optical signal and analog devices, such as photodiode and ADC, and a limitation in the interface speed between ADC and external processing devices. An advanced design would be required in real systems on the external interface to utilize this potentially available data outside the optical transceiver. These limitations need to be considered; nevertheless, this powerful digitization capability of digital coherent reception is an attractive feature, thus making digital coherent receivers one of the most viable options for providing a sufficient data volume for DNNs in fiber-optic communication.

Next, we extend the discussion about the data captured by digital coherent receivers to its informative aspect. The data captured by the digital coherent receiver are not a simple bit sequence; rather, they contain a complex optical electric field that experiences various events through the transmission link. A digital coherent receiver acquires data that contain all physical quantities of an incoming optical field within the bandwidth of the digital coherent receiver in terms of classical representation, i.e., phase and amplitude of each polarization. The received signal passes through optical fibers, optical amplifiers, and other optical devices over the transmission line before it reaches the receiver. A detailed analysis at the receiver can help estimate the impact of these optical devices on the received waveform.

The data captured by digital coherent receivers have advantages on the volume and informative aspects as indicated by the above discussion and can enable representation learning by DNNs for DL-based optical monitors. Therefore, combining digital coherent reception with DL for estimating the status of fiber-optic networks is promising direction.

6.3. Deep learning-based optical monitors

This section describes DL-based optical monitors [16,32–34], which are a type of data analytic-based optical monitors. DL-based optical monitoring is achieved by combining the large amount of optical waveform data acquired by digital coherent receivers with representation learning by DNN, as discussed in the previous section.

To clarify the scheme of the DL-based optical monitor, we consider the scheme comprising two parts: (1) the optical-measurement-and-digitization part that converts optical signals into digital data and (2) the data-analytic part that extracts useful summary information from the measured data. An overview of the scheme is shown in Fig. 6.5.

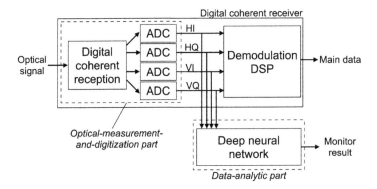

Figure 6.5 Overview of the deep learning-based optical monitor.

In Fig. 6.5, a digital coherent receiver and a DNN correspond to the optical-measurement-and-digitization part and the data-analytic part, respectively. First, as explained in Section 6.2.1, the optical signal input to the digital coherent receiver is converted into computable waveform data, which data are then fed to the DNN to extract summary information, which is useful for the network operator and manager. The unique aspect of DL-based optical monitoring is its use of end-to-end learning to minimize human intervention for feature engineering for maximizing flexibility during monitoring through DNN-enabled versatile information processing. This versatile information processing enables the extraction of a wide variety of useful information from the data.

The DL-based optical monitor has two operation modes that originate from DNN in the monitor: (1) training mode and (2) inference mode. The following sections discuss the basic and advanced topics of both operation modes.

6.3.1 Training mode of DL-based optical monitors

Fig. 6.6 shows the block diagram of the DL-based optical monitor in the training mode. The DL-based optical monitor trains the DNN inside before being deployed for monitoring. First, the reference optical signal is measured by the digital coherent receiver and stored in the raw waveform database. In the following data-augmentation block, the raw data can be augmented to improve model learning and tolerance against realistic conditions. (Its technical details are described in Section 6.3.2.1.) The data and the corresponding labels (expected DNN outputs in the supervised training dataset; see Section 6.2.2) are presented to the DNN in the DL-based optical monitor. The DNN is trained with supervised learning using backpropagation to minimize the loss, such as mean-squared error for regression or cross-entropy for classification. The minibatch stochastic gradient descent (SGD) algorithm with a controlled learning rate is generally used in the training phase. The learning rate is manually or automatically decayed using control algorithms such as Adam optimizer [35]. The trained DNN model, i.e., a trained set of weights of the DNN, is stored and used to scan and monitor the transmission line's status with the DL-based optical monitor in its inference mode.

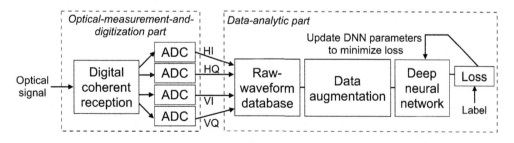

Figure 6.6 Overview of the training mode of the DL-based optical monitor.

6.3.2 Advanced topics for the training mode of DL-based optical monitors

In this section, additional advanced topics to enhance the training mode of DL-based optical monitors are described.

6.3.2.1 Data augmentation based on domain knowledge of optical communication

In this section, data augmentation by leveraging the domain knowledge of fiber-optic communication to train DNNs is discussed. One of the simplest ways to improve DNN model performance and generalization ability is to increase the amount of training data. However, doing so is not always feasible due to the expensive cost of data collection and labeling.

Data augmentation can solve this issue by creating "quasi-new" data from existing training data [15]. One of the most impressive examples of this process is data augmentation for an image recognition task, which recognizes objects in a given image data and classifies them into specific classes. Creating quasi-new images by modifying the input image data without changing its label is relatively easy in this case. The training image data can be translated, rotated, and flipped without changing their labels, e.g., "cat" or "dog." In contrast, the process of data augmentation for many other tasks, including optical monitoring, is not always easily understood.

The DL-based optical monitor is designed to utilize our domain knowledge of optical communication to augment input data. The input data for DNN of the DL-based optical monitor are data obtained via digital coherent reception. The input data can be straightforwardly treated as an optical electric field that comprises phase and amplitude of both horizontal and vertical polarization. Human experts in optical communication can utilize knowledge of the physical laws that govern the input optical waveform data. Thus, we can directly apply the knowledge to expand the input data. In contrast, if a direct detection system is used to collect the dataset, then the data captured by the direct detection system would contain only the amplitude of the signal. Thus, some data augmentation related to the phase information of the signal is not available due to information loss at the optical-measurement-and-digitization part of this system.

The data augmentation scheme based on the domain knowledge of optical communication [36] is shown in Fig. 6.7. Data augmentation based on domain knowledge has a cost advantage because no additional measurement cost and fixed computational cost are required to calculate an augmented data point, unlike in additional data collection with various measurement conditions. For highly reliable systems in particular, including optical fiber communication systems, data measurement cost corresponding to rare phenomena, such as system anomaly or failure, tends to be expensive because of their low probability of occurrence. In contrast, data augmentation based on domain knowledge can configure any transformation to create augmented data in principle, even if the transformation corresponds to rare phenomena. The following are simple examples of data augmentation for DL-based optical monitors.

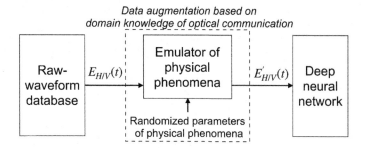

Figure 6.7 Data augmentation scheme for DL-based optical monitors.

6.3.2.1.1 Data augmentation on polarization state

Controlling the polarization states over transmission links is challenging. A training dataset for a DNN would have only a specific polarization state with a limited number of measurements. If the DNN model is trained using the data corresponding to a specific polarization state, then the model performance for real inputs that have various polarization states cannot be guaranteed. To solve this issue, we can use data augmentation on polarization states.

First, we construct a time-varying Jones vector from the measured HI, HQ, VI, and VQ data points sampled at a sampling rate of $F_s = 1/T_s$

$$
\begin{pmatrix} E_H(t_n) \\ E_V(t_n) \end{pmatrix} = \begin{pmatrix} HI(t_n) + jHQ(t_n) \\ VI(t_n) + jVQ(t_n) \end{pmatrix},
\tag{6.5}
$$

where j is an imaginary unit and $t_n = nT_s (n = 0, 1, 2, \cdots)$ denotes time. By multiplying the Jones vector by the 2×2 complex unitary matrix U, we can calculate the new Jones vector after polarization rotation.

$$
\begin{pmatrix} E'_H(t_n) \\ E'_V(t_n) \end{pmatrix} = U \begin{pmatrix} E_H(t_n) \\ E_V(t_n) \end{pmatrix} = \begin{pmatrix} a & b \\ -\bar{b} & \bar{a} \end{pmatrix} \begin{pmatrix} E_H(t_n) \\ E_V(t_n) \end{pmatrix}.
\tag{6.6}
$$

Note that $a\bar{a} + b\bar{b} = |a|^2 + |b|^2 = 1$ in U. Data augmentation on polarization states is equivalent to polarization randomization of the input signal for digital coherent transceivers. The virtual random polarization rotations by the data augmentation can help realize robust training against different polarization conditions between the training and test.

6.3.2.1.2 Data augmentation on the frequency offset

A typical digital coherent transceiver uses individual lasers for the signal and local light, thereby making the process of aligning the emitting frequencies of the two lasers a challenging task. In other words, there is a frequency difference between the signal and local lasers at the receiver side. The frequency offset is superimposed on the digitized optical electric field sampled from the optical hybrid's output, as described in Section 6.2.1. The frequency offset δf is different for each measurement due to the drifting of the emitting frequency of lasers. Thus, in principle, a mismatch of the frequency offset exists between the inference and training data, which may impair the model performance.

Data augmentation on the frequency offset is beneficial to cope with the frequency offset imposed on the inference data, adding a random additional frequency offset to the measured training data as follows:

$$E'_{H/V}(t_n) = E_{H/V}(t_n)exp(j2\pi f_{max}ut_n), \tag{6.7}$$

where j is an imaginary unit, $t_n = nT_s = n/F_s (n = 0, 1, 2, \cdots)$ denotes time, f_{max} is the maximum value of the frequency offset to be added, u is a uniform random value from -1 to $+1$, $E_H(t_n) = HI(t_n) + jHQ(t_n)$, and $E_V(t_n) = VI(t_n) + jVQ(t_n)$. This data augmentation increases the data diversity of the training data in terms of frequency offset. The DNN trained using the augmented training data is harnessed to perform a task correctly for various data inputs having frequency offsets ranging from $-f_{max}$ to $+f_{max}$.

The virtual frequency offsets by the data augmentation can help realize robust training against different sets of signal and LO lasers between the training and test. Section 6.5.5 describes experimental validation of this data augmentation.

6.3.2.2 Transfer learning for adaptation of DNNs

In this section, the issue of trained model performance degradation caused by a changing environment is considered. As already discussed in Section 6.3.2.1, model robustness against various environmental changes is one of the most important features required for maintaining the accuracy of DL-based optical monitors after deployment. Transfer learning [15] is an approach to tackle this problem through model adaptation after the initial training.

In the previous section, we simply assumed that the operator deploys the DNN model trained before its deployment or shipment. Here, for example, let us consider a situation in which several optical transceivers have different characteristics, such as different analog bandwidths. When a DNN model that is trained using a particular optical transceiver is applied to the other transceivers, it may not work well due to the mismatch/inconsistency of transceiver characteristics between different transceivers. To solve the issue, we can perform in-field additional training to overcome the mismatch/inconsistency, which is added to the factory's preshipment initial training. This process is called transfer learning [15].

Transfer learning uses the knowledge learned by a model in a certain domain (e.g., in a particular digital coherent receiver) for training in another domain (e.g., in another digital coherent receiver), thus allowing us to achieve highly accurate learning with a fewer number of additional training data points by transferring knowledge that can be used in different domains. Feature extraction and fine-tuning are the widely used implementations of transfer learning (Fig. 6.8).

1. **Feature extraction**: A pretrained DNN is used to extract meaningful features from new data. Newly added layers on top of the pretrained DNN are trained from scratch. In this case, the pretrained DNN that has already been trained in the previous domain does not need to be retrained.

2. **Fine-tuning**: The parameters in the early layers of the pretrained DNN are retained, and a few of the latter layers of the pretrained DNN are unlocked for training in the new domain. The unlocked latter layers and newly added layers (initialized by random values) are jointly trained for the new domain's new task. This joint learning makes the training more relevant for the new domain and task. This approach is advantageous because the early layers of the pretrained DNN can extract basic features that can be shared with many tasks. The features specific to the current task are extracted on the latter layers.

Transfer learning adjusts the model to different systems with a smaller amount of training data while ensuring a shared learning process. In optical communication, this method has been demonstrated in [37–40].

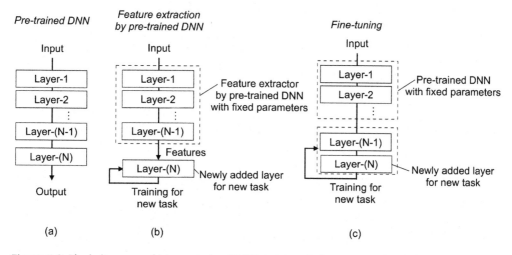

Figure 6.8 Block diagrams of (a) a pretrained DNN and transfer learning via (b) feature extraction and (c) fine-tuning.

6.3.2.3 Federated learning for collaborative DNN training over multiple operators

The transfer learning technique explained in Section 6.3.2.2 is used for sharing learned information among several domains. By implementing the information-sharing idea, we can design a framework of collaborative learning that shares information among models placed at different sites. In this section, an implementation of the collaborative learning approach called federated learning (FL) [40] is described. Pioneering work to introduce FL into optical networks was investigated in [43,44].

One of the practical applications of FL in fiber-optical networks is collaborative model training for multidomain networks operated by multiple network operators [43,44]. In this situation, the amount of data collected by one network operator is usually limited; this data limitation has a negative impact on the accuracy of the model trained using the local dataset collected by one operator (Fig. 6.9(a)). A straightforward solution overcome the data limitation is for several network operators to bring their data and jointly develop a shared model trained using the jointed dataset. The accuracy of the trained shared model is expected to be high because of the increased amount and diversity of training data (Fig. 6.9(b)). However, this straightforward solution is sometimes challenging in terms of the restrictions on the sharing of data owned by each network operator with other network operators because of data confidentiality as the collected data are likely to be sensitive. This issue becomes especially important when the data are shared across different countries or regions.

A learning framework of FL [41,42] can address this issue. An example that uses the FL framework for multidomain optical networks is shown in Fig. 6.9(c) [43,44]. First, each network operator first trains its own DNN model using its own collected data. Next, each network operator shares the parameters of the trained DNN model instead of the raw data. A central server aggregates the different models sent from several network operators to construct the shared model; then, the parameters of the shared model are sent back to each operator.

In the FL framework, each network operator can take advantage of joint model development while protecting its own measured raw data. This type of learning method will become more critical as the use of ML in fiber-optic networks becomes more common. Early demonstrations of the FL in optical networks were performed in [43, 44].

6.3.3 Inference mode of DL-based optical monitors

In this section, the DL-based optical monitor's inference mode to scan the fiber transmission link after its training mode is described. Fig. 6.10 shows the block diagram of the inference mode operation of the DL-based optical monitors. The DNN inside the monitor utilizes the DNN parameters trained in the training mode. The DNN then performs a forward propagation operation for the input data and outputs its result as a monitoring result.

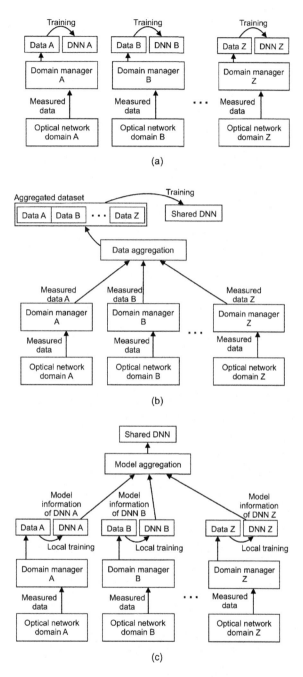

Figure 6.9 Federated learning for cooperative DNN training over multidomain optical networks. Model training and inference with (a) local training and local inference, (b) training by aggregated database and inference by local data, (c) federated training by model weight aggregation and inference by local data.

The DL-based optical monitor has a simple inference operation; however, this basic inference mode can be enhanced for advanced topics. The following section discusses this topic.

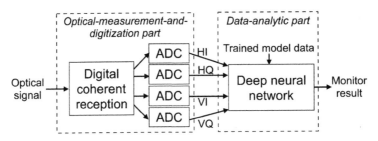

Figure 6.10 Overview of inference modes of DL-based optical monitors.

6.3.4 Advanced topics for inference modes of DL-based optical monitors

In this section, advanced discussions on the inference modes of DL-based monitors are presented.

6.3.4.1 Cloud-based vs. edge-based implementations

In this section, two implementation scenarios of the DL-based optical monitor, cloud- and edge-based implementation scenarios, are explained and their advantages and limitations are discussed. In the cloud-based implementation scenario, digital coherent receiver's measured waveform data are immediately sent to the centralized data center located at a different place, e.g., cloud data center. In the edge-based implementation scenario, the measured waveform is processed on edge-side device, e.g., digital coherent receivers inside or their embedded peripherals such as FPGA.

6.3.4.1.1 Cloud-based implementation of inference mode

Fig. 6.11(a) shows a block diagram of the first implementation scenario, i.e., cloud-based implementation. In the scenario, the digital coherent receiver's measured waveform data are sent to the centralized data center located at a different place for processing to realize a DL-based optical monitor.

One advantage of cloud-based implementation is that rich computational resources are available at the centralized data center. Well-equipped centralized cloud data centers can provide better computational efficiency than typical edge devices can.

Other advantages of this implementation are high flexibility and upgradability. All the processing parts for monitoring are centralized, thus ensuring that the monitoring algorithm can be easily deployed and upgraded using existing cloud-based tools, such as virtualization/container technologies.

The other possible advantage is high data reusability. Datasets collected at multiple digital coherent transceivers can be utilized for advanced multiinput and multimodal data analysis, such as network–wide fault isolation and prediction.

However, cloud-based implementation has limitations in terms of data security and latency when transmitting waveform data between the data measurement location (digital coherent receiver) and centralized datacenter. Transmitting raw waveform data may pose a security risk because data payload can be reconstructed from the waveform data. In addition, this scenario leads to a relatively large latency due to propagation between the digital coherent receiver and cloud data center, which may hinder real-time processing and feedback using the monitoring results.

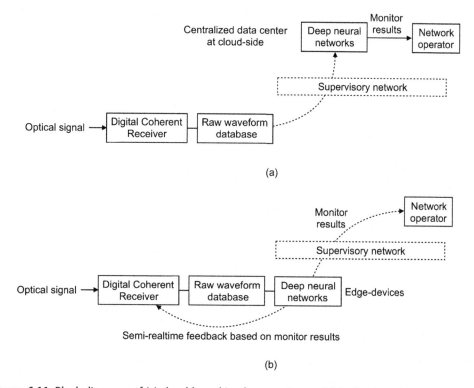

Figure 6.11 Block diagrams of (a) cloud-based implementation and (b) edge-based implementation of DL-based optical monitor.

6.3.4.1.2 Edge-based implementation of inference mode

Fig. 6.11(b) shows a block diagram of the second implementation scenario, i.e., edge-based implementation. Unlike the cloud-based implementation, the measured waveform is processed on the edge-side device, e.g., digital coherent receivers inside or in their embedded peripherals, such as embedded GPU and FPGA.

The main advantage of edge-based processing is reduced latency, which is achieved by reducing communication between the measurement and processing sites. A prompt response according to the monitoring results can be obtained because of the low latency. This feature is valuable when the transceiver or other network equipment needs real-time feedback. In addition, the edge-based implementation can minimize the risk in terms of the security aspect by processing and discarding data locally as communication paths no longer pose a risk of leaking sensitive information.

However, most edge-based implementations tend to suffer from limited available computation resources. Addressing this issue requires lightweight neural network models and entails additional discussion points related to a tradeoff between performance and the computational complexity of the neural network model, for example, lower bit accuracy to represent model parameters.

A positive aspect of the computational complexity issue at edge-side devices is that less computational resources are required for DNN inference than for DNN training. Thus, DNN inference can be available in optical transceivers or in peripheral devices.

Another positive aspect of implementing a DL-based optical monitor at the edge is the gap between the sampling rate of optical waveform sampling (e.g., tens or hundreds Gsample/s at the digital coherent transceiver) and monitoring result sampling. Most optical monitoring applications do not require an ultrahigh-speed (e.g., several tens of GHz) refresh rate for monitoring purposes. Thus, a real-time streaming DNN for the monitoring that can provide monitoring results at the sampling rate of the optical waveform at the digital coherent transceiver is not necessary in real-world applications. The typical refresh rate for monitoring applications is often several orders of magnitude lower than the ADC sampling rate for waveform data. Thus, we can utilize batched and time-stretched DNN processing instead of real-time streaming DNN processing for monitoring purposes.

Batched and time-stretched DNNs access and process only a small portion of the measured data. For example, we assume that we have a DNN that outputs a monitoring result from 100 input data points. When the input data for the DNN are sampled at 100 Gsample/s, we can compute the monitoring result using the DNN with a refresh rate of up to 1 GHz (= 100 GHz/100). If the required refresh rate for the monitoring is 10 MHz, then extracting one output out of every 100 monitoring outputs calculated at the maximum rate is sufficient. In other words, 99% of the data and processing will not be used in this specific case. The batched and time-stretched DNNs use the surplus time to reduce the circuit size for the DNN. A functional circuit block, such as multiply-accumulate operation and/or activation function in the neural network, can be reused for many calculations within the surplus time, thereby leading to smaller circuit sizes for DNN inferences.

6.3.4.2 Estimating the model uncertainty in inference mode

In this section, the DNN model uncertainty for the DL-based optical monitor to automate network operation is discussed. As shown in Fig. 6.1, the DL-based monitoring results are used to control the programmable optical network in the network automation scenario. From a social viewpoint, fiber-optic networks are a fundamental part of today's information society. Thus, poor control of the network caused by the monitor's incorrect estimation can lead to a critical problem. Although DL-based optical monitors show high accuracy because of representation learning by DNN, the model learns using a limited amount of training data and the same performance level is not guaranteed in the future.

The issue depends on the limited functionality of the DL-based optical monitors that provide only point-estimated monitoring results, e.g., a single OSNR value for a single input. If the DNN of the DL-based optical monitor can provide the model uncertainty information (or confidence in the current model's output), then we can use this uncertainty information to operate the optical networks. The uncertainty information can be provided to both human network operators and network controllers. The basic idea is as following: (1) Network operation is automated in a typical situation in which the trained model performs well with low uncertainty. (2) The automated operation is suspended in nontypical situations, i.e., high-uncertainty situations, where the trained model's performance deteriorates. (3) The operation moves to human decision-making (Fig. 6.1). The cause of the nontypical behavior indicated by a high uncertainty alert is identified and repaired by analyzing the system, including the DNN model, by employing human experts. This is a crucial function in operating mission-critical systems, such as optical networks.

Gaussian process regression (GPR) [14,45,46], a mathematically grounded tool to determine model uncertainty, has been widely used in optical monitoring to estimate model uncertainty [47,48]. However, GPRs have an issue on computational complexity at the inference phase with a large amount of training data points.

Dropout in the inference phase technique, which is an alternative model uncertainty estimation technique that is more suitable for trained DNN models, was recently proposed and developed by Gal *et al.* in Ref. [49]. This technique is a powerful tool to enhance the DL-based optical monitors, allowing monitors to provide model uncertainty information as an auxiliary result. Unlike GPR, the computation complexity of DNNs with dropout at the inference does not depend on the number of training data points. Thus, it is suitable for models trained using a large number of data points.

To illustrate the dropout at the inference, we first explain the existing standard dropout [50] that is widely used in training (not inferencing) a DNN to avoid overfitting. As shown in Fig. 6.12, the (standard) dropout samples a partial DNN from a full DNN by probabilistically deleting the full DNN's neurons with a probability of p for every input point and every forward and backward pass through the model training. We use all neurons scaled by $1/(1-p)$ for the inference phase. The standard dropout can improve the performance of a DNN but provides only point estimation results.

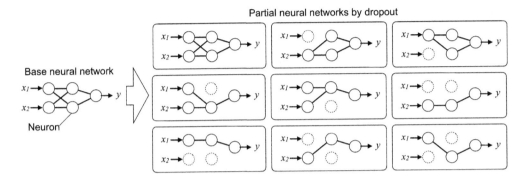

Figure 6.12 Partial neural networks sampled by dropout.

Unlike the standard dropout, the dropout technique is used at the inference phase to estimate model uncertainty, as shown in Fig. 6.13. The method for estimating model uncertainty includes performing a forward pass N-times through a DNN for a single input data using the dropout at the inference phase. The scaling of $1/(1-p)$ used in the standard dropout is not used in this inference to predict uncertainty. The N results generated from N stochastic partial networks by dropout are averaged to predict the point estimation results. To predict uncertainty, we collect the N results calculated from a single input data and take their standard deviation.

The idea behind this scheme is to emulate an ensemble of models. DNN training typically has a high computational cost, and the training of many different DNN models for an ensemble is sometimes challenging in terms of computational complexity. To mitigate this limitation in the framework with dropout, the model parameters (e.g., weights) are partially shared among the partial DNNs sampled by dropout. Parameter sharing allows a large number of different models to be represented with a limited computational complexity.

The dropout at the inference phase can be considered as an approximation of the Bayesian inference in the variational Bayesian learning framework [49]. In the variational Bayesian learning framework, weights in the neural network are not simple values but have a probabilistic distribution. Thus, the output also has a probabilistic distribution and can evaluate the uncertainty of the output. The dropout at the inference can approximate the Bayesian inference even in a framework of the standard DNN because a DNN sampled using the dropout at the inference phase can be treated as a DNN sampled from a probabilistic distribution in the variational Bayesian learning framework.

The dropout at the inference phase can be used to extract uncertainty information from existing trained DNNs without changing their implementations. Thus, the technique can be used to upgrade an existing DL-based optical monitor to an enhanced one that provides both point-estimated monitoring results and their uncertainties [51,52].

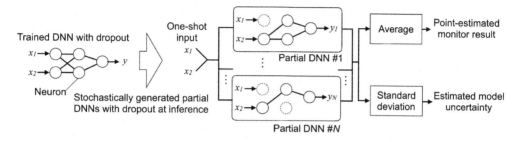

Figure 6.13 Dropout at inference to output uncertainty information for DL-based optical monitors.

6.4. Tips for designing DNNs for DL-based optical monitoring

In this section, the architecture of DNN for DL-based optical monitors is discussed. Many research activities have been performed to provide a basis for designing an optimal DNN for a given task; however, a comprehensive study has not been performed yet. Thus, a trial-and-error approach needs to be adopted in designing and developing DNNs for specific tasks. With this limitation, some useful heuristic guidelines and topics for designing DNNs for DL-based optical monitors are described in this section.

6.4.1 Shallow vs. deep network

According to the universal approximation theorem [53], a neural network with a sufficient number of hidden neurons with a nonlinear activation function can approximate any continuous function. However, we cannot use an infinite number of neurons in practical cases, thus requiring guidelines for constructing a sufficiently large network for a given task.

Two approaches can be used to construct a DNN: increasing the number of neurons in a single layer or stacking many layers with a relatively limited number of neurons. A DNN with a rectified linear unit (ReLU) activation function can represent the same mapping function with fewer neurons as compared with a shallow neural network [54]. This finding suggests that obtaining a powerful model by increasing the depth of the DNN layers is exponentially easier than obtaining one by increasing the number of units in a shallow network, thereby implying a significant advantage of deeper networks. However, a large capacity network does not always mean successful model training. Generally, increasing the number of network layers leads to difficulties during training, such as the vanishing gradient problem. Several techniques, such as dropout [50], batch normalization [55], and residual connection [56], have been developed to mitigate the training issues.

A method currently being used for constructing a DNN involves increasing the model capacity by increasing the number of layers and mitigating training difficulties by utilizing general techniques (e.g., dropout and batch normalization) and domain-specific techniques (e.g., data augmentation and utilizing implicit bias of data).[2]

6.4.2 DNN architecture for optical monitoring

In this section, a guideline for designing a network architecture for optical monitoring is discussed. We explore suitable model architectures by analyzing the structure of the input data to DNN. To show the impact of neural network architecture on optical monitoring tasks, we analyze two typical DNN architectures: fully connected DNN (FC-DNN) and convolutional neural network (CNN).

6.4.2.1 Fully-connected DNNs

In this section, we discuss a basic form of DNN: FC-DNN [15]. An FC-DNN comprises layered neurons that are connected only between adjacent layers. Each connected layer uses the previous layer's output as input. A neuron is a nonlinear unit that computes a single output from multiple inputs. A detailed explanation of the behavior of the neuron is provided in the introductory chapter. Fig. 6.14 shows an illustration of a single FC layer of the FC-DNN.

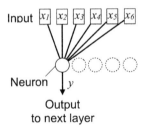

Figure 6.14 Data processing in FC layer.

---

[2] In-depth research on the learning of DNNs is still pending. One well-known paradox is as follows. Most of large-scale DNNs are overparametrized: the number of parameters of a DNN is usually much larger than that the amount of training data. In a traditional framework, the overparametrized model is likely to learn incorrect noise-induced patterns, resulting in a so-called overfitting problem, where the model works well with training data but not with test data. To avoid this problem, a regularization [14] that suppresses the model capacity during training is widely used. However, DNNs tend to generalize without overfitting even if the model is overparameterized. This paradox was unsolved when this chapter was written; however, recently, a new hypothesis called the lottery ticket hypothesis [57] has been proposed and has attracted much attention. Paying attention to such theoretical insights on DL is worthwhile for designing effective networks.

6.4.2.2 Convolutional neural networks

In this section, we briefly review CNNs [15] from the viewpoint of DL-based optical monitors. A CNN is a form of DNN that is used to process grid-like formatted data inspired by the brain's primary visual cortex and has been widely used in image processing because pixel images are equivalent to a two-dimensional (2D) grid-like array fitted to the CNN input. CNNs can be applied for DL-based optical monitors because a waveform sampled by constant time duration can be treated as a one-dimensional (1D) grid-like array that fits the CNN input.

In CNNs used in image processing, typical input data are in the form of a 2D grid with three channels, e.g., RGB channels. Considering the DL-based monitoring discussed in this chapter, the input data can be in the form of four channels and 1D data from a digital coherent receiver. The real and imaginary parts of each horizontal or vertical polarization of the complex optical field (HI, HQ, VI, and VQ) represent four channels similar to RGB channels in image data. Each channel has 1D values characterized by equal sampling intervals, similar to image data pixels.

A CNN comprises the convolutional layer(s), activation functions, and pooling layer(s). For 1D input data, discrete convolution, shown as the following equation, is performed in the convolutional layer:

$$s_i = \sum_{k=0}^{K-1} \sum_{j=0}^{L-1} x_{i-j,k} w_{j,k}, \tag{6.8}$$

where x is the input, w is the kernel, s is the feature map, and i, j, and k are indexed as integers. L is the length of the kernel, K is the number of input data channels, and the kernel's overall size is LK. In the convolutional layer of the CNN, the parameters learned by training are the coefficients of the kernel. The same kernel is used for the entire data, which is why the number of trainable parameters is dramatically reduced compared with the FC layer. Fig. 6.15 illustrates data processing on the CNN when $K = 1$ and $L = 3$. From the view of signal processing, this kernel can be considered a digital filter. For example, a low-pass, high-pass, or edge-detection filter can be constructed by training the kernel coefficients. Although these types of filters and their parameters have been implemented manually in conventional methods, the filter coefficients are automatically learned from the training data in the CNN.

6.4.2.3 DNN architecture for the optical monitoring

In this section, we discuss which DNN architecture—FC-DNN or CNN—is more suitable for DL-based optical monitors. We first analyze the connectivity between neurons in FC-DNN and CNN. Unlike the widely known impression of the architecture of FC-DNN and CNN, the neurons' functionality, i.e., activation function after dot prod-

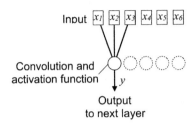

Figure 6.15 Data processing in CNN.

ucts of input and weights, in both FC-DNN and CNN can be considered identical. The difference between FC-DNN and CNN is in their connectivity among neurons.

The neurons in the FC layer are connected to all the neurons in the previous layer (Fig. 6.14), and each neuron in the FC layer has an independent set of parameters. The neurons in the convolutional layer are connected only to a local region in the input (Fig. 6.15), and each neuron in the convolutional layer shares the same set of parameters [15]. The parameter sharing, e.g., kernel in the convolutional layer, dramatically reduces the number of trainable parameters compared with the FC layer case. This reduction usually leads to better performance with the mitigation of model overfitting by reducing an excessive degree of freedom in the model if the local connectivity fits the nature of input data.

Next, we review the input data for the FC-DNN and CNN and analyze its nature. The input data to the DNNs (FC-DNN and CNN) of the DL-based optical monitor are an optical electric field sampled at equal time intervals.

$$\boldsymbol{HI} = (HI(t_1), HI(t_2), \cdots HI(t_N)), \tag{6.9}$$

$$\boldsymbol{HQ} = (HQ(t_1), HQ(t_2), \cdots HQ(t_N)), \tag{6.10}$$

$$\boldsymbol{VI} = (VI(t_1), VI(t_2), \cdots VI(t_N)), \tag{6.11}$$

$$\boldsymbol{VQ} = (VQ(t_1), VQ(t_2), \cdots VQ(t_N)), \tag{6.12}$$

where HI, HQ, VI, and VQ are the real (I) and imaginary (Q) part of each horizontal (H) or vertical (V) polarization of the complex optical field, and t_k indicates the time index. Fig. 6.16 shows an overview of the input data and data processing on the DNN.

First, the DL-based optical monitor scheme uses the data of the waveforms sampled asynchronously, meaning that we do not assume the presence of a specific frame pattern in the measured data. Thus, all data points characterized by a specific time in the measured dataset are essentially equivalent, e.g., both sets of data points (HI, HQ, VI, and VQ) at time of t_k and t_l are essentially equivalent for data processing purpose. With this nature of the input data, we can use the same kernel (parameters) for all input data points.

Next, we analyze the nature of the input data from the viewpoint of the use of the local connectivity on CNN. The axis of 1D convolution in the CNN designed for the DL-based optical monitor is a time axis of the input waveform. Thus, the local connectivity on the CNN can be justified if each waveform at a particular instance of time depends only on the waveform attained in the neighboring instance of time. We analyze this nature of the input data using the domain knowledge of optical communication. If the transmission line has no memory, then this condition is naturally satisfied. If the transmission line has memory, then we have to consider the spreading optical waveform on the time axis. The optical waveform is usually spread over the time axis through CD due to a propagation over a fiber transmission link. In this case, each data point strongly correlates to nearby data points and weakly correlates to farther data points in the input data for the DL-based optical monitor. This nature of the input data fits the CNN having local connectivity if the kernel size is large enough to support waveform spreading on the data.

According to the above discussion, the CNN should fit the DL-based optical monitors' input data, compared with the FC-DNN. The local connectivity in CNN fits the input data structure with local distortion in the optical waveform. We show experimental examples of DNN architectures in Section 6.5 through DL-based OSNR monitor's experimental results using FC-DNN and CNN.

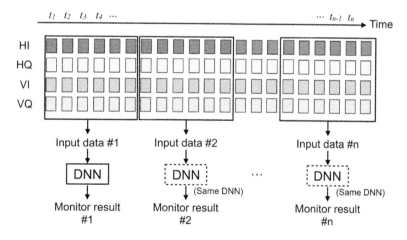

Figure 6.16 Input data structure for DL-based optical monitor.

6.5. Experimental verifications

In this section, we present experimental verifications for the features of DL-based optical monitors described in the preceding sections: DNN architecture for optical monitoring, CNN-based OSNR monitoring, CNN-based modulation format and baud rate classification, data augmentation using prior domain knowledge of the optical system, and estimating model uncertainty using the dropout at the inference phase.

6.5.1 Experimental setup for data collection

In this section, an experimental setup for acquiring training and test datasets of the DL-based optical monitor is described. Fig. 6.17 shows the details of the experimental setup that was explained in our previous work in [16,32–34,36,51,52].

Nyquist-filtered (roll-off factor = 0.01) 14- and 16-Gbaud (GBd) DP-QPSK, 16QAM, or 64QAM optical signals were received by the coherent receiver and digitized using ADCs with a sample rate of 40 or 80 Gsample/s, after amplified spontaneous emission (ASE) noise was loaded to vary the OSNR. Each signal was modulated using different random bit sequences generated using the Mersenne Twister method with a different seed for each training and test/inference of the DNN to avoid an overfitted evaluation [58]. The digitized samples, which are virtually reconstructed optical fields corresponding to HI, HQ, VI, and VQ, were presented to the DNN trained using the TensorFlow library [59] on a desktop computer equipped with GPUs.

Actual OSNRs were measured using an optical spectrum analyzer (OSA) to label the training dataset. This label information is not necessary for the test/inference. We confirmed the correct signal reception by evaluating the bit error ratio after demodulation DSP (not shown in this figure).

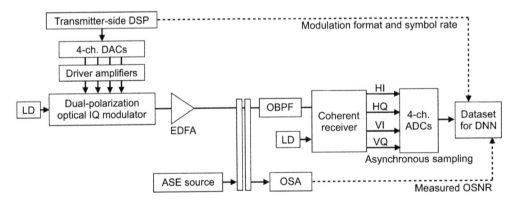

Figure 6.17 Experimental and simulation setup. DAC: digital-to-analog converter, IQ: in-phase and quadrature-phase, LD: laser diode, EDFA: erbium-doped fiber amplifier, ASE: amplified spontaneous emission, OBPF: optical bandpass filter, ADC: analog-to-digital converter, OSA: optical spectrum analyzer.

6.5.2 Neural network architecture for OSNR estimation task

In this section, we discuss which neural network architecture is suitable as monitor architecture on the basis of an experimental evaluation of OSNR estimation. Two types of DNN, i.e., FC-DNN and CNN, were trained and evaluated to find the difference due to neural network architecture.

6.5.2.1 DNN used in this experiment

Fig. 6.18 shows FC-DNN and CNN designed to estimate the OSNR. All DNNs were trained using supervised learning using backpropagation and the minibatch stochastic gradient descent algorithm with a controlled learning rate by the Adam optimizer. The loss was defined by mean-squared error between DNN output and actual OSNR measured by OSA and minimized through the training phase. To prevent overfitting, dropout (probability of 0.5) and batch normalization techniques were used for the networks.

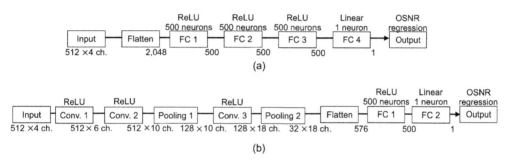

Figure 6.18 (a) Architecture of DNN #1 (FC-DNN) and (b) DNN #2 (CNN-1).

DNN #1 (FC-DNN):

Fig. 6.18(a) shows the network architecture of the FC-DNN used in this experiment. The FC-DNN had three hidden layers, which were determined according to our previous work [16]. Four channelized optical electric fields were provided as inputs to the FC-DNN: 512 samples × 4 channels (HI, HQ, VI, and VQ) of the digital coherent receiver outputs asynchronously sampled at 40 Gsample/s. The four-channel data were flattened before the first FC layer, going from 512 samples × 4 channels to 2,048 samples. The flattened data were provided to the FC 1 layer. The three hidden layers (FC 1, FC 2, and FC 3) had the same structure, each with 500 neurons and a ReLU [60] as activation function [15]. For the output in FC4, linear regression was used for OSNR estimation.

DNN #2 (CNN-1):

Fig. 6.18(b) shows the network architecture of the CNN used in this experiment. The CNN's input is exactly the same as that for DNN #1: four channelized optical electric fields asynchronously sampled at 40 Gsample/s. This sampled time-series data over the time axis was convolved in the convolutional layers, i.e., 1D convolution with trainable filter weights. This is equivalent to multiple filtering in the convolutional layers; thus, the number of output and input channels differed through the convolution layer (e.g., it increased from 512 samples × 4 channels to 512 samples × 6 channels through the convolutional layer 1 in Fig. 6.18(b). All convolutional layers in this work used ReLU as the activation function. The pooling layer was a max-pooling layer with a stride of 4. These pooling layers reduced the amount of data to one-fourth, e.g., from 512 samples to 128 samples, through pooling 1 in Fig. 6.18(b). The data were flattened before the first FC layer, going from 32 samples × 18 channels to 576 samples. The flattened data were used for the FC layers. For output, linear regression was used for OSNR estimation.

6.5.2.2 Results and discussion

The model performance of the FC-DNN (DNN #1) was evaluated using different numbers of records in the training datasets. The training and test datasets were a mixture of 14- and 16-GBd DP-QPSK signals with OSNRs ranging from 8 to 28 dB. The number of records of 14- and 16-GBd DP-QPSK signals in both training and test datasets was adjusted to be in the ratio of 1:1.

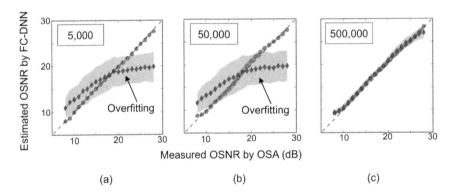

Figure 6.19 OSNR estimation by FC-DNN (DNN #1) with number of training data of (a) 5,000, (b) 50,000, and (c) 500,000.

The OSNR values estimated by the DNN #1 trained using 5,000 training data points are shown in Fig. 6.19(a). The black dashed line shows the line where the output of the DNN #1 is equal to the OSNR measured by the OSA. If the DNN #1 were trained correctly, then the measured points would agree with this line. The red (mid gray in print version) circles and blue (dark gray in print version) diamonds indicate the output of DNN #1 using the training and test datasets, respectively. The transparent area indicates the standard deviation of the DNN #1 outputs. The DNN #1 trained using 5,000 training data points estimated correct OSNR values for the training dataset but yielded inaccurate values for the test dataset. Thus, the DNN #1 trained using 5,000 data fell to an overfitting region.

Fig. 6.19(b) shows the results when the number of training data points increased tenfold (50,000 data points). Overfitting was observed even with 50,000 data points. As shown in Fig. 6.19(c), training with 10 times more data (500,000 data points) mitigated the overfitting and DNN #1 almost correctly estimated OSNR values for both training and test datasets. Although a large number of training datasets (e.g., 500,000) can help avoid overfitting and enable FC-DNN to estimate OSNR for unknown test datasets as known training datasets, this situation suggests that the small amount of training data (e.g., 5,000) may induce overfitting in the FC-DNN.

Next, we performed the same OSNR estimation task using a CNN to compare the FC-DNN- and CNN-based OSNR monitors. We trained the CNN (DNN #2) shown in Fig. 6.18(b) to estimate the OSNR. The training and test data were the same as those used for the FC-DNN (DNN #1). The results with 5,000, 50,000, and 500,000 training data points are shown in Figs. 6.20(a), 6.20(b), and 6.20(c), respectively. Notably, as shown in Fig. 6.20(a), the CNN (DNN #2) outputs almost correct values for the test data even with a smaller amount of training data (e.g., 5,000). In other words, the CNN (DNN #2) was successfully trained while avoiding overfitting even with a small amount of training data. This result supports the discussion in Section 6.4.2.

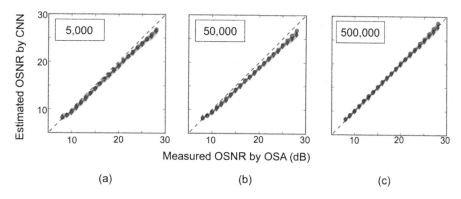

Figure 6.20 OSNR estimation by CNN (DNN #2) with number of training data points of (a) 5000, (b) 50,000, and (c) 500,000.

6.5.3 Detailed experimental evaluation of CNN-based OSNR estimators

In this section, we discuss the details of the experimental characterization of a CNN-based OSNR monitor for multiple signal types, i.e., 14- and 16-GBd DP-QPSK, 16QAM, and 64QAM signals. The discussion in this section is based on our previous work presented in [33,34].

6.5.3.1 DNN used in this section

DNN #3 (CNN-2):

Fig. 6.21 shows the network architecture of the CNN used in this section. The CNN's general structure is the same as that of the DNN #2, but the network parameters are different, as shown in Fig. 6.21.

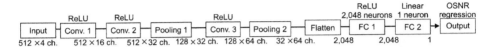

Figure 6.21 Convolutional neural network (DNN #3) architecture for OSNR estimation.

6.5.3.2 Results and discussion

First, the estimated OSNR values were obtained by the DNN #3 trained using 16-GBd DP-QPSK signal, with OSNR varying from 11 to 33 dB. Fig. 6.22 shows the mean value of the OSNR values estimated by the DNN #3 as a function of the actual OSNRs measured by OSA. The test dataset that contained 10,000 records of 16-GBd DP-QPSK signal was used for the evaluation. Figs. 6.22(a) and 6.22(b) show the DNN-estimated OSNR obtained by the training and test datasets, respectively. The results show that the DNN #3 was successfully trained to estimate OSNR even for unknown test datasets while avoiding the overfitting.

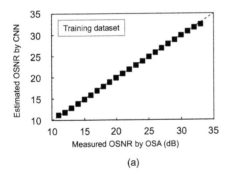 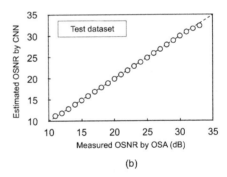

Figure 6.22 Evaluation results of CNN-based OSNR estimator (DNN #3) by using 16-GBd DP-QPSK (a) training dataset and (b) test dataset. These figures are based on the results in Ref. [33].

Next, to investigate multiformats coverage by the CNN, we trained DNN #3 using different datasets having mixed signal types, comprising 14- and 16-GBd DP-QPSK, 16QAM, and 64QAM signals with different OSNR values. The total number of training data points was 1,000,000. The trained CNN was evaluated over multiple modulation formats and symbol rates. We used six test datasets that contained each modulation format and symbol rate. For this test, 10,000 data records were used for each test dataset. The single CNN trained with the mixed training dataset was evaluated using the six test datasets. Fig. 6.23(a) shows the bias error of the OSNR values estimated using the trained CNN averaged over OSNRs from 11 to 33 dB. The measured bias errors were around 0.3 dB across six different test datasets, corresponding to different signal types.

Next, we also calculated the standard deviation of the OSNRs estimated by the CNN for the same six test datasets. As shown in Fig. 6.23(b), the measured standard deviation remained within 0.4 dB across the six test datasets. The results show that the trained CNN (DNN #3) learned the functionality to accurately estimate OSNRs over multiple modulation formats and symbol rates.

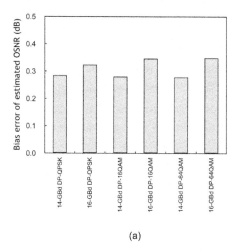 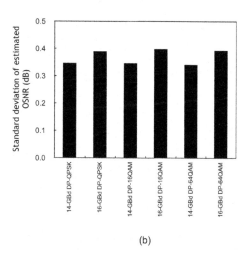

(a)   (b)

Figure 6.23 (a) Bias error and (b) standard deviation of CNN-based OSNR estimator (DNN #3) using test datasets with 14- and 16-GBd DP-QPSK/16QAM/64QAM. These figures are based on the results in Ref. [33].

In addition, to check the applicability of the CNN-based OSNR monitor against transmission, we evaluated the tolerance against residual CD using the CNN trained with the mixed dataset. Note that verification under more comprehensive fiber transmission situations, including nonlinearity, is left for future work. In this experiment, CD was added via post-digital signal processing. Fig. 6.24 shows the tolerance against residual CD for the 16-GB DP-QPSK test dataset. The bias error and standard deviation of the CNN–estimated OSNRs are plotted on the actual OSNR of 20 dB in Figs. 6.24(a) and 6.24(b), respectively. The trained CNN was insensitive to the residual CD up to 50,000 ps/nm, and the result suggests that the trained CNN-based OSNR monitor can be used even after transmission with a large amount of uncompensated CD.

6.5.4 Versatile monitoring using DNN

In principle, the DNN of DL-based monitoring can provide versatile transformations for monitoring. DL-based optical monitors have the potential to estimate various op-

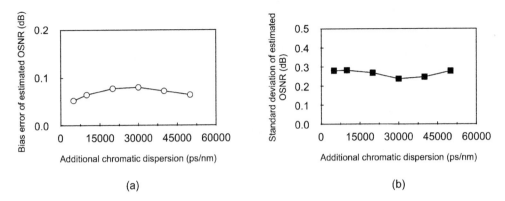

Figure 6.24 (a) Bias error and (b) standard deviation of CNN-based OSNR estimator (DNN #3) as a function of residual chromatic dispersion. These figures are based on the results in Ref. [33].

tical parameters that are not only OSNR but also other physical parameters, meaning that the inherent capability of DL-based optical monitors can extract useful information unexpected in the design stage but turned out to be important later. This section describes such a potential by showing another type of monitoring function, i.e., modulation format and symbol rate identification, using DL-based optical monitors. The discussion in this section is based on our previous work presented in [34].

6.5.4.1 DNN architecture used in this experiment

DNN #4 (CNN-3):

Fig. 6.25 shows the CNN network architecture to classify the modulation and symbol rate used in this section. The CNN's general structure is the same as that of DNN #2, but the network parameters are different, as shown in Fig. 6.25. We used softmax in the last layer for output instead of the linear activation function used in DNN #2.

Figure 6.25 CNN for modulation format/symbol rate classification (DNN #4).

6.5.4.2 Results and discussion

With the use of DNN #4, the classification of incoming optical signal types by CNN was experimentally demonstrated. The input data format of DNN #4 was the same as that of the other DNNs (e.g., DNN #1, #2, and #3) for OSNR estimation. The incoming signal was classified by DNN #4 into six signal types: 14-GBd DP-QPSK, 16-GBd DP-QPSK, 14-GBd DP-16QAM, 16-GBd DP-16QAM, 14-GBd DP-64QAM, and 16-GBd DP-64QAM.

Figs. 6.26 (a) and 6.26(b) show the classification accuracy for each combination of the signal type on the received OSNRs ranging from 10 to 29 dB. Fig. 6.26(a) shows the accuracy evaluated on the training dataset for each signal type. The results indicate >95% accuracy with OSNRs >10 dB for all six signal types. Fig. 6.26(b) shows the evaluated accuracy on the test dataset. The results also indicate an accuracy of >95% in terms of averaged value over six signal types with >17-dB OSNRs. As shown in Fig. 6.26(b), a gap in the classification accuracy between the training and test datasets can be found with the low-OSNR region. Even in low-OSNR cases (e.g., an OSNR of 10 dB), the trained DNN #4 shows a classification accuracy of >80%, thereby indicating that DNN #4 successfully learned how to classify signal types of received signals. The accuracy shown in Fig. 6.26 was calculated from a single input of DNN #4, i.e., only 512 samples at 40 Gsamples/s. This detail poses a possibility to further improve the classification accuracy using multiple inputs of DNN if necessary.

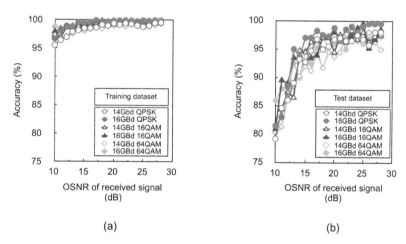

Figure 6.26 Accuracy of modulation format/symbol rate classification for (a) training and (b) test datasets by DNN #4. These figures are based on the results in Ref. [34].

6.5.5 Data augmentation based on domain knowledge of optical transceivers

As discussed in Section 6.3.2.1, DL-based optical monitors have to be robust against various operational conditions including drifting or switching characteristics of the optical device. Thus, the DL-based optical monitor trained with a specific training dataset has to be operated in more general situations, including a situation not covered by the original training dataset. Data augmentation based on domain knowledge of the optical transceiver is a promising way of building a robust DNN model for DL-based monitors over varying operational conditions. Data augmentation can be realized as a preprocessing block before DNN in the training phase. We experimentally verified this concept using the CNN-based OSNR estimator with data augmentation to increase its robustness against a laser frequency offset between a signal and a LO in a digital coherent receiver. The discussion in this section is based on our previous work presented in [36].

6.5.5.1 DNN used in this section

DNN #5 (CNN-4):

Fig. 6.27 shows the network architecture of the CNN used in this section. The CNN's general structure was the same as that of DNN #2, but the detailed network parameters were different, as shown in Fig. 6.27. The CNN's input is four channelized optical electric fields: 512 samples × 4 channels (HI, HQ, VI, and VQ) of the digital coherent receiver outputs asynchronously sampled at 80 Gsample/s.

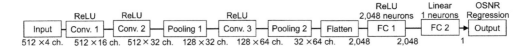

Figure 6.27 CNN for the data-augmentation experiment (DNN #5).

6.5.5.2 Results and discussion

DNN #5 was trained to estimate the OSNR of the incoming optical signal. As shown in Fig. 6.6, the data-augmentation block was inserted before the DNN in this experiment. The data-augmentation block was activated in the training mode of the DL-based OSNR monitor.

The data-augmentation block used in this experiment was designed to enhance the tolerance against frequency offset between the signal and the local laser of the digital coherent receiver. The practical processing of the data-augmentation block for frequency offset is given by Eq. (6.7). DNN #5 was trained with and without the data-augmentation block to highlight the impact of the data augmentation. We used the same number of records in the training dataset for both training scenarios for a fair comparison. In other words, the data-augmentation block only enhanced the diversity of the training dataset but did not increase the number of records in the training dataset. f_{max} in Eq. (6.7) is a parameter to specify the expected frequency offset tolerance enhanced by data augmentation and was set to 1.25 and 2.5 GHz in this experiment, which is based on the specification of the Integrable Tunable Laser Assembly (iTLA) by the Optical Internetworking Forum (OIF) [61].

First, a sample case study was investigated through a simulation to illustrate the impact of data augmentation. The DNN of the DL-based OSNR monitor was trained with a simulation dataset that had an optical waveform without frequency offset. The data-augmentation block was skipped in this training scenario. Fig. 6.28(a) shows the trained DNN-estimated OSNR values averaged over a test dataset with frequency offset

$\delta f = 0$. The condition of the test dataset is the same as that of the training dataset in terms of frequency offset; thus, the trained DNN provided accurate estimated values of OSNR.

Next, the same DNN trained with the no-frequency-offset training dataset was evaluated using different test dataset that had 1-GHz frequency offset. Fig. 6.28(b) shows the average value of estimated OSNRs over this test dataset. The result shows that the DNN of the DL-based OSNR monitor cannot estimate OSNR values accurately because the frequency offset in the input data was out of range of the training dataset.

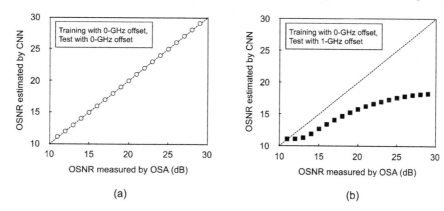

(a)   (b)

Figure 6.28 Evaluation results of CNN-based OSNR estimator (DNN #5) trained with simulation data with frequency offset = 0. Using test dataset with frequency offset = (a) 0 and (b) 1 GHz. These figures are based on the results in Ref. [36].

To solve the inaccurate OSNR estimation as shown in Fig. 6.28(b), the data-augmentation block for frequency offset was used in training mode. f_{max} in the data-augmentation block was set to 0, 1.25, and 2.5 GHz. Fig. 6.29 shows the simulation results of the estimated OSNR values obtained by the trained DNNs. Figs. 6.29(a), 6.29(b), and 6.29(c) show the results of the DNN trained together with the data-augmentation block with the f_{max} of 0, 1.25, and 2.5 GHz, respectively. Each test dataset comprises data with a specific frequency offset of 0, 1, 2, and 3 GHz.

As shown in Fig. 6.29(a), the case without the data-augmentation block, i.e., the case with f_{max} of 0 Hz, led to inaccurate OSNR estimation except for the input data with 0-Hz frequency offset. Fig. 6.29(b) shows the data-augmentation block with f_{max} of 1.25 GHz ensured to accurately estimate OSNR for the input data with a maximum frequency offset of 1 GHz. When the f_{max} was increased up to 2.5 GHz, the operational range to ensure accurate OSNR estimation was expanded up to 2 GHz, as shown in Fig. 6.29(c).

Finally, we experimentally evaluated the bias error and standard deviation of estimated OSNR values with different frequency offsets on test datasets for a more detailed discussion. A baseline dataset equivalent to no residual frequency offset was measured

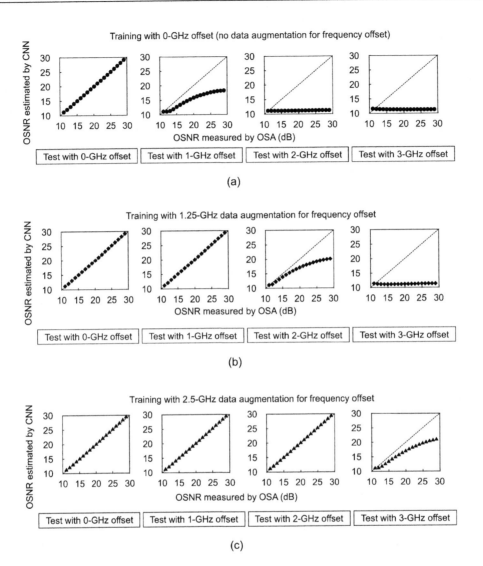

Figure 6.29 Evaluation results of CNN-based OSNR estimators (DNN #5) trained with data augmentation for frequency offsets of (a) 0, (b) 1.25, and (c) 2.5 GHz. These figures are based on the results in Ref. [36].

during the experiment using a finely frequency-tuned LO laser. The residual frequency offset between signal and LO was within 150 MHz. DNN #5 was trained with this baseline dataset with the data augmentation block in this experiment. The f_{max} in the data–augmentation block was set to 0, 1.25, and 2.5 GHz. Test datasets with different frequency offsets were measured in the experiments with varying actual emitting frequencies of the LO laser.

Figs. 6.30(a) and 6.30(b) show the experimental results of bias errors and standard deviations as a function of frequency offset on test datasets. The plotted bias errors and standard deviations were the averaged values over each test dataset, including various OSNRs ranging from 11 to 29 dB. For both bias error and standard deviation, the experimental results show that the data–augmentation block enhanced the operating range where the DL–based OSNR monitor can provide accurate OSNR values. The operational ranges were almost equivalent to the f_{max} used in the data–augmentation block. However, a slight degradation was observed even within the range of the f_{max}, i.e., +/−1.25 or +/−2.5 GHz, because of the limited analog bandwidth of ADCs in this experimental setup.

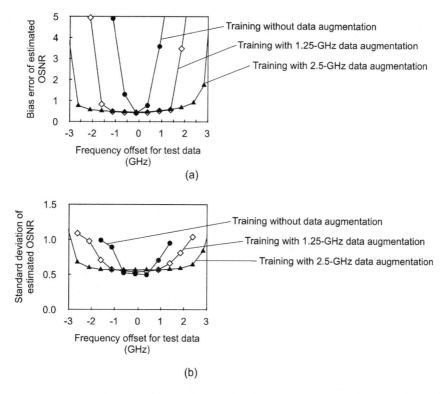

Figure 6.30 Experimental results of (a) bias error and (b) standard deviation of OSNR estimated by the CNN-based optical monitor (DNN #5) trained with data augmentation for frequency offsets of 0, 1.25, and 2.5 GHz. These figures are based on the results in Ref. [36].

6.5.6 Estimating uncertainty by dropout at inference

The DL–based optical monitors were trained and tested to satisfy the accuracy criteria discussed in previous sections. However, this approach still does not guarantee sufficient estimation accuracy under unexpected test case conditions, e.g., a newly developed signal that is not included in the test dataset. Thus, as discussed in Section 6.3.4.2, a technique needs to be developed to assess the current accuracy of a DL-based optical monitor toward robust and continuous evolution of automated optical networks.

As discussed in Section 6.3.4.2, a DL-based optical monitor enhanced by N-times inference using the dropout at the inference phase can simultaneously provide a point-estimated monitoring result (e.g., OSNR) and its uncertainty information by taking the average and standard deviation of N-outputs obtained by the N-times inference for a single input. This section shows the feasibility of the DL-based monitor enhanced by the dropout at the inference by evaluating a DL-based OSNR monitor in cases where the DNNs of the monitor were trained with either a limited number of records or partially missing records in a training dataset. The experimental results indicate that the monitor successfully showed that its own output has large uncertainties due to limited amount of training data or missing records in the training dataset. The discussion and results in this section are based on our previous work presented in [50,51].

6.5.6.1 DNN used in this experiment

DNN #6 (CNN-5):

Fig. 6.31 shows the network architecture of the CNN used in this section. The CNN's general structure was almost the same as that of the DNN #2, but the detailed parameters were different, as shown in Fig. 6.31. In this case, four channelized optical electric fields were provided as input to the CNN: 512 samples × 4 channels (HI, HQ, VI, and VQ) asynchronously sampled at 80 Gsample/s. The pooling layer was a max-pooling layer with a stride of 2.

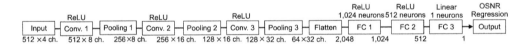

Figure 6.31 CNN for estimating the model uncertainty by dropout at inference (DNN #6).

6.5.6.2 Results and discussion

To investigate the relationship between the actual standard deviation of point-estimated outputs and estimated model uncertainties obtained using the dropout at the inference, we first trained DNN #6 with datasets having different numbers of points: 10,000, 50,000, 100,000, 300,000, and 375,700. Subsequently, we calculated the estimated uncertainty of each trained model using the dropout at inference phase technique. 200 partial networks were generated by the dropout at inference phase for each evaluation. A test dataset with 10,000 records, including different OSNR data ranging from 11 to 30 dB, was used for the inference. The actual standard deviation plotted in Fig. 6.32(a) decreased with increasing number of records in the training datasets. The model uncertainty estimated using the dropout at the inference phase showed almost the same tendency of the actual standard deviation, as shown in Fig. 6.32(b).

The results suggest that the actual deviation (i.e., the confidence level of the DNN's current output) can be estimated using DNN based on the dropout at the inference phase. The estimated uncertainty can be provided to network operators and can trigger an alert when the estimated uncertainty exceeds the given threshold, as discussed in Section 6.3.4.2.

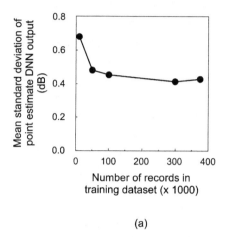 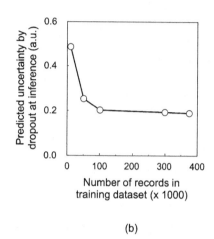

(a)   (b)

Figure 6.32 (a) Actual and (b) predicted model uncertainty as a function of the number of records in training datasets. These figures are based on the results in Ref. [51].

Next, we investigate another case where the model uncertainty is caused by a lack of specific training data points. To highlight this effect, we prepared artificial training datasets by deleting data corresponding to OSNR of 17, 18, and 19 dB from the complete training dataset. To ensure the same total number, i.e., 375,700 records, the other data points (w.r.t. OSNR in the ranges of 11–16 and 20–30 dB) were added in place of the deleted data points. After training DNN #6 using this incomplete dataset or the reference complete dataset that has full data points, we evaluated the bias errors and standard deviations of the point-estimated outputs on the two trained DNNs.

In this case study, we first investigated the effect using the different number of partial neural networks generated by the dropout at inference. We evaluated the estimated model uncertainty in two cases: (1) a case with no corresponding training data points, i.e., the OSNR of 18 dB in the DNN trained with the incomplete dataset, and (2) a case with corresponding training data points, i.e., the OSNR of 23 dB in the DNN trained with the incomplete dataset.

Fig. 6.33 shows the standard deviation of the estimated uncertainties normalized from the mean value of estimated uncertainties for each OSNR. The fluctuation in the estimated uncertainty decreased with increasing number of partial neural networks generated using the dropout at the inference phase. This graph also suggests that 500 partial neural networks would be enough to estimate uncertainty by the dropout at inference in this specific case. A large number of partial neural networks means that an estimated uncertainty was calculated from a more diverse ensemble of partial neural networks. The fluctuation in the estimated uncertainties can be reduced by increasing the number of partial neural networks generated using the dropout at the inference phase.

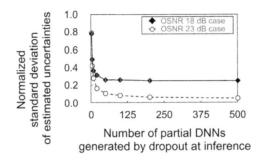

Figure 6.33 Normalized standard deviation of estimated uncertainties as a function of the number of partial DNNs at inference. These figures are based on the results in Ref. [51].

On the basis of the investigation in the previous paragraph, we performed the inference with 500 partial neural networks by the dropout at inference. In Figs. 6.34(a) and 6.34(b), the actual and estimated model uncertainties were flat and low because of the complete dataset used for training this model. Moreover, Figs. 6.34(c) and 6.34(d) show that the model's outputs have large standard deviations for corresponding OSNRs due to the lack of corresponding training data, which is due to the incomplete training dataset, i.e., no data points corresponding to OSNRs of 17, 18, and 19 dB. The whiskers in Figs. 6.34(b) and 6.34(d) show the standard deviation over the multiple estimated uncertainties from multiple test data. The results show that the dropout at the inference phase has the potential to predict the deviations of estimated OSNRs.

6.6. Future direction of data-analytic-based optical monitoring

In this section, a possible evolutionary scenario of the DL-based optical monitors is discussed. The preceding sections in this chapter discussed how monitoring functionalities can be learned from a dataset without prior knowledge by utilizing representation learning of DNNs. Although this approach was successfully demonstrated for many optical system indicators, such as OSNR, PMD, and modulation formats, we can further discuss monitoring capability for more-sophisticated system indicators.

As an example of monitoring such a sophisticated indicator, an algorithm that estimates a fiber-longitudinal optical power profile over a multispan optical transmission

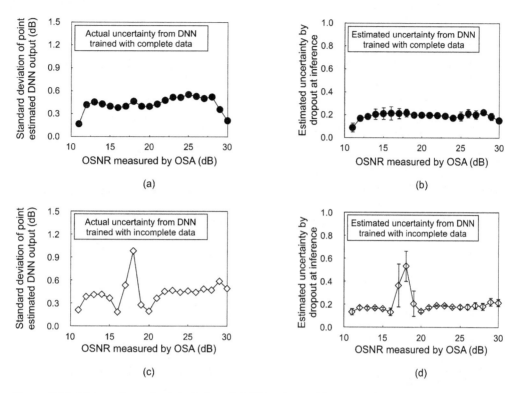

Figure 6.34 (a) Actual standard deviation of OSNR estimated by a DNN trained with the complete dataset, (b) estimated uncertainty by a DNN trained with a complete dataset (N = 500), (c) actual standard deviation of OSNR estimated by a DNN trained with an incomplete dataset, (d) estimated uncertainty by a DNN trained with an incomplete dataset (N = 500). These figures are based on the results in Ref. [51].

line from only the receiver end information was proposed and demonstrated [62–65]. This fiber-longitudinal monitor is a variant of the data analytic-based optical monitors derived from a study on automatic optimization of nonlinear equalizers called digital backpropagation in fiber-optic communications [66]. The algorithm explicitly utilizes prior domain knowledge of the fiber-optic transmission line that has a distributed nonlinearity due to the Kerr effect. This concept can extend the existing data analytic-based optical monitor that can estimate only cumulative quantities (e.g., accumulated OSNR and CD) over the entire fiber transmission line to an enhanced family of optical monitoring methods that can estimate distance-wise quantities. An example of the monitoring results is shown in Fig. 6.35, on the basis of the result shown in [62]. This distance-wise monitoring allows the network operator to identify where the network's physical layer anomaly is occurring and respond quickly to the anomaly.

The method shown in [62–65] is not based on DL but rather on the latest rule-based algorithm. Nevertheless, the authors believe that the concept that utilizes prior domain knowledge of optical transmission line, including distributed nonlinearity, can also enhance the DL-based optical monitor by unifying the prior domain knowledge into the DNN for the DL-based monitor. One possible way to embed the prior knowledge into the DNN is to design the DNN architecture to fit the prior knowledge, similar to the CNN discussed in the previous section. The other approach for including the prior domain knowledge is to use the data-augmentation technique discussed in Section 6.3.2.1. Although not explicitly discussed in this chapter, reflecting the system's knowledge through learning methodology, such as modifying the loss function during the training phase, may be possible.

A fusion of learning and prior knowledge would be a promising direction to realize advanced perception for optical networks in the future as evolved DL-based optical monitors.

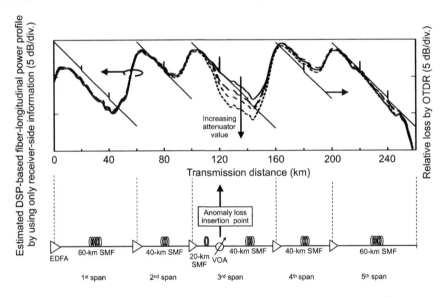

Figure 6.35 Example of experimental measurement using an advanced rule-based fiber-longitudinal monitor calculated from only receiver-end information [62] in normal (no excess loss, solid line) and anomaly cases with three different insertion losses (dotted, dashed, and long dashed lines) with measured relative loss by OTDR corresponding to the solid line. This figure is based on the results in Ref. [62] and [63].

6.7. Summary

Automated operation of optical networks is a promising way to treat the ever-increasing operational complexity in a cost-effective manner. As an essential part of optical network automation, we showed the recent advances and experimental verification results of a DL-based optical monitor that comprising a digital coherent receiver and a DNN. We described this DL-based optical monitor in terms of its training and inference mode and discussed advanced techniques to enhance it, including data augmentation, transfer learning, federated learning, and model uncertainty estimation. We also discussed guidelines for designing DNNs for optical monitoring tasks. Using multiple modulation formats (DP-QPSK, 16QAM, and 64QAM), we performed experimental verifications, proving the DL-based optical monitor concept. Results show the feasibility of learning-based monitoring using DL-based monitors. Over the next decade, significant progress is expected in end-to-end learning using DNNs and the design of DNNs is also expected to include prior domain knowledge of optical communication systems.

Acknowledgment

This work was partly supported by the National Institute of Information and Communications Technology (NICT), Japan, Grant Number 20501.

References

[1] R.J. Essiambre, G. Kramer, P.J. Winzer, G.J. Foschini, B. Goebel, Capacity limits of optical fiber networks, J. Lightwave Technol. 28 (4) (2010) 662–701.

[2] S.J. Savory, Digital coherent optical receivers: algorithms and subsystems, IEEE J. Sel. Top. Quantum Electron. 16 (5) (2010) 1164–1179.

[3] Y. Pointurier, Design of low-margin optical networks, J. Opt. Commun. Netw. 9 (1) (2017) A9–A17.

[4] S. Yan, A. Aguado, Y. Ou, R. Wang, R. Nejabati, D. Simeonidou, Multi-layer network analytics with SDN-based monitoring framework, J. Opt. Commun. Netw. 9 (2) (2017) A271–A279.

[5] F. Meng, Y. Ou, S. Yan, K. Sideris, M.D.G. Pascual, R. Nejabati, D. Simeonidou, Field trial of a novel SDN enabled network restoration utilizing in-depth optical performance monitoring assisted network re-planning, in: Proc. of Optical Fiber Communications Conference and Exhibition (OFC), 2017, 2017, Th1J.8.

[6] S. Yan, F.N. Khan, A. Mavromatis, D. Gkounis, Q. Fan, F. Ntavou, K. Nikolovgenis, F. Meng, E.H. Salas, C. Guo, C. Lu, A.P.T. Lau, R. Nejabati, D. Simeonidou, Field trial of machine-learning-assisted and SDN-based optical network planning with network-scale monitoring database, in: Proc. of European Conf. Optical Communication (ECOC), 2017, PDP Th.PDP.B.4.

[7] R. Elschner, F. Frey, C. Meuer, J.K. Fischer, S. Alreesh, C. Schmidt-Langhorst, L. Molle, T. Tanimura, C. Schubert, Experimental demonstration of a format-flexible single carrier coherent receiver using data-aided digital signal processing, Opt. Express 20 (27) (2012) 28786–28791.

[8] J.K. Fischer, S. Alreesh, R. Elschner, F. Frey, M. Nolle, C. Schmidt-Langhorst, C. Schubert, Bandwidth-variable transceivers based on four-dimensional modulation formats, J. Lightwave Technol. 32 (16) (2014) 2886–2895.

[9] M. Eiselt, B. Teipen, K. Grobe, A. Autenrieth, J.P. Elbers, Programmable modulation for high-capacity networks, in: Proc. of European Conference and Exhibition on Optical Communication (ECOC), 2011, Tu.5.A.5.

[10] T. Tanimura, T. Hoshida, T. Kato, S. Watanabe, M. Suzuki, H. Morikawa, Throughput and latency programmable optical transceiver by using DSP and FEC control, Opt. Express 25 (10) (2017) 10815–10827.

[11] M. Jinno, H. Takara, K. Yonenaga, A. Hirano, Virtualization in optical networks from network level to hardware level, J. Opt. Commun. Netw. 5 (10) (2012) A46–A56.

[12] C.C.K. Chan, Optical Performance Monitoring: Advanced Techniques for Next-Generation Photonic Network, Academic Press, 2010.

[13] F.N. Hauske, M. Kuschnerov, B. Spinnler, B. Lankl, Optical performance monitoring in digital coherent receivers, J. Lightwave Technol. 27 (16) (2009) 3623–3631.

[14] C.M. Bishop, Pattern Recognition and Machine Learning, Springer, 2006.

[15] I. Goodfellow, Y. Bengio, A. Courville, Deep Learning, MIT Press, 2016.

[16] T. Tanimura, T. Hoshida, J.C. Rasmussen, M. Suzuki, H. Morikawa, OSNR monitoring by deep neural networks trained with asynchronously sampled data, in: Proc. of 21st OptoElectronics and Communications Conference (OECC), 2016, TuB3-5.

[17] K. Zhang, Y. Fan, T. Ye, Z. Tao, S. Oda, T. Tanimura, Y. Akiyama, T. Hoshida, Fiber nonlinear noise-to-signal ratio estimation by machine learning, in: Proc. of Optical Fiber Communications Conference and Exhibition (OFC), 2019, Th2A.45.

[18] D. Zibar, M. Piels, R. Jones, C.G. Schaeffer, Machine learning techniques in optical communication, J. Lightwave Technol. 34 (6) (2016) 1442–1452.

[19] F.N. Khan, K. Zhong, X. Zhou, W.H. Al-Arashi, C. Yu, C. Lu, A.P.T. Lau, Joint OSNR monitoring and modulation format identification in digital coherent receivers using deep neural networks, Opt. Express 25 (15) (2017) 17767–17776.

[20] D. Wang, M. Zhang, Z. Li, J. Li, M. Fu, Y. Cui, X. Chen, Modulation format recognition and OSNR estimation using CNN-based deep learning, IEEE Photonics Technol. Lett. 29 (19) (2017) 1667–1670.

[21] A.S. Kashi, Q. Zhuge, J.C. Cartledge, S. Ali Etemad, A. Borowiec, D.W. Charlton, C. Laperle, M. O'Sullivan, Nonlinear signal-to-noise ratio estimation in coherent optical fiber transmission systems using artificial neural networks, J. Lightwave Technol. 36 (23) (2018) 5424–5431.

[22] J. Estaran, R. Rios-Muller, M.A. Mestre, F. Jorge, H. Mardoyan, A. Konczykowska, J.-Y. Dupuy, S. Bigo, Artificial neural networks for linear and non-linear impairment mitigation in high-baudrate IM/DD systems, in: Proc. of European Conf. Optical Communication (ECOC), 2016, M2B.2.

[23] D. Zibar, L.H.H. de Carvalho, M. Piels, A. Doberstein, J. Diniz, B. Nebendahl, C. Franciscangelis, J. Estaran, H. Haisch, N.G. Gonzalez, J.C.R.F. de Oliveira, I.T. Monroy, Application of machine learning techniques for amplitude and phase noise characterization, J. Lightwave Technol. 33 (7) (2015) 1333–1343.

[24] X. Wu, J.A. Jargon, R.A. Skoog, L. Paraschis, A.E. Willner, Applications of artificial neural networks in optical performance monitoring, J. Lightwave Technol. 27 (16) (2009) 3580–3589.

[25] J.A. Jargon, X. Wu, H.Y. Choi, Y.C. Chung, A.E. Willner, Optical performance monitoring of QPSK data channels by use of neural networks trained with parameters derived from asynchronous constellation diagrams, Opt. Express 18 (5) (2010) 4931–4938.

[26] X. Wu, J.A. Jargon, L. Paraschis, A.E. Willner, ANN-based optical performance monitoring of QPSK signals using parameters derived from balanced-detected asynchronous diagrams, IEEE Photonics Technol. Lett. 23 (4) (2011) 248–250.

[27] F.N. Khan, T.S.R. Shen, Y. Zhou, A.P.T. Lau, C. Lu, Optical performance monitoring using artificial neural networks trained with empirical moments of asynchronously sampled signal amplitudes, IEEE Photonics Technol. Lett. 24 (12) (2012) 982–984.

[28] J. Thrane, J. Wass, M. Piels, J.C.M. Diniz, R. Jones, D. Zibar, Machine learning techniques for optical performance monitoring from directly detected PDM-QAM signals, J. Lightwave Technol. 35 (4) (2017) 868–875.

[29] D-S. Ly-Gagnon, S. Tsukamoto, K. Katoh, K. Kikuchi, Coherent detection of optical quadrature phase-shift keying signals with carrier phase estimation, J. Lightwave Technol. 24 (1) (2006) 12–21.

[30] K. Kikuchi, Fundamentals of coherent optical fiber communications, J. Lightwave Technol. 34 (1) (2016) 157–179.

[31] Y. LeCun, Y. Bengio, G. Hinton, Deep learning, Nature 521 (2015) 436–444.

[32] T. Tanimura, T. Hoshida, T. Kato, S. Watanabe, J.C. Rasmussen, M. Suzuki, H. Morikawa, Deep learning based OSNR monitoring independent of modulation format, symbol rate and chromatic dispersion, in: Proc. of European Conf. Optical Communication (ECOC), 2016, 2016, Tu.2.C.2.

[33] T. Tanimura, T. Hoshida, T. Kato, S. Watanabe, H. Morikawa, Convolutional neural network-based optical performance monitoring for optical transport networks, J. Opt. Commun. Netw. 11 (2019) A52–A59.

[34] T. Tanimura, T. Hoshida, T. Kato, S. Watanabe, H. Morikawa, Data-analytics-based optical performance monitoring technique for optical transport networks (invited), in: Proc. of Optical Fiber Communications Conference and Exhibition (OFC), 2018, 2018, Tu3E.3.

[35] J. Ba, D. Kingma, Adam: a method for stochastic optimization, in: Proc. of 3rd International Conference on Learning Representations (ICLR), 2015.

[36] T. Tanimura, T. Hoshida, T. Kato, S. Watanabe, H. Morikawa, Simple learning method to guarantee operational range of optical monitors, J. Opt. Commun. Netw. 10 (2018) D63–D71.

[37] W. Mo, Y. Huang, S. Zhang, E. Ip, D.C. Kilper, Y. Aono, T. Tajima, ANN-based transfer learning for QoT prediction in real-time mixed line-rate systems, in: Proc. of Optical Fiber Communications Conference and Exhibition (OFC), 2018, 2018, paper W4F.3.

[38] L. Xia, J. Zhang, S. Hu, M. Zhu, Y. Song, K. Qiu, Transfer learning assisted deep neural network for OSNR estimation, Opt. Express 27 (2019) 19398–19406.

[39] J. Yu, W. Mo, Y. Huang, E. Ip, D.C. Kilper, Model transfer of QoT prediction in optical networks based on artificial neural networks, J. Opt. Commun. Netw. 11 (10) (2019) C48–C57.

[40] J. Pesic, M. Lonardi, E. Seve, N. Rossi, T. Zami, Transfer learning from unbiased training data sets for QoT estimation in WDM networks, in: Proc. of European Conf. Optical Communication (ECOC), 2020, 2020, Mo2K-2.

[41] H. Brendan McMahan, Eider Moore, Daniel Ramage, Seth Hampson, Blaise Agueray Arcas, Communication-efficient learning of deep networks from decentralized data, in: Proc. of 20th International Conference on Artificial Intelligence and Statistics (AISTATS), 2017.

[42] R. Shokri, V. Shmatikov, Privacy-preserving deep learning, in: Proc. of 22nd ACM SIGSAC Conference on Computer and Communications Security, CCS'15, 2015.

[43] B. Shariati, P. Safari, A. Mitrovska, N. Hashemi, J.K. Fischer, R. Freund, Demonstration of federated learning over edge-computing enabled metro optical networks, in: Proc. of European Conf. Optical Communication (ECOC), 2020, 2020, Tu5A-2.

[44] L. Ruan, S. Mondal, I. Dias, E. Wong, Low-latency federated reinforcement learning-based resource allocation in converged access networks, in: Proc. of Optical Fiber Communications Conference and Exhibition (OFC), 2020, 2020, W2A.28.

[45] C.E. Rasmussen, C.K.I. Williams, Gaussian Processes for Machine Learning (Adaptive Computation and Machine Learning), MIT Press, 2006.

[46] J. Lee, J. Sohl-Dickstein, J. Pennington, R. Novak, S. Schoenholz, Y. Bahri, Deep neural networks as Gaussian processes, in: Proc. of Int. Conf. Learn. Represent., 2018.

[47] J. Wass, J. Thrane, M. Piels, R. Jones, D. Zibar, Gaussian process regression for WDM system performance prediction, in: Proc. of Optical Fiber Communications Conference and Exhibition (OFC), 2017, 2017, Tu3D.7.

[48] F. Meng, S. Yan, K. Nikolovgenis, Y. Ou, R. Wang, Y. Bi, E.H. Salas, R. Nejabati, D.E. Simeonidou, Field trial of Gaussian process learning of function-agnostic channel performance under uncertainty, in: Proc. of Optical Fiber Communications Conference and Exhibition (OFC), 2018, 2018, W4F.5.

[49] Y. Gal, Z. Ghahramani, Dropout as a Bayesian approximation: representing model uncertainty in deep learning, in: Proc. of 33rd Int. Conf. Mach. Learn., 2016, 2016, pp. 1651–1660.

[50] N. Srivastava, G. Hinton, A. Krizhevsky, I. Sutskever, R. Salakhutdinov, Dropout: a simple way to prevent neural networks from overfitting, J. Mach. Learn. Res. 15 (2014) 1929–1958.

[51] T. Tanimura, T. Hoshida, T. Kato, S. Watanabe, OSNR estimation providing self-confidence level as auxiliary output from neural networks, J. Lightwave Technol. 37 (7) (2019) 1717–1723.

[52] T. Tanimura, T. Kato, S. Watanabe, T. Hoshida, Deep neural network based optical monitor providing self-confidence as auxiliary output, in: Proc. of European Conf. Optical Communication (ECOC), 2018, 2018, We1D.5.

[53] G. Cybenko, Approximation by superpositions of a sigmoidal function, Math. Control Signals Syst. (MCSS) 2 (4) (1989) 303–314.

[54] G. Montufar, R. Pascanu, K. Cho, Y. Bengio, On the number of linear regions of deep neural networks, in: Proc. of 27th International Conference on Neural Information Processing Systems, vol. 2, NIPS'14, 2014, pp. 2924–2932.

[55] S. Ioffe, C. Szegedy, Batch normalization: accelerating deep network training by reducing internal covariate shift, in: Proc. of 32nd International Conference on Machine Learning (ICML), 2015.

[56] K. He, X. Zhang, S. Ren, J. Sun, Deep residual learning for image recognition, in: Proc. of IEEE Conference on Computer Vision and Pattern Recognition (CVPR), 2016, 2016, pp. 770–778.

[57] J. Frankle, M. Carbin, The lottery ticket hypothesis: finding sparse, trainable neural networks, in: Proc. of International Conference on Learning Representations (ICLR), 2019, p. 2019.

[58] T.A. Eriksson, H. Bülow, A. Leven, Applying neural networks in optical communication systems: possible pitfalls, IEEE Photonics Technol. Lett. 29 (23) (2017) 2091–2094.

[59] M. Abadi, P. Barham, J. Chen, Z. Chen, A. Davis, J. Dean, M. Devin, S. Ghemawat, G. Irving, M. Isard, M. Kudlur, J. Levenberg, R. Monga, S. Moore, D.G. Murray, B. Steiner, P. Tucker, V. Vasudevan, P. Warden, M. Wicke, Y. Yu, X. Zheng, Tensorflow: a system for large-scale machine learning, in: Proc. of 12th USENIX Symposium on Operating Systems Design and Implementation (OSDI), 2016.

[60] X. Glorot, A. Bordes, Y. Bengui, Deep sparse rectifier neural networks, in: Proc. of 14th International Conference on Artificial Intelligence and Statistics (AISTATS), 2011, pp. 315–323.

[61] Integrable tunable laser assembly MSA, June 26, 2008.

[62] T. Tanimura, S. Yoshida, K. Tajima, S. Oda, T. Hoshida, Fiber-longitudinal anomaly position identification over multi-span transmission link out of receiver-end signals, J. Lightwave Technol. 38 (9) (2020) 2726–2733.

[63] T. Tanimura, K. Tajima, S. Yoshida, S. Oda, T. Hoshida, Experimental demonstration of a coherent receiver that visualizes longitudinal signal power profile over multiple spans out of its incoming signal, in: Proc. of European Conf. Optical Communication (ECOC), 2019, 2019, post-deadline paper PD.3.4.

[64] T. Tanimura, S. Yoshida, S. Oda, K. Tajima, T. Hoshida, Advanced data-analytics-based fiber-longitudinal monitoring for optical transport networks (invited), in: Proc. of European Conf. Optical Communication (ECOC), 2020, 2020, We1H-3.

[65] T. Tanimura, S. Yoshida, K. Tajima, S. Oda, T. Hoshida, Concept and implementation study of advanced DSP-based fiber-longitudinal optical power profile monitoring toward optical network tomography [Invited], J. Opt. Commun. Netw. 13 (2021) E132–E141.

[66] T. Tanimura, T. Hoshida, T. Tanaka, L. Li, S. Oda, H. Nakashima, Z. Tao, J.C. Rasmussen, Semi-blind nonlinear equalization in coherent multi-span transmission system with inhomogeneous span parameters, in: Proc. of Optical Fiber Communications Conference and Exhibition (OFC), 2010, 2010, paper OMR6.

## CHAPTER SEVEN

Machine Learning methods for Quality-of-Transmission estimation

Memedhe Ibrahimi[a], Cristina Rottondi[b], and Massimo Tornatore[a]
[a]Department of Electronics, Information and Bioengineering, Politecnico di Milano, Milan, Italy
[b]Department of Electronics and Telecommunications, Politecnico di Torino, Torino, Italy

7.1. Introduction

The continuous growth in the complexity of optical networks has stimulated the investigation of Machine Learning (ML) as an enabling methodology to accomplish several tasks in network control, design and management. In particular, the emergence of coherent technologies has introduced a large number of tunable design parameters in optical transponders as variable modulation formats, symbol rates, forward error correction (FEC) coding rate, and adaptive channels spacing in flex-grid networks. Such plethora of design parameters offers system engineers numerous possible configurations for deploying lightpaths [1], but it also increases the complexity of design and operation of optical networks, thus making it difficult to rely on accurate modeling of the system behavior through analytical models. A crucial challenge for system engineers is related to the non-linear nature of signal propagation in an optical fiber. Several analytical models can provide satisfactory solutions through closed-form formulas. However, even when such solutions are available, design margins must be adopted to account for uncertainties in the values of the parameters provided as inputs to the models, leading to under-utilization of networks resources and undesirable increase of network management costs.

ML is expected to offer the capability to capture the complex non-linear behavior of optical signal propagation, e.g., by using supervised learning algorithms that can leverage the knowledge contained in historical network measurements. The adoption of ML approaches can be useful especially for cross-layer optimizations (i.e., optimizations jointly covering both the physical and the network layer), since knowledge extracted from data gathered at the physical layer, e.g., by monitoring the Bit Error Rate (BER) or Signal-to-Noise Ratio (SNR) of deployed lightpaths, can trigger changes at network layer, e.g., in routing, spectrum and modulation format assignments [2], with the final goal of supporting network self-configuration and fast decision-making. Indeed, the significant increase of traffic and the stringent requirements in terms of capacity and latency imposed by modern applications (e.g., those arising in 5G communications) make ML a promising candidate to provide to network engineers with new cost- and resource-efficient design and management tools for optical network optimization.

Among the design and management tasks that can be addressed using ML approaches, estimating the QoT of a candidate lightpath prior to its deployment plays a pivotal role. QoT may refer to one or several physical layer parameters, such as SNR or BER, which determine quantitatively the quality of the optical signal at the receiver.[1] Through metrics such as SNR and BER, network engineers are able to determine if the quality of the received optical-signal is below or above a pre-defined QoT threshold. QoT values are dependent on several design parameters, such as, e.g., modulation format, baud rate, channel spacing, physical path and others. Optimizing the choice of such parameters is not a trivial task for network engineers, but it is essential to achieve optimized design and planning of optical networks.

Before the recent introduction of ML as a tool for QoT estimation, QoT estimation was performed (and is yet, in most practical deployments) through either of the two following approaches: i) "exact" analytical models, e.g., those based on the split-step Fourier transform method [3], which simulate light propagation along the fiber and provide accurate results at the expense of a heavy computational burden and ii) margined formulas, e.g., those based on the Gaussian Noise model [4] for the estimation of non-linear impairments, which are computationally faster, but imply high margins to ensure lightpath feasibility and lead to higher cost and under–utilization of network resources [5].

In optical communications, there are two main sources of impairments which impact the QoT: i) amplified spontaneous emission (ASE) noise, introduced by the presence of optical amplifiers, also referred to as *linear noise* and ii) non-linear interference (NLI) noise due to Kerr effect, referred to as *non-linear noise*. For the sake of simplicity, and without loss of generality, let us concentrate on SNR as a possible QoT metric. SNR is defined as the ratio between the power of the optical signal and the power of the noise signal, given by the sum of ASE and NLI. Among the two noise components, the ASE noise is relatively easier to model as it can be estimated as a function of two main parameters of optical amplifiers, namely, amplifier gain (G) and noise figure (NF) (see, e.g., Eq. (1) in [6]), while the estimation of NLI noise component can be dealt with several models [4,7,8] that depend on a large set of parameters.

Despite the intensive research and significant advancements achieved in the last decade to provide accurate and computationally-efficient QoT modeling, ML is emerging as a useful methodology for QoT estimation and modeling of uncertainties that are due to not having an exact knowledge of input parameters and/or knowing the exact behavior of network elements. Some of these uncertainties are unavoidable, as, in

---

[1] Note that we interchangeably refer to SNR or BER as the QoT metric throughout the chapter. Moreover, note that there is a direct relation/conversion of BER to SNR values, given the bit rate and modulation format characterizing a connection request.

practical deployments, it is not always possible to precisely know and/or monitor the parameters of some network elements. Among the main sources of uncertainty we mention here three relevant examples: the noise figure and gain of optical amplifiers [9–11], fiber span losses [12–14] and filtering penalties [15,16].

Uncertainties on optical amplifier parameters impact the modeling of the linear noise, which is the main source of transmission impairments in optical communications. In practice, the gain of optical amplifiers is frequency-dependent and depends on the load of the channel and spectrum of the input power of the channels reaching the amplifier. Moreover, uncertainties in span losses lead to uncertainties in setting the amplifier gain and noise figure. Uncertainties due to span losses affect also the modeling of nonlinear effects. Filtering penalties are due to the presence of wavelength selective switches (WSSs) at reconfigurable optical add/drop multiplexer (ROADM) nodes. In practice, a lightpath traverses different ROADMs (e.g., from different vendors) and experiences different degrees of filtering penalty. The filtering penalty for a given lightpath does not show a linear dependence on the number of filters crossed and there are uncertainties due to the misalignment of filters to the grid and due to the fact that filters could come from multiple vendors with diverse characteristics.

ML is regarded as a tool capable of estimating all the above mentioned uncertainties (and possibly, several other sources of uncertainty, not reported here for the sake of conciseness), thanks to the knowledge extracted from monitoring data gathered from the network. Moreover, ML promises to overcome both the shortcomings of "exact" analytical models by providing computationally-efficient solutions, and the shortcomings of margined formulas, by allowing to set lower margins.

Several ML approaches have been recently proposed for the QoT estimation of unestablished lightpaths. We refrain here from exhaustively covering the recent and abundant literature on ML-based QoT estimation (interested readers are referred to [2,17]). In this chapter, instead, we describe some exemplificative recent ML-based methods for QoT estimation, guiding the reader through the main principles and approaches for ML-based QoT estimation. The chapter is organized as follows: in Section 7.2, we describe how to perform QoT estimation using two classical ML approaches, namely classification and regression. Then, in Section 7.3, we cover more recent ML applications for QoT estimation based on active learning and transfer learning. In Section 7.4 we discuss how to incorporate ML-based QoT estimation frameworks in optimization tools. Finally, the chapter is concluded in Section 7.5 by some illustrative numerical results.

7.2. Classification and regression models for QoT estimation

7.2.1 Classification approaches for QoT estimation

QoT parameters like SNR and BER are most commonly estimated based on measurements gathered by means of optical performance monitors (OPMs) installed at the receiver nodes. The estimation of the SNR/BER for an unestablished lightpath relies on measurements collected from already established lightpaths. An unestablished lightpath (yet to be deployed) can be characterized by a set of features: the traffic volume to be served is an example of a numerical feature while the modulation format used for transmission is a categorical feature. A set of instances, which are considered independent from each other, represents a dataset.

QoT estimation of unestablished lightpaths through ML has been commonly modeled/formulated as a binary classification problem. A class is described by a binary value associated to each instance: "1" if it satisfies a given rule (positive, or *True* instances), "0" otherwise (negative, or *False* instances). For example, considering the BER as QoT metric, 1 is associated to an instance if the BER of the lightpath characterized by the features constituting the instance is below a pre-defined BER threshold T, 0 otherwise. Conversely, if SNR is considered, then 1 is associated to an instance where the SNR of the corresponding lightpath is above a pre-defined SNR threshold and 0 otherwise.

A classifier can be considered as a function that maps a point of the features space to a real number. Such a real number is the score of the instance, which can in turn be converted in the probability P_{pos} of belonging to class *True*. The choice of features should be done in such a way that they contain information useful to discriminate the class of an instance. Conversely, feature that show no correlation to the class of the instance, i.e., non-informative features, may reduce the performance of the classifiers. Hence, it is of utmost importance to identify features which are informative for the QoT classification problem.

For a classifier to be used, it should initially be trained by means of a training dataset, i.e., a set of instances whose class is known. During training, the classifier learns to map the features space and the class. Several classification algorithms have been proposed in the literature [18]. Among the most frequently used ML-based classification approaches for QoT estimation are: Neural Networks (NN), Support Vector Machines (SVM), Random Forests (RF), k-Nearest Neighbors (kNN), Logistic Classifiers. The effectiveness of various ML classification methods for QoT estimation of unestablished lightpaths is discussed in [19].

Once the training phase is over, the classifier can be used to test instances that were not part of the training set. Given a test instance characterized by a set of features, the classifier predicts the probability that the instance under test belongs to the positive class. This probability is obtained from the output score of the classifier: the score is close to "1" for instances that are very likely to be positive, and close to "0" for instances that are very likely to be negative. Note that, for instances that are difficult to classify, the classifier may return a score close to 0.5 as the instance may belong to either class.

In most practical applications, a single well-defined output is needed, therefore, the output of the classifier is a binary value such that a test instance is classified as positive only if the output score is greater or equal than a threshold γ (a typical value of γ is

0.5). Note that, practically, this means that a test instance will be classified as positive if the classifier produces a score equal to 0.9 (the instance is very likely positive) and if the classifier produces a score equal to 0.51, indicating a large uncertainty in the prediction. To avoid, or at least decrease, this uncertainty one might consider to explore the application of regression for QoT estimation, as it will be discussed in Section 7.2.2.

7.2.1.1 Performance evaluation metrics - ML classification

Given a trained classifier and a testing set, according to the threshold γ, the testing instances are divided in four groups:

- True Positive (TP) samples, i.e., positive samples that are correctly classified;
- True Negative (TN) samples, i.e., negative samples that are correctly classified;
- False Positive (FP) samples, i.e., negative samples that are incorrectly classified as positive;
- False Negative (FN) samples, i.e., positive samples that are incorrectly classified as negative.

The True Positive Rate (TPR) is defined as the fraction of all positive instances that are classified as such, i.e., $TPR = \frac{TP}{TP+FN}$. Conversely, the false positive rate is the fraction of all negative instances that are incorrectly classified as positive: $FPR = \frac{FP}{FP+TN}$. Note that both the TPR and FPR are in the $[0, 1]$ range. An ideal classifier has $TPR = 1$ and $FPR = 0$. By increasing the value of γ, the number of instances that are classified as positive is reduced, and the number of samples that are classified as negative is increased.

The performance of a classifier is most commonly measured through two metrics: *accuracy* and *Area Under the ROC Curve (AUC)* [18], where ROC stands for Receiver Operating Characteristic. The accuracy corresponds to the fraction of correctly classified test instances, i.e., $\frac{TP+TN}{TP+TN+FP+FN}$. Accuracy is easy to interpret and understand, but, has the following drawbacks:

- Accuracy is affected by the relative frequency of the two classes in the testing set.
- Accuracy depends on the choice of the threshold γ used to determine the classifier outputs.
- Accuracy does not capture the ability of a classifier to identify difficult (ambiguous) instances as such.

The AUC is another effective and robust metric for the performance of binary classifiers, which does not depend on the specific choice of γ. AUC ranges from 0.5 (for a "dummy" classifier) to 1 (for an ideal classifier). According to [20], the value of the AUC is preferable to accuracy when evaluating the quality of classifiers, and has a very useful intuitive interpretation as follows: pick a negative and a positive sample at random from the testing dataset and score both samples with the trained classifier. The AUC of the classifier can be interpreted as the probability that the classifier returns a larger score for the positive sample than for the negative sample. Therefore, for any choice of a negative and a positive sample, a classifier with AUC $= 1$ will score lower the negative

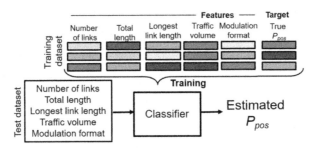

Figure 7.1 The classifier structure. (Adapted with permission from [1]. © The Optical Society)

sample rather than the positive sample and thus implying that there exists a threshold γ which perfectly separates negative and positive samples. Conversely, a classifier that returns random scores will have an AUC close to 0.5.

7.2.1.2 Illustrative description of a classifier for QoT estimation

We now report the exemplifying description of a classifier for QoT estimation proposed in [1]. As shown in Fig. 7.1, the set of features adopted in [1] to train the classifier are: the number of links of the lightpath; the lightpath total length (in km); the length of its longest link (in km); the traffic volume it serves (in Gb/s); the modulation format used for transmission. Note that these features do not account for cross-channel non-linear effects. If complete knowledge of the lightpaths already deployed in the networks is available, the following additional features may be included: the smallest left/right guardband sizes separating the considered channel from the nearest left/right neighboring channels (i.e., a worst case over all links traversed by the perspective lightpath is considered); the traffic volume and modulation format of the left/right nearest neighboring channels (i.e., the neighboring channels separated by the smallest guardband, among all the left/right neighbors over every link traversed by the perspective lightpath). These six additional features capture information on cross-channel nonlinear effects.

The target variable that the classifier predicts is a binary variable, which is *True* if and only if the lightpath BER is lower than a given reference (a typical BER reference value is, e.g., $4 \cdot 10^{-3}$). Note that the BER value is affected by other factors than those captured by the classification features (e.g., time-varying penalties, uncertainties in optical amplifier parameters). Therefore, it may occur that two lightpaths with identical sets of features exhibit different BER values and in turn different values of the target variable, i.e., the association between feature values and BER value is not deterministic. The classifier is trained on a training dataset and quantitatively evaluated on a separate testing dataset (see Fig. 7.1). Since the classifier requires feature values that are numeric and that have comparable ranges to avoid numerical instability, the features are pre-processed as follows: *i*) the modulation format feature, which can take one of six possible categorical values, is replaced by six distinct binary features (one for each possible format); for each instance, the feature corresponding to the modulation format in use will take value 1, whereas the other five will take value 0; *ii*) the values of each feature are offset and re-scaled to ensure that their distribution in the whole training set has mean equal to 0 and standard deviation equal to 1. At training time, the offset and scaling parameters for each feature are estimated. When the classifier is applied to test samples, the feature values are re-scaled using such parameters. Given the set of features, the classifier outputs the predicted probability that the instance under test belongs to the positive class. If the predicted probability is equal or greater than a threshold (e.g., $\gamma = 0.5$), the instance is classified as positive.

7.2.2 Regression approaches for QoT estimation

Even though several studies have addressed ML-based QoT estimation as a classification problem [1,19,21–23], classification-based approaches have three main drawbacks: *i*) they do not convey how close to the system threshold the predicted BER/SNR is; *ii*) they do not return the predicted distribution of the BER/SNR value; *iii*) during training, there is no distinction between a training sample with a SNR (BER) slightly above (below) the threshold, and a training sample associated to a SNR (BER) that is way above (below) the threshold, thus leading to a loss of information.

To tackle these issues, ML–based QoT estimation approaches based on *regression* have also been investigated. Analogously to classification, a lightpath is characterized by a set of features (e.g., traffic volume, modulation format, total length of the lightpath, longest link length, number of links in the path). However, since the BER/SNR depends on several other factors not captured by the set of features considered, e.g., amplifier noise figure changes, it may still exhibit variations. Hence, the BER/SNR associated to a lightpath configuration can be modeled as a random variable and thus be characterized by a probability distribution function (PDF). The regressor estimates the parameters that characterize the PDF of such random variable.

An illustrative example of possible regression approaches is shown in Fig. 7.2, for the case of SNR estimation. Given an unestablished lightpath described by a set of features, the associated SNR may exhibit different values due to varying network conditions, e.g., the overall channel load, guardband size, configuration of spectrally-adjacent lightpaths, noise figure characterization of optical amplifiers. Such values could be measured only a-posteriori (i.e., after the lightpath establishment) by means of optical performance monitors [24] at the receiver node, and constitute the *ground truth* empirical PDF (see Fig. 7.2.a). A scalar value of the SNR can be estimated by means of a standard ML-regression model (Fig. 7.2.b). However, a standard ML-regression approach does not capture the uncertainties in the SNR value. Instead, if a regressor is used to estimate the parameters of the distribution of SNR (e.g., mean and variance of a Gaussian distribution as in Fig. 7.2.c or the first four moments of a Gamma distribution as in Fig. 7.2.d), it is then possible to assess how well the estimated PDF fits the observed ground truth samples. This approach was initially introduced in [25] and generalized in [26], and more details will be provided in Section 7.2.2.1.

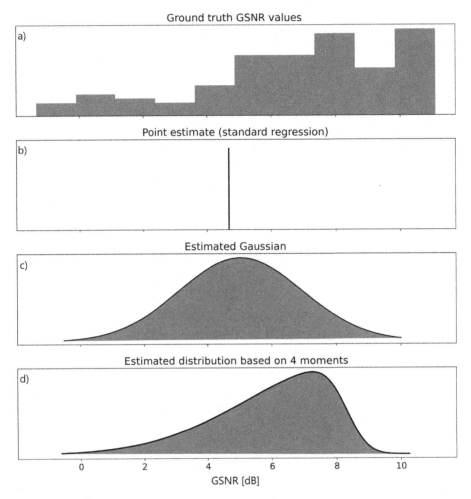

Figure 7.2 Illustrative example of SNR estimation: a) ground truth SNR values, b) SNR point estimate through a standard regressor, c) estimated Gaussian distribution of the SNR, d) and estimated distribution of the SNR based on 4 moments in dB units. (Reprinted with permission from [26]. © The Optical Society)

As ML regression approaches allow a network operator to make more informed decisions for lightpath deployment in comparison to classification approaches, a number of studies have recently focused on ML regression models for QoT estimation. In [27], authors investigate the performance of a data-driven QoT model on a dynamic metro optical network supporting both unicast and multicast connections. An accuracy of 95% is achieved in deciding if a lightpath has a sufficient QoT for the connection to be established. A methodology to accurately and rapidly evaluate optical transmission performance using ML is proposed in [28]. It involves the training of an optical performance calibration model through ML and the use of such a model in a routing engine. In [13], authors investigate a neural–network–based QoT regressor in presence of uncertainty on span lengths to decrease or eliminate this uncertainty. Authors in [29] propose a framework to improve the accuracy by which non–linear impairments are modeled on a per link basis. A significant improvement compared to previously proposed models is shown in terms of accuracy and RMSE. A ML–regression model to estimate the penalties due to filter spectral shape uncertainties at ROADM nodes and Erbium Doped Fiber Amplifier (EDFA) gain ripple effect is given in [30]. As a result in capturing the uncertainties introduced by the amplifier gain ripple and filtering at ROADM nodes, the proposed approach ensures a better estimation of QoT of light-paths and a reduction in design margins. Uncertainties in the estimation of SNR due to the variation of fiber attenuation coefficient, dispersion and non–linear coefficients as well as amplifier noise figure in each span are studied in [14]. In the following, we describe ML regression approaches which estimate the parameters of the distribution of the QoT metric, e.g., SNR [26].

7.2.2.1 Regression models for QoT estimation

In order to statistically model the SNR distribution, two approaches can be adopted: *i)* considering well-known (branded) distributions and make use of the mean and variance of the ground truth (GT) distribution of SNR to reconstruct a statistically well-known distribution of the SNR, and *ii)* removing the assumption of using well-known distributions and instead estimate the four moments and quantiles of a generic distribution of the SNR. Since scenario *i)*, i.e., considering well-known distributions, limits the ability to reconstruct the distribution of the SNR, the analysis is extended by removing the assumption of a pre-determined distribution and by estimating the moments or quantiles of the SNR distribution. Hence, three ML–regression based approaches for performing SNR's PDF estimation for candidate lightpaths are considered:

- a *Matched Gaussian Distribution Regressor (MD-R)*: given the set of features of the perspective lightpath, estimates the values of mean and variance of the Gaussian distribution approximating the SNR distribution.
- a *Moments Estimation Regressor (ME-R)*: enhances *MD-R* to estimate the first four moments of the distribution, i.e., mean, variance, skew and kurtosis.
- a *Quantile Estimation Regressor (QE-R)*: further generalizes the *MD-R* approach by removing any underlying assumption on the Gaussianity of the SNR distribution and leveraging the regressor to estimate n equally spaced distribution quantiles.

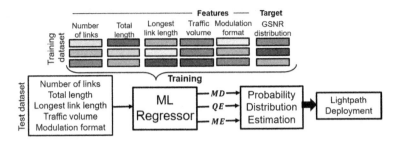

Figure 7.3 The main building blocks of the ML regression approach for estimating the parameters of the SNR distribution. (Reprinted with permission from [26]. © The Optical Society)

Given a dataset, the *n* quantiles are the *(n-1)* values that divide the empirical distribution in *n* ranges with the same number of samples.

In Fig. 7.3, the main building blocks of the SNR distribution estimation approach are shown. The figure is analogous to Fig. 7.1, in which a classification approach is considered for QoT estimation. Given the set of features of a perspective lightpath, the corresponding parameters of the probability distribution are estimated depending on the ML estimation model being considered. The estimated SNR distribution is then leveraged to take a decision about the deployment of the candidate lightpath.

For each estimation approach, a numerical analysis of the estimated parameters of the distribution is performed in terms of R2 score and Root Mean Square Error (RMSE) (see Section 7.5). The R2 score defines the proportion of the variance in the dependent variable (i.e., each of the considered distribution parameters) that is predictable by observing the independent variables (i.e., the lightpath's features) [31]. R2 is always less or equal than one: a perfect prediction achieves an R2 equal to one; a model with R2= 0.9 explains 90% of the variance of the data, and is unable to capture the remaining 10%. The RMSE measures the standard deviation of the errors that the system makes in its predictions [32]. The RMSE is non-negative, and a perfect prediction achieves RMSE= 0. RMSE is more difficult to relate to the quality of the model since it is dependent on the scale of the target variable.

Since RMSE and R2 are more difficult to interpret from a network operation point of view, the Kullback–Leibler divergence (d_{KL}) is adopted as a performance metric to compare the capability of the various considered estimation approaches in reconstructing the actual PDF of the SNR. The d_{KL} measures the difference between a given probability distribution, e.g., the reconstructed SNR distribution based on the moments by *ME-R*, and the ground truth empirical SNR distribution. A d_{KL} of zero indicates that the two considered distributions are identical, while a divergence d_{KL} equal to one indicates that the distributions are nowhere similar. The definition of d_{KL} can be found in Eq. (48) in [33].

A complete discussion on *MD-R QE-R* and *ME-R* can be found in [26], while a further numerical analysis is provided in Section 7.5.

7.3. Active and transfer learning approaches for QoT estimation

The vast majority of recently-proposed ML-based QoT-estimation approaches adopt offline supervised learning, i.e., the ML algorithms are trained using large sets of historical training samples. In most cases, the SNR and/or BER are assumed to be measured at the receiver and it is also assumed that the set of training samples contains thousands of values of QoT parameters of already deployed lightpaths. Indeed, to obtain high estimation accuracy, the training set needs to be sufficiently large and to contain samples that explore the whole feature space, but such samples might not always be available in a real operating network. In general, SNR/BER data is expensive to be acquired, e.g., due to lack of telemetry equipment, which makes the measurement collection difficult. Moreover, lightpaths with above-threshold BER (i.e., exhibiting faults or malfunctions) are unlikely to be observed in real deployments due to the conservative system-design strategies (i.e., high margins) typically adopted to guarantee transmission quality. Therefore, to complement the training set with above-BER-threshold samples, probe lightpaths [1,34] (i.e., lightpaths which do not carry user traffic that are deployed with the objective of QoT monitoring) can be used to acquire data associated with critical transmission configurations that would not be normally adopted for customer traffic. However, installing these probes incurs in additional operational costs and higher occupation of spectral resources.

Therefore, when the number of training samples is limited, two approaches can be applied to enlarge the training dataset and improve the prediction accuracy of a ML-based QoT estimator: *i) Active Learning* and *ii) Transfer Learning*.

Active Learning (AL) is based on the principle of acquiring a limited number of additional samples via dedicated probe lightpaths. AL can be used to select training instances to be acquired in order to improve the model prediction performance while minimizing the number of installed probes.

Transfer Learning (TL) consists in exploiting "external" training data collected from a different network [35]. TL allows to improve the performance of the ML QoT estimator by training it with data (samples) gathered from a *source domain* which is different from the *target domain* where the QoT estimator is used.

7.3.1 Active learning

How to provide accurate ML-based QoT predictions in presence of small/ incomplete training sets is an important and scarcely explored research issue. A possible approach is the application of an *Active Learning* (AL) method that works on top of a ML predictor based on Gaussian Processes (GP). After an initial training with a limited number of instances,

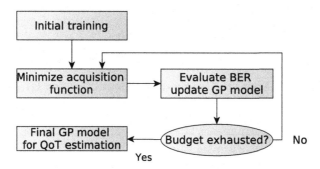

Figure 7.4 The active learning framework. (Reprinted with permission from [37]. © The Optical Society)

the AL algorithm iteratively asks to collect only a few selected training samples with specific characteristics, with the intent of minimizing the number of required samples. In particular, samples that minimize a specifically tailored *acquisition function* are sought. The objective of the acquisition function is to maximize the increase of prediction accuracy at each iteration.

It follows that an AL algorithm requires two components: a ML model that, given a training set, returns predictions for unobserved inputs, and an acquisition function that guides the selection of new instances to be added to enlarge the training set. These two components form the core of an iterative procedure where: *i*) the ML model is fitted to the training set; *ii*) the model predictions are used to build an acquisition function; *iii*) the minimizer of the acquisition function determines the next data point to be added to the training set; *iv*) the procedure is repeated until a predefined budget of added samples is reached. Fig. 7.4 represents the block diagram summarizing the above mentioned phases. In the context of QoT estimation, the ML model could either predict whether a certain instance is above or below the critical system threshold with a standard classifier, or estimate the BER/SNR value with a regression model and then threshold the prediction. In the framework depicted in Fig. 7.4, predictions are obtained by means of Gaussian Process (GP), which are briefly introduced in the next paragraphs with application to regression problems.

7.3.1.1 Gaussian Processes for QoT estimation

Gaussian-Processes (GPs) [36] are a probabilistic non-parametric learning algorithm which provide both a prediction and a quantification of its uncertainty. GPs ensure a reliable and fast uncertainty computation, thus allowing for an efficient exploration of the input space. As such, they are preferred in comparison to other models, e.g., neural networks [37].

Given the features of a candidate lightpath $\mathbf{x}^* \in \mathbb{X} \subset \mathbb{R}^L$ (where L is the size of the feature space) in input, a GP model will output a prediction of the BER/SNR value.

In particular, we observe a training set of ℓ points $\mathbf{X}_\ell = \{\mathbf{x}_1, \ldots, \mathbf{x}_\ell\} \subset \mathbb{X}$, coupled with ℓ response values $\mathbf{y} = (y_1, \ldots, y_\ell)^T \in \mathbb{R}^\ell$ where

$$y_i = f(\mathbf{x}_i) + \varepsilon \qquad (7.1)$$

for $\mathbf{x}_i \in \mathbb{X}$, $i = 1, \ldots, \ell$ with a measurement error $\varepsilon \sim N(0, \sigma_{\text{noise}}^2)$. Eq. (7.1) means that we assume that BER/SNR observations are generated by a latent function f and corrupted by a small measurement noise ε. This observation model allows to handle data points where two observations exhibit different BER/SNR values, while having the same input features. We denote by $\mathbf{f} = (f(\mathbf{x}_1), \ldots, f(\mathbf{x}_\ell)) \in \mathbb{R}^\ell$ the latent function values. The observation model described in Eq. (7.1) can be summarized as $p(\mathbf{y} \mid \mathbf{f}) = N(\mathbf{f}, \sigma_{\text{noise}}^2 I_\ell)$, where $I_\ell \in \mathbb{R}^{\ell \times \ell}$ is the identity matrix. Note that for the QoT estimation problem at hand, $\mathbf{x}_i$ is a vector describing the ith lightpath's features in the training set, coupled with y_i, i.e., the (noisy) BER/SNR value observed for that lightpath. The vector $\mathbf{x}_i$ is considered here as a vector of real values in $[0, 1]^L$. The j-th ordinal feature is thus transformed with a linear transformation that maps each $x^j \in [min_j, max_j]$ into $\tilde{x}^j \in [0, 1]$ as $\tilde{x}^j = (x^j - min_j)/(max_j - min_j)$, where min_j, max_j are the minimum and the maximum values for the feature j.

GP regression assumes that the latent function f is a realization of a Gaussian process. Thus, in a Bayesian sense, we assume that the latent vector $\mathbf{f}$ has a prior distribution given by $p(\mathbf{f}) = N(m(\mathbf{X}_\ell), K)$, with $m(\mathbf{X}_\ell) = (m(\mathbf{x}_1), \ldots, m(\mathbf{x}_\ell))^T \in \mathbb{R}^\ell$ and K a positive definite matrix with elements $K_{i,j} = k(\mathbf{x}_i, \mathbf{x}_j)$. The function k is a positive definite kernel which, along with m, the mean function, defines the Gaussian process.

The GP mean function m and kernel k are chosen before observing the data, and therefore encode our prior knowledge. The prior mean is an arbitrary function that encodes trends of the latent function f known before observing any data. A zero mean function can be used for the QoT estimation, as no known trend is available for the BER/SNR. The kernel determines the smoothness of the GP regression fit and can be used to encode prior knowledge on f. For example, if the latent function is known to be periodic, we can choose a periodic kernel and all prior realizations of the GP will be periodic functions. However, if we assume that the BER/SNR function does not have, a priori, specific properties, then we can choose a kernel from a stationary family, so as to have a dependency on a few hyper-parameters tuned from data, e.g., the Matérn family.

The prior and the observation model in Eq. (7.1) can be combined with the Bayes theorem to obtain the posterior distribution.

$$p(\mathbf{f} \mid \mathbf{y}) = \frac{p(\mathbf{f})p(\mathbf{y} \mid \mathbf{f})}{p(\mathbf{y})}.$$

In GP regression, the posterior distribution $p(\mathbf{f} \mid \mathbf{y})$ has the remarkable property of being normally distributed, with analytical expressions for its mean and covariance (see [36], chapter 2). This implies that no sampling is needed to compute the posterior distribution. However, because the analytical formulas have a complexity of $O(\ell^3)$, they become practically infeasible for large datasets. This issue is often not problematic in AL applications because the size of the training set is intrinsically limited by the cost of acquiring new samples.

7.3.2 Transfer learning

In a Transfer Learning (TL) scenario, the main assumption is that it is possible to rely on a large training dataset collected from a *source domain*, which can be used to train a model that predicts the QoT of lightpaths to be established in a different *target domain*, for which only a small labeled training dataset is available [38–40]. In a TL scenario, the aim is to make use of the data from the source domain to adapt a good model to the target domain. While most of the TL-based approaches rely on learning certain parameters of the model, e.g., weights and biases of a neural network, another possible approach relies on the manipulation of the dataset gathered from the source domain to make it suitable for re-use to train the ML-model to be applied to the target domain. Such approach is known as *Domain Adaptation* (DA) [41] and is based on the principle that the samples from one domain carry useful information regarding the QoT estimation problem in the other domain.

In principle, the more the source and target domain differ, the less accurate the QoT estimation is expected to be. For example, the two domains may differ in terms of type of optical fibers and transmission equipment or of length of the links. Indeed, the mapping between features and target variable somewhat differs between the two domains, as the joint probability distribution of features and target variable is not the same in the two domains. As a result, samples from the target domain need to be added to the training data from the source domain in order for the ML model to achieve a sufficient estimation performance. To apply DA techniques in practical applications, it is crucial to quantify the samples needed from the target domain.

TL is a ML-based approach that has received considerable interest in the last few years. For further reading on the use of TL for QoT estimation on optical networks the reader is addressed to the following references [38,39,42–44].

An Artificial Neural Network (ANN) - based TL framework has been adopted in [38] to predict the Q-factor in a real-time optical system with mixed line rates. By applying TL, it is shown that an accurate QoT prediction can be achieved for three domains and this by making use of only a few dozens of training samples for the tuning of ANN weights. TL-based approaches also enable fast re-modeling in case of variation of system parameters, e.g., optical launch power, bit-rate and chromatic dispersion. A Deep Neural Network (DNN) approach for SNR estimation in [39] shows that a significant reduction in the number of samples required from the target domain can be obtained by applying TL, in comparison to a randomly initialized weights scheme, without previously acquiring knowledge from the source domain. By transferring the hyper-parameters of DNN at the initial stage, a fast adaptation to the channel variation is achieved with smaller training set size and fewer calculations. The performance of TL-based approaches has been experimentally tested in [42], where a joint prediction of SNR and identification of modulation format is considered. A TL methodology for QoT estimation in multi-domain networks with broker orchestration is discussed in [43,44]. QoT monitoring data of intra-domain paths collected by each network domain manager are associated to a set of encoded features (to avoid disclosure of confidential domain information) and provided as samples to the broker plane. The broker leverages the samples acquired from various domains to train a DNN-based QoT estimator for each inter-domain path.

In the following subsection, three DA techniques are discussed. A preliminary analysis of existing DA approaches for QoT estimation is presented in [45]. Such analysis is generalized in [46,47] by considering various topologies, re-scaling their link length by different factors, to obtain a wider set of domains. The performance of DA techniques is assessed as a function of the number of available training instances from the target domain and on the degree of dissimilarity between the two domains.

7.3.2.1 Domain adaptation techniques

In the following, three DA techniques applied in [45] are presented: Bayesian Updating (BU), Feature Augmentation (FA) [48] and CORrelation ALignment (CORAL) [49].

Bayesian Updating (BU) is a supervised DA technique which consists in using a model trained on the source dataset $\mathbf{S}$ as prior model and updating it with the data from the target domain $\mathbf{T}$. BU is computationally reasonable with GP models because they allow for analytical updates of the posterior distribution (see, e.g., [50]), so the distribution can be updated exactly at affordable computational costs.

FA is a supervised DA technique that implements a simple approach which encodes the domain of a sample by augmenting its feature vector. In particular, the length of the original feature vector $\mathbf{x}$ is tripled, with a rule that depends on the domain: if the sample belongs to $\mathbf{S}$, the resulting feature vector is computed as $\mathbf{x}' = \langle \mathbf{x}, \mathbf{x}, \mathbf{0} \rangle$; otherwise, for a sample in $\mathbf{T}$, the feature vector is redefined as $\mathbf{x}' = \langle \mathbf{x}, \mathbf{0}, \mathbf{x} \rangle$. This augmentation transformation is applied to all samples, both in the training and test phases.

CORAL is an unsupervised DA technique that transforms the features in $\mathbf{S}$ to match the second-order statistics of the features in $\mathbf{T}$. Because of the difference in the domains, the instances in $\mathbf{S}$ are contained in a different manifold of the space of features than the ones in $\mathbf{T}$: a model learned on $\mathbf{S}$ will therefore underperform on the target domain. CORAL applies a transformation ϕ that re-colors the whitened features of the samples in $\mathbf{S}$ with the covariance matrix estimated from the feature distribution of the samples in $\mathbf{T}$; the model is learned on the transformed data. Because estimating such covariance matrix does not require information about the labels of samples in $\mathbf{T}$, the approach is unsupervised and we use the notation $\mathbf{T}_{\text{unlabeled}}$. Note that, in our application scenarios, generating $\mathbf{T}_{\text{unlabeled}}$ simply requires to select the routes and transmission configurations of a large set of potential lightpaths, without measuring their BER/SNR. Since generating feature vectors associated to perspective lightpath configurations come at no cost, the cardinality of set $\mathbf{T}_{\text{unlabeled}}$ is assumed to be large. The method estimates the transformation ϕ from $\mathbf{S}$ to $\mathbf{T}_{\text{unlabeled}}$, then trains the prediction model on $\phi(\mathbf{S})$.

For illustrative purposes, the performance of BU, FA and CORAL in terms of RMSE and R2 score is shown in Section 7.5 and compared with three benchmark approaches:

- *Source Domain Baseline* (SDB) that trains the regressor only on $\mathbf{S}$;
- *Reduced Target Domain Baseline* (RTDB) that trains the regressor only on $\mathbf{T}$;
- *Large Target Domain Baseline* (LTDB) that trains the regressor on $\mathbf{T}'$ containing a larger number of samples from the target domain (i.e., $|\mathbf{T}'| \gg |\mathbf{T}|$).

7.3.3 When to apply AL/DA during network lifecycle

The AL and DA approaches discussed in the previous sections may be applied in the early life-stages of optical networks. Consider a newly deployed optical network, which is assumed as not populated by any lightpath at time t_0. When the first lightpath request has to be provisioned (i.e., $\mathbf{T} = \varnothing$), the viable ML options for QoT estimation are:

- acquire a dataset $\mathbf{S}$ from a different network domain and use it to train the BER/SNR estimator (SDB);
- acquire a dataset $\mathbf{S}$ from a different domain and a dataset $\mathbf{T}_{\text{unlabeled}}$ of unlabeled data from the current network domain and apply CORAL to train the SNR predictor (note that, the acquisition of $\mathbf{T}_{\text{unlabeled}}$ is straightforward and comes at no cost, because it consists of a collection of lightpath configurations, without need of assessing their BER/SNR).

Moreover, the estimation obtained by SDB and CORAL can be improved by AL, given that the installation of probe lightpaths is viable. The samples indicated by the AL algorithms and acquired by dedicated probes will enlarge the set $\mathbf{S}$. This combination of AL and DA is indicated as SDB + AL or CORAL + AL. Conversely, in case an external dataset can not be acquired, the first deployment should be done through traditional BER/SNR estimation methods (e.g., the Gaussian Noise model [4]).

Once the network starts to be populated with lightpaths, the BER/SNR can be monitored and measured and the set $\mathbf{T}$ starts being populated ($0 < |\mathbf{T}| < |\mathbf{T}_0|$). At this stage $t_1 > t_0$, two additional DA options become applicable to improve the performance of the BER/SNR predictor:

- Bayesian Update (BU) is applied on the model previously trained on dataset $\mathbf{S}$ by making use of the new samples from dataset $\mathbf{T}$;
- Feature Augmentation (FA) technique is applied on the datasets $\mathbf{S}$ and $\mathbf{T}$ and uses both datasets to train the predictor.

Table 7.1 Taxonomy of the use of DA/AL techniques, depending on the size of the source and target domain datasets. (Adapted with permission from [47]. © The Optical Society)

Network lifetime	Target domain dataset	$|S| = 0$		$|S| > 0$	
		Probes available	Probes not available	Probes available	Probes not available
t_0	$|\mathbf{T}| = 0$	–	–	SDB CORAL	SDB+AL CORAL+AL
$t_1 > t_0$	$0 < |\mathbf{T}| < |\mathbf{T}_0|$	–	–	BU, FA	BU, FA
	$|\mathbf{T}| \geq |\mathbf{T}_0|$	RTDB	RTDB+AL	BU, FA	BU, FA
$t_2 > t_1$	$|\mathbf{T}| \gg |\mathbf{T}_0|$	LTDB	LTDB	LTDB	LTDB

The adoption of AL becomes feasible when set $\mathbf{T}$ contains a sufficient number of samples ($|\mathbf{T}| \geq |\mathbf{T}_0|$) to boost an initial model training, regardless that an "external" dataset might not be acquired. This scenario is indicated as RTDB+AL.

Finally, at time $t_2 > t_1$, when the size of $\mathbf{T}$ becomes large ($|\mathbf{T}| \gg |\mathbf{T}_0|$), it can be expected that training the BER/SNR predictor with dataset $\mathbf{T}$ yields to a sufficiently good estimation performance (LTDB) and AL/DA techniques are no longer necessary. Table 7.1 summarizes the above mentioned scenarios.

7.4. On the integration of ML in optimization tools

The main optimization problem underlying the planning of Elastic Optical Networks (EONs) is the so-called Routing, Modulation Format and Spectrum Assignment (RMSA), which consists in assigning a lightpath with a certain modulation format to every traffic request connecting a source to a destination node over a designated spectrum portion. To obtain a feasible RMSA solution, the QoT of each lightpath must be estimated before deployment to verify which modulation formats can be safely used, considering that physical-layer impairments can affect signal propagation. Hence, QoT estimation is an essential preliminary step for achieving effective RMSA. Most commonly, RMSA and QoT estimation are addressed as separate problems, however it is of utmost importance to consider approaches which address the RMSA and QoT estimation jointly.

RMSA in EON has been broadly investigated in the past decade [51,52], however there number of studies attempting to integrate the outputs of QoT estimators in RMSA is still limited. A few ML-based approaches for QoT estimation have been integrated in resource allocation tools. For example, a neural network estimator trained using the total length of a lightpath, the maximum link length, the number of traversed amplifiers, the degree of the nodes and the channel wavelength as features can be utilized by a heuristic which performs dynamic routing and spectrum allocation [27]. Other illustrative examples of integrating ML-based approaches in optimization tools may be having a ML-based regressor queried by a heuristic algorithm with the objective of performing core, route and spectrum assignment in a multi-core fibers scenario [53]. A fuzzy C-means clustering algorithm has been adopted to evaluate candidate lightpath configurations in an optical network with multi-core fibers. For an incoming traffic request, once a set of feasible routes, spectrum and core allocations are identified, the ML algorithm is used to identify the most suitable option for the services' transmission needs, depending on its associated service level [54]. A QoT estimator making use of deep neural networks can be interfaced with a heuristic algorithm performing Routing, Modulation format and Spectrum Assignment (RMSA) [55].

Several recent works make use of ML to perform joint routing and spectrum assignment [56–59]. The standard routing and wavelength assignment (RWA) problem is considered as a supervised classification problem in [56], while in [57] the RSA strategy selection for a given traffic demand is implemented using deep neural networks. In [58,59] a reinforcement learning framework is proposed for performing routing, modulation format and spectrum assignment for dynamic network scenarios. In the following, we describe an exemplificative framework that integrates the ML-based QoT estimation with RMSA in EONs [60].

7.4.1 RMSA integrating ML-based QoT estimation in EONs

A possible approach to jointly solve the RMSA and QoT estimation problems consists in integrating the probabilistic output of a ML classifier for QoT estimation in an Integer Linear Program (ILP) formulation for RMSA, considering a static traffic scenario. Once an initial solution is found, the QoT estimator is called iteratively to further refine it by including features about the neighbor channels of each lightpath, to account for inter-channel interference. Based on the QoT estimation output and on the optimal network configuration calculated at the previous round, additional constraints are then inserted in the ILP formulation to prevent that the deployment of neighboring channels causes excessive QoT degradation at the receiver nodes [60]. The classifier is trained with a dataset of input instances consisting in the same list of features introduced in Section 7.2.1.2 (i.e., lightpath total length and maximum link length, the number of traversed links, the amount of traffic to be transmitted and the modulation format to be adopted for transmission). Optionally, the classifier may consider additional features characterizing the neighbor channels of lightpath: guardband size, traffic volume and modulation format of the spectrally nearest left and right channels (co-propagating with the considered lightpath along at least one of its links).

7.4.1.1 Integrated network planning framework

As a baseline network design approach, reach constraints of the ILP model are set based on pre-computed reach values obtained through margined analytical formulas (see

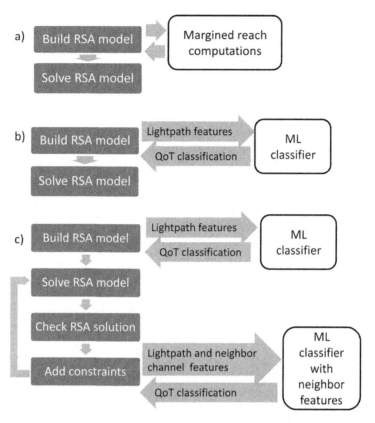

Figure 7.5 Margined (a), ML-based (b) and iterative (c) RMSA frameworks. (Reprinted with permission from [60]. © IEEE)

Fig. 7.5a). System margins are needed to account for the uncertainties due to physical layer parameters' characterization (e.g., amplifier characteristics, span losses or filtering penalties) and in the actual network configuration (e.g., inter-channel crosstalk due to neighbor lightpaths).

ML-based approaches for QoT estimation can be integrated with RMSA by replacing the margined formulas with a ML-based QoT classifier as the one described in Section 7.2.1 which outputs the probability that a lightpath configuration is feasible. Based on such output, reach constraints can then be defined and inserted in the RMSA model (see Fig. 7.5b). The classifier is expected to capture the behavior of the physical layer more accurately than margined formulas by making use of the monitored data gathered from already deployed lightpaths.

A challenge in the development of ML-based QoT estimators is to reduce the number of erroneous classifications of a lightpath configuration as feasible, despite its actual BER exceeding the threshold. To achieve such an objective, an alternative can be to consider additional information on the spectrally adjacent channels co-propagating with the candidate lightpath as additional input to the classifier.

However, in a realistic scenario, an a-priori ML-based estimation for all the possible configurations of a lightpath and its neighbors (e.g., left and right nearest neighbor channels) is not feasible. For example, assuming traffic volume of ten different sizes, six available modulation formats and ten different widths for guardbands, there is a total of $6^3 \cdot 8^3 \cdot 10^2 = 11,059,200$ configurations to be pre-computed for each lightpath. Hence, integrating information on neighboring channels in RMSA models introduces huge scalability issues.

An iterative procedure (see Fig. 7.5c) may be considered to overcome such limitations. After finding a solution to the initial RMSA formulation, for each lightpath in the solution, a query to the ML classifier is issued containing the exact features of that lightpath and of its neighboring channels (hence, there is no need to account for all possible cases as in the previous example). If the classifier returns a negative outcome (i.e., the lightpath QoT is not feasible), an additional constraint is added that excludes the unacceptable deployment of the lightpath and its neighbors from the RMSA solution. The process is iterative, until either a feasible solution for all lightpaths is found, or a maximum number of iterations is reached. For a more detailed description of this approach, including the ILP formulation, the reader is referred to [60].

7.5. Illustrative numerical results

We now report a set of illustrative numerical results referring to the ML-based approaches discussed in the previous sections. Initially, we briefly introduce the data generation tools used in these studies and then show results for ML-based classification, regression, active and transfer learning approaches.

7.5.1 Data generation

The two following network topologies are considered for performing the numerical assessment: Japan (Jnet) and NSF (NSFnet) network topologies.

The dataset used for training the ML prediction models should be large enough to obtain a good estimation and avoid biases. As field data are hard to acquire and, even if acquired, it is highly unlikely that lightpaths with an unfeasible QoT are deployed, synthetic data are often utilized to train ML models. In this chapter, we consider two well-known transmission simulation tools for the purpose of data generation: E-tool [1] and GNPy [6].

E-tool assumes optical channels multiplexed in an EON with standard slice width of 12.5 GHz, 4 THz bandwidth per link, and elastic transceivers operating at 28 Gbaud with optical channel bandwidth of 37.5 GHz (i.e., 3 slices). For the transmission channel, transparent links of dispersion uncompensated standard single mode fiber (SSMF)

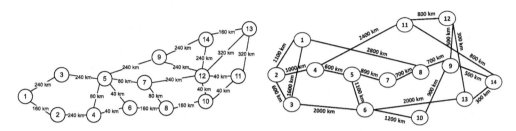

Figure 7.6 Japan (Jnet) and NSF (NSFnet) network topologies. (Adapted with permission from [1]. © The Optical Society)

with attenuation coefficient of 0.2 dB/km are considered, in which the signal power restoration is achieved through using identical optical amplifiers with equal spacing of 100 km, gain of 20 dB and noise figure of 5 dB. A system margin is added as a per-link random penalty parameter according to an exponential distribution with a 1 dB average. The system margin emulates physical impairments not captured by the model and is chosen to have an exponential distribution because it ensures the maximum entropy distribution and provides a conservative estimation of the margin. Adding a per-link stochastic penalty allows to account for the number of nodes along the path and, e.g., it helps differentiating between lightpaths with similar lengths but crossing different amounts of nodes.

In order to model the creation of a balanced training dataset representative of the whole feature space, three possible approaches can be adopted during the training phase: *i)* using historical data, derived from deployed lightpaths; *ii)* using random probes, i.e. provisioning additional probe traffic requests over unoccupied spectrum portions and *iii)* leveraging selective probes: probe traffic is momentarily transmitted over the same route that would be chosen for the perspective lightpath, using the least spectrally efficient modulation format with reach (computed according to margined formulas) below the lightpath length. Random probes are generated by randomly selecting a source-destination pair and traffic volume in the range [50, 500] Gbps (50 Gbps granularity is assumed) and creating a triplet scenario (source, destination, traffic volume). A route among w-shortest paths (e.g., $w=3$) and one out of B possible modulation formats (e.g., among the following 6 alternatives: dual polarization (DP)-BPSK, DP-QPSK and DP-n-QAM, with n = 8, 16, 32, 64) is chosen. The left/right guardbands separating the lightpath from its neighbor channels over each link (with uniform distribution in the range [12.5, 112.5] GHz) are also randomly selected, as well as their modulation format and traffic volume (with the same procedure described above). Regarding the generation of the testing dataset, M scenarios are randomly selected (e.g., $M=50$), and for each scenario, all the $w \cdot B$ possible combinations (3 routes · 6 modulation formats) are considered to test the feasibility of all deployment options, and for each combination a prediction is provided. In total there are $M \cdot w \cdot B$ settings (50 · 3 · 6 = 900) and for each setting, the BER/SNR calculation is repeated $k = 100$ times to obtain 100 instances of the exponentially distributed random variable that emulates fast time-varying impairments.

GNPy is a vendor-agnostic software tool experimentally validated in [61,62], which, given the resource allocation of a fully coherent wavelength division multiplexed (WDM) network, calculates the SNR of a lightpath over a given route. The data generated through GNPy assumes a fixed 50 GHz bandwidth allocation for optical channels, transceivers operating at 32 GBaud, 80 km fiber spans and optical channels with 0 dBm launch power. The *advanced noise model* of amplifiers is selected, which uses a 3rd order polynomial where the amplifier's noise figure is a function of the amplifier gain. Moreover, gain and noise figure ripples are modeled as functions of frequency. The motivation behind the utilization of the *advanced noise model* is to introduce randomization on the amplifier gain and noise figure, to simulate uncertainties on these parameters. GNPy accounts for the uncertainties in the noise figure of amplifiers by adding randomly generated values with a uniform distribution to the noise figure ripple: [-0.1, +0.1] dB/THz for the tilt and [-1, +1] dB for the offset, while maintaining the frequency correlation. Conversely, in the case of E-tool, the parameter uncertainties are modeled through the system margin which is accounted as a per-link random penalty calculated according to an exponential distribution. To generate synthetic datasets, given a set of features that characterize a lightpath, the SNR calculation is performed k times in order to obtain multiple instances of the random penalties, due to the uncertainties in the noise figure of amplifiers. Considering M scenarios (e.g., M = 90) and k instances (e.g., $k = 500$) for each scenario, a total of $M \cdot k$ samples (e.g., 45,000) are generated and for each sample the target variable, i.e., the associated SNR is computed. Note that the traffic volume is assumed to be of constant value, thus it is excluded from the set of features since it carries no information. Likewise, modulation format is not included in the feature set because the NLI contribution calculated through the GN-model is not dependent on the modulation format.

7.5.2 Classification

A numerical assessment of the classifier discussed in Section 7.2.1 is performed over the Jnet and NSFnet topologies. Three datasets (A, B and C) are generated using the E-tool (see Table 7.2) and the classification performance is evaluated in terms of accuracy and AUC. Using dataset A, three k-Nearest Neighbors (kNN) classifiers with k = 1, 5, 25 and five Random Forest classifiers with 1, 5, 25, 100 and 500 estimators are considered. The classifiers are trained with instances including all the 11 features reported in Subsection 7.2.1.2. With the exception of kNN with k = 1 and RF with 1 estimator, all others perform similarly well. Therefore, a RF classifier with 25 estimators is adopted for the following analysis [1].

Table 7.2 Description of datasets A, B and C. (Adapted with permission from [1]. © The Optical Society)

Dataset	A	B	C
Topology	Japan	NSF	Japan
Probing approach	Random	Random	Selective
Training dataset size	90,000	90,000	10,639
Percentage of probe instances (in training dataset)	94.5%	94.5%	20.3%
Testing dataset size	90,000	90,000	–
Positive instances (training)	53,121	9081	9845
Positive instances (testing)	52,120	8515	–

Table 7.3 Subsets of the considered features. (Adapted with permission from [1]. © The Optical Society)

No.	Feature \Subset	S1	S2	S3	S4	S5	S6	S7
1	Number of links	✓	✓	✓	✓			
2	Lightpath length	✓	✓	✓	✓	✓	✓	
3	Length of the longest link	✓	✓	✓	✓			
4	Traffic volume	✓	✓	✓		✓		✓
5	Modulation format	✓	✓			✓	✓	✓
6-11	Guardband, modulation format and traffic volume of the nearest left and right neighbor	✓						

A question of practical importance is: which features are more important to achieve a good accuracy and AUC? This question derives from the fact that considering a higher or lower number of features implies a higher or lower burden in terms of monitor deployment and control complexity. Note that, initially, eleven features are considered. The usefulness of each feature is then evaluated by comparing the performance after training the classifier over datasets A and B, considering the subsets of all the features as given in Table 7.3.

For each subset of features, the classifier is tested over two sets: *i)* the full testing dataset and *ii)* a subset of instances with BER in range $[4 \cdot 10^{-4}, 4 \cdot 10^{-2}]$, i.e., focusing on test samples which are near the system threshold ($T = 4 \cdot 10^{-3}$) and thus more "difficult" to classify. The obtained results are reported in Fig. 7.7. In the case of dataset B, when focusing on test instances "near to threshold", both metrics decrease in comparison to the values obtained over the full testing dataset, whereas for dataset A such decrease is not so significant. This implies that the performance of the classifier is still acceptable even for test instances with a BER close to the threshold T.

Results for both topologies show that training the classifier with the feature sets S1, S2 and S5 leads to the highest and comparable AUC values. Note that S1 includes all

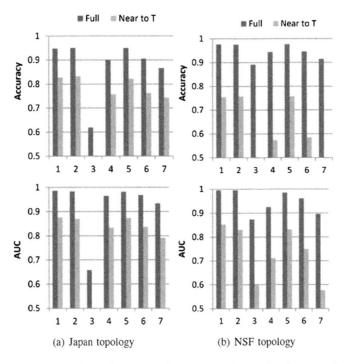

Figure 7.7 Accuracy and AUC depending on the feature set selection for dataset A (left) and B (right). (Reprinted with permission from [1]. © The Optical Society)

the 11 features, whereas S2 excludes the features characterizing the nearest neighbor lightpaths. However, focusing on the AUC of "near to threshold" instances, results obtained in scenario S1 are slightly higher, which leads to conclude that information on the closest neighbors does provide some insights for classification of the instances with BER close to T, as intuition would suggest. In particular, S5 includes only three attributes (total lightpath length, traffic volume and modulation format) which suggests that information on the number of links and length of the longest link are not very useful if the previous three features are used. Similarly, knowing the length of the longest link of the lightpath does not bring much additional information on system impairments, once the number of links and the lightpath length are known: this is due to the fact that both linear and nonlinear penalties are mainly determined by the two latter attributes. However, if either the traffic volume or the lightpath length are removed from set S5 (as in subsets S6 and S7, respectively), classification performance degrades both in terms of AUC and accuracy, especially for "near to threshold" instances. Results similar to those achieved with feature set S6 are obtained also for the feature set S4, which includes also the number of links and the length of the longest link to the features already included in S6. Performance degradation becomes extremely severe when eliminating the modulation format from the feature set, as done in S3 (which contains only the

Table 7.4 AUC comparison of probing approaches. (Adapted with permission from [1]. © The Optical Society)

Training set	AUC (full testing dataset)	AUC (testing dataset near to T)
C (historical)	0.77	0.74
C (selective, 5% probes)	0.85	0.76
C (selective, 10% probes)	0.87	0.77
C (selective, 25% probes)	0.89	0.78
C (selective, 50% probes)	0.89	0.77
A (random)	0.98	0.87

traffic volume and the lightpath characteristics). With such training features, the AUC is slightly higher than 0.6, meaning that the improvement w.r.t. a random classification (which would return 0.5) is scarce.

Regarding dataset C, the classification performance is considered after all the instances obtained by probing lightpaths carrying dummy traffic are removed from the training set and randomly sampling a set of 1000 instances. The experiment is repeated by replacing either 50, 100 or 500 instances with randomly chosen instances among the ones obtained via selective probing. AUC results averaged over 50 trials are reported in Table 7.4 and compared to results obtained by training the classifier over 1000 randomly chosen instances from dataset A. Results show that relying exclusively on historical data leads to low AUC values, as the vast majority of them belongs to the class of positive instances. Including instances obtained from selective probes in the training dataset (which mostly belong to the class of negative instances) improves the AUC: the highest improvement is obtained with 250 probes (increase by more than 0.12 when evaluating the classifier over all the instances of the testing set, and of almost 0.05 when focusing on test instances with BER in the range $[4 \cdot 10^{-4}, 4 \cdot 10^{-2}]$). Note that, increasing the probing instances up to 500 did not lead to further improvement in the performance and it even degrades the performance if probing instances are above 500. However, the overall performance is still lower than the one obtained when using training instances drawn from dataset A, which is constituted by almost 95% of instances obtained by random probes, thus ensuring a more exhaustive coverage of the whole feature space. It follows that a good classification performance can be achieved only at the price of an extensive deployment of random probe lightpaths. If probe lightpaths are deployed, the performance of the prediction model can be improved by applying Active Learning, as discussed in Section 7.3.1.

7.5.3 Regression

Numerical results for the estimation approaches presented in Section 7.2.2, i.e., *MD-R*, *ME-R* and *QE-R*, are provided in terms of RMSE and R2 score. Moreover, a *cost*

Table 7.5 RMSE (lower is better) and R2 score (higher is better) for *MD-R* and *ME-R* regressors. (Adapted with permission from [26]. © The Optical Society)

Regressor	Dataset / Target variable	RMSE	RMSE Baseline	R2
MD-R and ME-R	GNPy/ Mean	0.07	2.45	0.99
	GNPy/ Variance	0.02	0.06	0.91
ME-R	GNPy/ Skew	0.11	0.15	0.47
	GNPy/ Kurtosis	0.11	0.14	0.43
MD-R and ME-R	E-tool/ Mean	1.14	6.37	0.97
	E-tool/ Variance	2.26	8.51	0.93
ME-R	E-tool/ Skew	0.25	0.60	0.83
	E-tool/ Kurtosis	0.63	1.14	0.69

Table 7.6 RMSE (lower is better) and R2 score (higher is better) for *QE-R* regressor. (Adapted with permission from [26]. © The Optical Society)

Quantile	Dataset	RMSE	RMSE Baseline	R2
Q1	GNPy	0.07	2.41	0.99
	E-tool	2.39	7.07	0.89
Q2	GNPy	0.07	2.43	0.99
	E-tool	1.85	6.52	0.92
Q3	GNPy	0.08	2.46	0.99
	E-tool	1.54	6.27	0.94
Q4	GNPy	0.10	2.49	0.99
	E-tool	1.37	5.99	0.95

analysis which quantifies the penalties in the context of wrong lightpath deployment decisions is provided. Such a cost analysis allows to quantify the advantage of performing distribution estimation with *ME-R* compared to *QE-R* and *MD-R*. Moreover, the advantage of *QE-R* and *MD-R* is provided in comparison to several baseline approaches.

Jnet topology is considered and data are generated considering both tools, i.e., E-tool and GNPy as reported in Subsection 7.5.1.

The RMSE of each of the proposed regression approaches (*MD-R*, *ME-R* and *QE-R*) is compared to that of a baseline regressor that always predicts the mean value of each target variable calculated on the whole training set (RMSE Baseline). The values for RMSE and R2 score are reported in Table 7.5 and 7.6 for *MD-R*, *ME-R*, *QE-R* and RMSE-Baseline for both GNPy and E-tool generated data. Results show that the three proposed estimators (*MD-R*, *QE-R* and *ME-R*) perform much better than the baseline in terms of RMSE and achieve comparable performance in terms of RMSE and R2. Note that, the low values of R2 for *Skew* and *Kurtosis* are expected, since in principle they are much noisier compared to the first two moments and therefore more difficult to predict.

Next, we report the cost analysis that refers to the decision making penalties concerning the deployment of candidate lightpaths. Several scenarios are compared, in which decisions are made based on the estimated SNR probability distributions obtained by *MD-R*, *QE-R* and *ME-R*, the point estimation of the mean SNR value, and baseline approaches.

Given a set of features, let us consider a lightpath j which belongs to the set J of candidate lightpaths. The question to be addressed is: is SNR_j lower than a system defined threshold SNR_T? Given the set of features, let F_{G_i} be the estimated CDF of the random variable G_i that models the SNR, according to estimator i, where $i \in \{MD\text{-}R, QE\text{-}R, ME\text{-}R\}$. The probability that the SNR is below the threshold SNR_T, according to estimator i, can be computed as $p_i = F_{G_i}(SNR_T)$. Each estimator will estimate a different probability and the lightpath deployment decision is to be made based on the estimated probability. Two penalty costs are considered, associated to the two ways that a decision can be wrong:

- an underestimation cost (C_u), that is paid when SNR_j is estimated to be lower (Below) than SNR_T, but is in fact higher (Above) than SNR_T,
- an overestimation cost (C_o), that is paid when SNR_j is estimated to be higher (Above) than SNR_T, but is in fact lower (Below) than SNR_T.

The expected penalty for a deployment decision (Above or Below) is the probability that such decision is wrong, times the cost of taking such a wrong decision. Therefore, if the decision is that SNR_j is *below* SNR_T, the estimated probability of being wrong is equal to $(1 - p_i)$. Hence, the expected cost of deciding that $SNR_j < SNR_T$, according to estimator i is:

$$C_{i,\text{Below}} = (1 - p_i) \cdot C_u$$

while if the decision is that SNR_j is *above* SNR_T, the probability of being wrong is equal to p_i. Hence, the expected cost of deciding that $SNR_j > SNR_T$, according to estimator i is:

$$C_{i,\text{Above}} = p_i \cdot C_o$$

For each estimator i, a decision D_i is taken according to the following rules:

$$D_i = \begin{cases} \text{Below} & \text{if } C_{i,\text{Below}} < C_{i,\text{Above}} \\ \text{Above} & \text{otherwise} \end{cases}$$

The decision D_i is compared to the decision that would be made by leveraging the ground truth (D_{GT}), i.e., the one computed based on actual SNR measurements. The ground truth decision is defined as follows:

$$D_{GT} = \begin{cases} \text{Below} & \text{if } SNR < SNR_T \\ \text{Above} & \text{otherwise} \end{cases}$$

where SNR is the actual sampled value. If $D_i \neq D_{GT}$, then the respective cost associated to the decision is added to the total penalty cost of estimator i (PC_i).

$$PC_i = (\sum_{j \in J} min(C_{i,\text{Below}}, C_{i,\text{Above}}))/|J|$$

where $|J|$ is the total number of lightpaths. Please note that $min(C_{i,\text{Below}}, C_{i,\text{Above}})$ is added to PC_i since the estimator makes the decision based on the comparison between $C_{i,\text{Below}}$ and $C_{i,\text{Above}}$, so the minimum cost reflects the cost of the wrong decision of the estimator.

The performance of MD-R, QE-R and ME-R is compared with four baselines in terms of lightpath deployment decision penalties:

- *ADB*: always decide Below;
- *ADA*: always decide Above;
- *RD*: random decision;
- *CI*: cost–insensitive decision, i.e., make the decision Below (Above) if the SNR mean value estimated by the Gaussian regressor, i.e., MD-R, is below (above) the threshold.

Additionally, a lower bound of the obtainable cost which reflects the penalty cost incurred by an "ideal" estimator (noted as IE) is considered, which always returns as output an estimated CDF identical to GT.

A numerical value is assigned to each cost type: $C_u = 1$ cu (cost unit) and $C_o = 10$ cu. These values can be interpreted as follows: if the lightpath's SNR is underestimated, a lightpath configuration may be erroneously considered as infeasible while in fact it is feasible. Therefore, a lightpath with a less spectrally-efficient modulation format will be deployed, leading to the unnecessary occupation of some spectral resources, yet no service disruption will be incurred. Conversely, in the case of overestimation, a lightpath with a modulation format which will lead to a below-threshold SNR will erroneously be deployed, eventually resulting in service disruption. From a network operator's point of view, the penalty in case of service disruption is higher compared to the penalty of deploying a lightpath that does not adopt the most spectrally-efficient modulation format. This disparity is captured by the cost values, though they may not exactly reflect the actual economic losses experienced by a network operator.

A set of 100 sequences of deployment decisions is performed, each one including 500 candidate lightpaths. Therefore, the total penalty cost for each estimator is averaged over 100 simulations. The E-tool has been used for generating the data samples.

Table 7.7 Penalty Cost (PC) in cost units (cu) for each estimator. (Adapted with permission from [26]. © The Optical Society)

	IE	ME-R	QE-R	MD-R	ADB	CIB	RD	ADA
PC_i [cu]	0.059	0.069	0.074	0.087	0.275	1.527	2.532	7.248

The average penalty per instance for each estimator is reported in Table 7.7, ordered from best to worst performance. Note that *MD-R*, *QE-R* and *ME-R* all perform much better than the baseline approaches. Regarding IE, as expected it provides the lowest cost penalty, as it serves as lower bound for the penalty cost analysis.

The advantages of estimating the probability distribution instead of a point estimate are confirmed in the results reported in Table 7.7. The cost penalty of the estimation that assumed a Gaussian distribution (*MD-R*) is significantly lower in comparison to the standard regressor which only estimated the mean value of SNR (*CIB*). Additionally, the more sophisticated estimators (e.g., *ME-R*) achieve an even better performance in terms of cost penalty ($0.069 < 0.087$). Note that, since the ratio between lightpath configurations with an SNR below threshold to lightpath configurations with an SNR above threshold is 70% to 30%, *ADB* provides a better cost result compared to *ADA*.

7.5.4 Active learning and transfer learning

For each topology given in Fig. 7.6 (Jnet and NSFnet), a dataset $\mathcal{R}$ is generated through the E-tool (see Subsection 7.5.1). The performance of the AL and DA regression approaches is evaluated in terms of RMSE and R2 on a 6000 test sample chosen at random from the original full dataset $\mathcal{R}_{target}$ before any training. The remaining part of $\mathcal{R}_{target}$, denoted $\mathcal{R}_{target}^{train}$ (12,000 points), is used to select points for the training sets and as the pool of pre-selected points for the acquisition function optimization. The AL and DA methods are validated against the RTDB and LTDB baselines (see Section 7.3.2.1), where LTDB assumes $|\mathbf{T}'| = 1000$ and RTDB assumes $|\mathbf{T}| = 50$ (where $\mathbf{T}, \mathbf{T}' \subset \mathcal{R}_{target}^{train}$), whereas $|\mathbf{S}| = 75, 125, 250, 500, 750, 1000$ (where $\mathbf{S} \subset \mathcal{R}_{source}$). For CORAL only, we set $\mathbf{T}_{unlabeled} = 1000$ (where $\mathbf{T}_{unlabeled}$ contains the feature vectors of 1000 elements in $\mathcal{R}_{target}^{train}$). The AL algorithm starts from RTDB and up to 750 new samples are added. Every 100 iterations, RMSE and R2 are evaluated. Experiments are repeated 10 times, with random extraction of elements of $\mathbf{S}$ (resp. $\mathbf{T}$) from set $\mathcal{R}_{source}$ (resp. $\mathcal{R}_{target}^{train}$). The feature vector of each sample includes five features associated to the lightpath (i.e., total lightpath length, longest link length, number of traversed link, traffic volume and modulation format).

While the complete set of numerical analysis can be found in [47], here we report the numerical results in Fig. 7.8 in terms of RMSE values (bottom) and R2 values (top) obtained by DA, AL and DA+AL. When DA techniques are applied, the source domain is the Jnet and the target domain is the NSFnet. For a correct interpretation of Fig. 7.8 (top left), keep in mind that: *i*) RTDB (on the extreme left) represents a lower

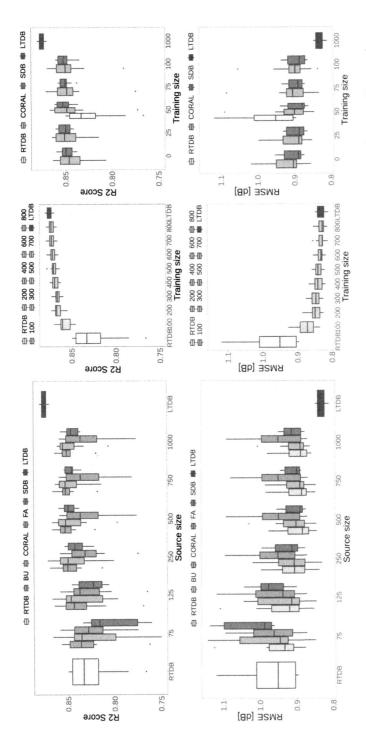

Figure 7.8 R2 (top) and RMSE (bottom) obtained by DA (left), AL (center) and DA+AL (right) approaches with target domain NSFnet and source domain Jnet. (Reprinted with permission from [47]. © The Optical Society)

performance bound, as it is obtained with limited data from target domain; *ii)* LTDB (on the extreme right) represents an upper performance bound, as it is obtained with a large training dataset from target domain; *iii)* all the approaches in the middle represent the possible increase in performance obtained thanks to DA (BU, FA and CORAL approaches) in comparison to the SDB baseline. Note that, CORAL achieves median R2 score values between [0.835-0.856] and outperforms SDB which achieves R2 values between [0.816-0.847] when $|S|=75-|S|=1000$. When $|S|$ is low, BU shows performance comparable to RTDB but for higher cardinalities of $|S|$, the performance improves and is comparable to that of CORAL. Conversely, FA is always below CORAL and slightly outperforms RTDB only when $|S| \geq 500$.

Results obtained through AL are given on the top center of Fig. 7.8 (the x-axis reports the size of **T**). AL achieves a R2 of 0.859 after 100 iterations, i.e., for $|\mathbf{T}| = 100$, which is comparable to the performance of CORAL with 1000 source domain samples. Increasing $|\mathbf{T}|$ to 200, AL outperforms all DA techniques. However, as the generation of an initial training set can be more costly, a pure AL approach might not always be the lowest-cost solution.

Results on the top right of Fig. 7.8 are obtained by combining AL and SDB or CORAL under the assumption that $|\mathbf{S}| = 500$. As a reference, the performance of RTDB for $|\mathbf{T}| = 50$ is considered. Note that AL+SDB has a median ranging between 0.849 and 0.852, which is slightly higher compared to AL+CORAL, which achieves a median ranging between 0.844 and 0.850 for $(|\mathbf{T}| = 0)$ and $(|\mathbf{T}| = 100)$, respectively.

Finally, the results at the bottom left in Fig. 7.8 report the RMSE when comparing the predictions to the ground truth values. Note that both CORAL and BU reduce the gap from the reference value reached by LTDB. Specifically, the median RMSE ranges between $[0.889 - 0.946]$ dB and is lower compared to SDB, which ranges between $[0.913 - 0.988]$. On the other hand, RTDB reaches a median RMSE equal to 0.952 dB. AL achieves the most consistent RMSE reduction as low as 0.872 using 100 probes (bottom center). AL+SDB and AL+CORAL achieve a RMSE which ranges between $[0.887 - 0.910]$ dB, thus reducing the values obtained by SDB and CORAL.

To provide a practical relevance of the results shown, in Table 7.8 the percentage (averaged over 10 repetitions) of test data that recorded an error below 0.5 dB, 1 dB and 2 dB in the best scenario for each method is given. Note that for AL with 800 iterations more that 50% of the errors are below 0.5 dB, and more than 97% are below 2 dB.

7.6. Future research directions and challenges

In the past few years, Machine Learning has become the *go-to* tool for solving problems regarding the estimation of quality of transmission in optical networks. A list of surveys, workshops and dedicated sessions in various conferences regarding the

Table 7.8 Target NSFnet. Percentage of absolute errors, ϵ, smaller than thresholds in dB. (Adapted with permission from [47]. © The Optical Society)

Method	RMSE	$\epsilon < 0.5$	$0.5 \leq \epsilon < 1$	$1 \leq \epsilon < 2$	$\epsilon \geq 2$
RTDB	0.9522	0.4598	0.2910	0.2087	0.0405
BU (1000)	0.8885	0.4164	0.3343	0.2239	0.0254
CORAL (1000)	0.8899	0.4632	0.3059	0.1939	0.0370
FA (1000)	0.9501	0.4618	0.2932	0.1989	0.0461
SDB (500)	0.8905	0.4018	0.3352	0.2449	0.0181
LTDB	0.8367	0.5152	0.3184	0.1448	0.0216
AL (800 its)	0.8260	0.5090	0.3143	0.1492	0.0275

applications of ML in optical communications can be found in [17]. We list below some possible research directions and challenges for ML applied to QoT estimation:

Data availability: while most works employing ML for QoT estimation rely on synthetic generated data, the availability of non-proprietary data gathered from the field remains a challenge. As more players from the industry join forces with academia, the expectation of having field data available is slowly becoming a reality. However, the main challenge remains the fact that networks do not operate in faulty conditions, so it is not easy to gather a dataset which covers the whole search space. Generation of meaningful synthetic data using tools as GNPy can be a direction, but more investigation on the actual accuracies achievable through a mix of real and synthetic data during training is still an open research issue.

ML methodologies: most studies that adopt ML for QoT estimation rely on offline supervised learning approaches, i.e., based on historical data. However, optical networks evolve with time and are intrinsically dynamic. Therefore, employing active ML approaches for QoT estimation is a promising research direction, as shown in some of the studies introduced in previous sections. Active ML approaches allow to interactively ask the user to collect data with specific characteristics and are likely to open new ways to integrate human-to-machine interaction in ML approaches for QoT estimation. Similarly, transfer learning has been so far investigated for a specific version of the problem (estimation of SNR), but further investigation of transfer learning in the context of other versions of the QoT estimation problems (e.g., those aimed at estimating ripple or aging parameters) is needed.

ML approaches have proven to be useful in estimating the QoT of unestablished lightpaths and allowing a better usage of networks resources compared to several traditional QoT estimation methods. ML can be deployed as a *black box* or as a complementary tool to existing physical models. Given the high accuracy and efficiency of some of the physical models currently leveraged, the vision for the future of applying ML for QoT estimation lies more on it operating as a refinement tool, especially for capturing uncertainties in the parameters of physical components such as optical amplifiers.

Ultimately, the vision for future research opportunities is not to use ML as a stand-alone tool for QoT estimation and optimizing resource allocation, but rather as a complementary tool which operates synergistically with other optimization tools.

7.7. Conclusion

In this chapter we have reviewed some popular applications of Machine Learning in QoT estimation of unestablished lightpaths. We overviewed classical ML-based methodologies to perform QoT estimation, i.e., classification and regression, and also described the more recent active learning and transfer learning approaches. Finally, we provided illustrative numerical results to show the application of the ML approaches discussed in the previous sections.

Most of the early works on ML for QoT estimation refer to the application of classification approaches and achieve a high accuracy in estimating whether a lightpath configuration satisfies the QoT constraints. However, as classification-based approaches do not convey information about how far from the QoT threshold a lightpath configuration is, regression-based ML approaches are suggested: in addition to overcoming the drawbacks of classification-based approaches, regression-based approaches allow network operators to make informed decisions and exactly determine how near to/far from the system threshold the predicted QoT metric is. In the case that training data are scarce (as it happens in several practical scenarios), Active Learning and Transfer Learning, when properly used, can reach the estimation capabilities of the classical supervised approaches without requiring a large dataset.

References

[1] C. Rottondi, L. Barletta, A. Giusti, M. Tornatore, Machine-learning method for quality of transmission prediction of unestablished lightpaths, IEEE/OSA Journal of Optical Communications and Networking 10 (2) (2018) A286–A297.

[2] F. Musumeci, C. Rottondi, A. Nag, I. Macaluso, D. Zibar, M. Ruffini, M. Tornatore, An overview on application of machine learning techniques in optical networks, IEEE Communications Surveys Tutorials 21 (2) (2019) 1383–1408.

[3] J. Shao, X. Liang, S. Kumar, Comparison of split-step Fourier schemes for simulating fiber optic communication systems, IEEE Photonics Journal 6 (4) (2014) 1–15.

[4] P. Poggiolini, G. Bosco, A. Carena, V. Curri, Y. Jiang, F. Forghieri, The GN-model of fiber non-linear propagation and its applications, Journal of Lightwave Technology 32 (4) (2013) 694–721.

[5] Y. Pointurier, Design of low-margin optical networks, IEEE/OSA Journal of Optical Communications and Networking 9 (1) (2017) A9–A17.

[6] A. Ferrari, M. Filer, K. Balasubramanian, Y. Yin, E. Le Rouzic, J. Kundrat, G. Grammel, G. Galimberti, V. Curri, GNPy: an open source application for physical layer aware open optical networks, IEEE/OSA Journal of Optical Communications and Networking 12 (6) (2020) C31–C40, https://doi.org/10.1364/JOCN.382906.

[7] P. Serena, A. Bononi, A time-domain extended Gaussian noise model, Journal of Lightwave Technology 33 (7) (2015) 1459–1472, http://jlt.osa.org/abstract.cfm?URI=jlt-33-7-1459.

[8] R. Dar, M. Feder, A. Mecozzi, M. Shtaif, Properties of nonlinear noise in long, dispersion-uncompensated fiber links, Optics Express 21 (22) (2013) 25685–25699, https://doi.org/10.1364/OE.21.025685, http://www.opticsexpress.org/abstract.cfm?URI=oe-21-22-25685.

[9] E. Seve, J. Pesic, Y. Pointurier, Accurate QoT estimation by means of a reduction of EDFA characteristics uncertainties with machine learning, in: 2020 International Conference on Optical Network Design and Modeling (ONDM), 2020, pp. 1–3.

[10] A. D'Amico, S. Straullu, A. Nespola, I. Khan, E. London, E. Virgillito, S. Piciaccia, A. Tanzi, G. Galimberti, V. Curri, Using machine learning in an open optical line system controller, IEEE/OSA Journal of Optical Communications and Networking 12 (6) (2020) C1–C11, https://doi.org/10.1364/JOCN.382557.

[11] A. Mahajan, K. Christodoulopoulos, R. Martinez, S. Spadaro, R. Munoz, Machine learning assisted EDFA gain ripple modelling for accurate QoT estimation, in: 45th European Conference on Optical Communication (ECOC 2019), 2019, pp. 1–4.

[12] D.W. Boertjes, M. Reimer, D. Cote, Practical considerations for near-zero margin network design and deployment [invited], IEEE/OSA Journal of Optical Communications and Networking 11 (9) (2019) C25–C34, https://doi.org/10.1364/JOCN.11.000C25.

[13] J. Pesic, M. Lonardi, N. Rossi, T. Zami, E. Seve, Y. Pointurier, How uncertainty on the fiber span lengths influences QoT estimation using machine learning in WDM networks, in: 2020 Optical Fiber Communications Conference and Exhibition (OFC), 2020, pp. 1–3.

[14] I. Sartzetakis, K.K. Christodoulopoulos, E.M. Varvarigos, Accurate quality of transmission estimation with machine learning, IEEE/OSA Journal of Optical Communications and Networking 11 (3) (2019) 140–150.

[15] A. Mahajan, K. Christodoulopoulos, R. Martinez, S. Spadaro, R. Munoz, Modeling filtering penalties in ROADM-based networks with machine learning for QoT estimation, in: 2020 Optical Fiber Communications Conference and Exhibition (OFC), 2020, pp. 1–3.

[16] C. Delezoide, P. Ramantanis, P. Layec, Weighted filter penalty prediction for QoT estimation, in: 2018 Optical Fiber Communications Conference and Exposition (OFC), 2018, pp. 1–3.

[17] Y. Pointurier, Machine learning techniques for quality of transmission estimation in optical networks, Journal of Optical Communications and Networking 13 (4) (2021) B60–B71.

[18] C.M. Bishop, Pattern Recognition and Machine Learning, Springer, 2006.

[19] R.M. Morais, J. Pedro, Machine learning models for estimating quality of transmission in DWDM networks, Journal of Optical Communications and Networking 10 (10) (2018) D84–D99.

[20] P.A. Flach, J. Hernández-Orallo, C.F. Ramirez, A coherent interpretation of AUC as a measure of aggregated classification performance, in: ICML, 2011.

[21] T. Jimenez, J.C. Aguado, I. de Miguel, R.J. Duran, M. Angelou, N. Merayo, P. Fernandez, R.M. Lorenzo, I. Tomkos, E.J. Abril, A cognitive quality of transmission estimator for core optical networks, Journal of Lightwave Technology 31 (6) (2013) 942–951.

[22] S. Aladin, C. Tremblay, Cognitive tool for estimating the QoT of new lightpaths, in: 2018 Optical Fiber Communications Conference and Exposition (OFC), 2018, pp. 1–3.

[23] N. Sambo, Y. Pointurier, F. Cugini, L. Valcarenghi, P. Castoldi, I. Tomkos, Lightpath establishment assisted by offline QoT estimation in transparent optical networks, IEEE/OSA Journal of Optical Communications and Networking 2 (11) (2010) 928–937.

[24] K. Christodoulopoulos, P. Kokkinos, A. Di Giglio, A. Pagano, N. Argyris, C. Spatharakis, S. Dris, H. Avramopoulos, J. Antona, C. Delezoide, et al., Orchestra-optical performance monitoring enabling flexible networking, in: 2015 17th International Conference on Transparent Optical Networks (ICTON), IEEE, 2015, pp. 1–4.

[25] M. Ibrahimi, H. Abdollahi, A. Giusti, C. Rottondi, M. Tornatore, Machine learning regression vs. classification for QoT estimation of unestablished lightpaths, in: OSA Advanced Photonics Congress (AP) 2020, Optical Society of America, 2020, p. NeM3B.1.

[26] M. Ibrahimi, H. Abdollahi, C. Rottondi, A. Giusti, A. Ferrari, V. Curri, M. Tornatore, Machine learning regression for QoT estimation of unestablished lightpaths, Journal of Optical Communications and Networking 13 (4) (2021) B92–B101.

[27] T. Panayiotou, S.P. Chatzis, G. Ellinas, Performance analysis of a data-driven quality-of-transmission decision approach on a dynamic multicast- capable metro optical network, IEEE/OSA Journal of Optical Communications and Networking 9 (1) (2017) 98–108.

[28] R.M. Morais, B. Pereira, J. Pedro, Fast and high-precision optical performance evaluation for cognitive optical networks, in: 2020 Optical Fiber Communications Conference and Exhibition (OFC), 2020, pp. 1–3.

[29] X. Liu, H. Lun, M. Fu, Y. Fan, L. Yi, W. Hu, Q. Zhuge, A three-stage training framework for customizing link models for optical networks, in: 2020 Optical Fiber Communications Conference and Exhibition (OFC), 2020, pp. 1–3.

[30] A. Mahajan, K. Christodoulopoulos, R. Martínez, S. Spadaro, R. Muñoz, Modeling EDFA gain ripple and filter penalties with machine learning for accurate QoT estimation, Journal of Lightwave Technology 38 (9) (2020) 2616–2629.

[31] N.R. Draper, H. Smith, Applied Regression Analysis, vol. 326, John Wiley & Sons, 1998.

[32] A. Géron, Hands-on Machine Learning with Scikit-Learn, Keras, and TensorFlow: Concepts, Tools, and Techniques to Build Intelligent Systems, O'Reilly Media, 2019.

[33] S.-H. Cha, Comprehensive survey on distance/similarity measures between probability density functions, City 1 (2) (2007) 1.

[34] Y. Pointurier, M. Coates, M. Rabbat, Cross-layer monitoring in transparent optical networks, Journal of Optical Communications and Networking 3 (3) (2011) 189–198.

[35] K. Weiss, T.M. Khoshgoftaar, D. Wang, A survey of transfer learning, Journal of Big Data 3 (1) (2016) 9.

[36] C.K. Williams, C.E. Rasmussen, Gaussian Processes for Machine Learning, vol. 2 (3), the MIT Press, 2006, p. 4.

[37] D. Azzimonti, C. Rottondi, M. Tornatore, Reducing probes for quality of transmission estimation in optical networks with active learning, Journal of Optical Communications and Networking 12 (1) (2020) A38–A48.

[38] J. Yu, W. Mo, Y.-K. Huang, E. Ip, D.C. Kilper, Model transfer of QoT prediction in optical networks based on artificial neural networks, IEEE/OSA Journal of Optical Communications and Networking 11 (10) (2019) C48–C57.

[39] L. Xia, J. Zhang, S. Hu, M. Zhu, Y. Song, K. Qiu, Transfer learning assisted deep neural network for OSNR estimation, Optics Express 27 (14) (2019) 19398–19406.

[40] Q. Yao, H. Yang, A. Yu, J. Zhang, Transductive transfer learning-based spectrum optimization for resource reservation in seven-core elastic optical networks, Journal of Lightwave Technology 37 (16) (2019) 4164–4172.

[41] S. Sun, H. Shi, Y. Wu, A survey of multi-source domain adaptation, Information Fusion 24 (2015) 84–92, https://doi.org/10.1016/j.inffus.2014.12.003, http://www.sciencedirect.com/science/article/pii/S1566253514001316.

[42] Y. Cheng, W. Zhang, S. Fu, M. Tang, D. Liu, Transfer learning simplified multi-task deep neural network for PDM-64QAM optical performance monitoring, Optics Express 28 (5) (2020) 7607–7617.

[43] X. Chen, B. Li, R. Proietti, C.-Y. Liu, Z. Zhu, S.B. Yoo, Demonstration of distributed collaborative learning with end-to-end QoT estimation in multi-domain elastic optical networks, Optics Express 27 (24) (2019) 35700–35709.

[44] C.-Y. Liu, X. Chen, R. Proietti, S.B. Yoo, Evol-TL: evolutionary transfer learning for QoT estimation in multi-domain networks, in: 2020 Optical Fiber Communications Conference and Exhibition (OFC), Optical Society of America, 2020, p. Th3D.1.

[45] R. Di Marino, C. Rottondi, A. Giusti, A. Bianco, Assessment of domain adaptation approaches for QoT estimation in optical networks, in: 2020 Optical Fiber Communications Conference and Exhibition (OFC), Optical Society of America, 2020, p. Th3D.2.

[46] C. Rottondi, R. di Marino, M. Nava, A. Giusti, A. Bianco, On the benefits of domain adaptation techniques for quality of transmission estimation in optical networks, IEEE/OSA Journal of Optical Communications and Networking 13 (1) (2020) A34–A43.

[47] D. Azzimonti, C. Rottondi, A. Giusti, M. Tornatore, A. Bianco, Comparison of domain adaptation and active learning techniques for quality of transmission estimation with small-sized training datasets [invited], IEEE/OSA Journal of Optical Communications and Networking 13 (1) (2021) A56–A66, https://doi.org/10.1364/JOCN.401918.

[48] H. Daumé III, Frustratingly easy domain adaptation, arXiv preprint, arXiv:0907.1815, 2009.

[49] B. Sun, J. Feng, K. Saenko, Correlation alignment for unsupervised domain adaptation, in: Domain Adaptation in Computer Vision Applications, Springer, 2017, pp. 153–171.

[50] C.K. Williams, C.E. Rasmussen, Gaussian Processes for Machine Learning, vol. 2, MIT Press, Cambridge, MA, 2006.

[51] B.C. Chatterjee, N. Sarma, E. Oki, Routing and spectrum allocation in elastic optical networks: a tutorial, IEEE Communications Surveys Tutorials 17 (3) (2015) 1776–1800, https://doi.org/10.1109/COMST.2015.2431731.

[52] M. Klinkowski, P. Lechowicz, K. Walkowiak, Survey of resource allocation schemes and algorithms in spectrally-spatially flexible optical networking, Optical Switching and Networking 27 (2018) 58–78.

[53] Q. Yao, H. Yang, R. Zhu, A. Yu, W. Bai, Y. Tan, J. Zhang, H. Xiao, Core, mode, and spectrum assignment based on machine learning in space division multiplexing elastic optical networks, IEEE Access 6 (2018) 15898–15907, https://doi.org/10.1109/ACCESS.2018.2811724.

[54] H. Yang, Q. Yao, A. Yu, Y. Lee, J. Zhang, Resource assignment based on dynamic fuzzy clustering in elastic optical networks with multi-core fibers, IEEE Transactions on Communications 67 (5) (2019) 3457–3469, https://doi.org/10.1109/TCOMM.2019.2894711.

[55] X. Chen, R. Proietti, H. Lu, A. Castro, S.J.B. Yoo, Knowledge-based autonomous service provisioning in multi-domain elastic optical networks, IEEE Communications Magazine 56 (8) (2018) 152–158, https://doi.org/10.1109/MCOM.2018.1701191.

[56] I. Martín, S. Troia, J.A. Hernandez, A. Rodriguez, F. Musumeci, G. Maier, R. Alvizu, O. Gonzalez de Dios, Machine learning-based routing and wavelength assignment in software-defined optical networks, IEEE Transactions on Network and Service Management 16 (3) (2019) 871–883, https://doi.org/10.1109/TNSM.2019.2927867.

[57] J. Yu, B. Cheng, C. Hang, Y. Hu, S. Liu, Y. Wang, J. Shen, A deep learning based RSA strategy for elastic optical networks, in: 2019 18th International Conference on Optical Communications and Networks (ICOCN), 2019, pp. 1–3.

[58] X. Chen, B. Li, R. Proietti, H. Lu, Z. Zhu, S.J.B. Yoo, DeepRMSA: a deep reinforcement learning framework for routing, modulation and spectrum assignment in elastic optical networks, Journal of Lightwave Technology 37 (16) (2019) 4155–4163, https://doi.org/10.1109/JLT.2019.2923615.

[59] X. Luo, C. Shi, L. Wang, X. Chen, Y. Li, T. Yang, Leveraging double-agent-based deep reinforcement learning to global optimization of elastic optical networks with enhanced survivability, Optics Express 27 (6) (2019) 7896–7911, https://doi.org/10.1364/OE.27.007896, http://www.opticsexpress.org/abstract.cfm?URI=oe-27-6-7896.

[60] M. Salani, C. Rottondi, M. Tornatore, Routing and spectrum assignment integrating machine-learning-based QoT estimation in elastic optical networks, in: IEEE INFOCOM 2019 - IEEE Conference on Computer Communications, 2019, pp. 1738–1746.

[61] A. Ferrari, M. Filer, K. Balasubramanian, Y. Yin, E. Le Rouzic, J. Kundrat, G. Grammel, G. Galimberti, V. Curri, Experimental validation of an open source quality of transmission estimator for open optical networks, in: 2020 Optical Fiber Communications Conference and Exhibition (OFC), 2020, pp. 1–3.

[62] M. Filer, M. Cantono, A. Ferrari, G. Grammel, G. Galimberti, V. Curri, Multi-vendor experimental validation of an open source QoT estimator for optical networks, Journal of Lightwave Technology 36 (15) (2018) 3073–3082, https://doi.org/10.1109/JLT.2018.2818406.

CHAPTER EIGHT

Machine Learning for optical spectrum analysis

Luis Velasco[a], **Marc Ruiz**[a], **Behnam Shariati**[b], **and Alba P. Vela**[a]
[a]Universitat Politècnica de Catalunya, Barcelona, Spain
[b]Fraunhofer HHI, Berlin, Germany

In this Chapter, we propose Machine Learning (ML) based solutions for Optical Spectrum Analysis of Elastic Optical Networks (EON). Specifically, we present novel failure detection and identification solutions utilizing the optical spectrum traces captured by cost-effective coarse-granular Optical Spectrum Analyzers (OSA). We demonstrate the effectiveness of the developed solutions for detecting and identifying filter-related failures in the context of Spectrum-Switched Optical Networks (SSON), as well as transmitter-related laser failures in Filterless Optical Networks (FON). Such detection and identification can contribute to the cost reduction and lowering the required margin in optical networks.

8.1. Introduction

Optical Performance Monitoring (OPM) [4] can enable several important and advanced network functionalities including *i) adaptive impairments compensation, ii) reliable network operation, iii) efficient resource allocation with physical layer consideration*, and *iv) failure detection and identification*, which can greatly bring down both, the repair time and the operational cost of optical networks [8].

Nowadays, with the introduction of coherent transmission systems, OPM has proven to be a key functionality in the receiver side as most of the Digital Signal Processing (DSP) functions built-in the coherent receivers rely on the appropriate estimation of linear impairments, which requires channel parameter monitoring. Optical Signal to Noise Ratio (OSNR) monitoring is commonly realized by the DSP module of the coherent receivers. DSP also monitors pre-Forward Error Correction (FEC) Bit Error Rate (BER). Even though coherent receivers allow monitoring these parameters in the egress node, OSNR and power monitoring is still needed across the network to have full visibility of the network operation and performance.

One of the key features to be exploited in next-generation optical networks is the availability of monitoring parameters that can be used by data analytics applications, especially those based on ML [12]. This paves the way for future autonomic networking as defined in [13], including self-protection and self-healing.

8.1.1 Failure detection and localization

Many research efforts have been dedicated to developing failure localization techniques for *hard failures*, i.e., unexpected events that suddenly interrupt the established connections (see, e.g., [9]). Nonetheless, the identification and localization of *soft failures*, i.e., events that progressively degrade the QoT, remains rather unexplored. Owing to the fact that soft failures might eventually evolve to hard failures, it is of paramount importance not only to detect them a priori, before connections disruption, but also to localize their cause in order to take the proper action, e.g., finding a restoration path for the affected connections avoiding the element in failure [14]. In this chapter we focus on soft-failure failures impacting the optical spectrum of the signals, like Laser Drift (LD), Filter Tightening (FT), and Filter Shift (FS), which progressively degrade the Quality of Transmission (QoT) of optical connections.

Aiming at detecting traffic anomalies in packet networks, the authors in [1] proposed bringing data analytics toward the network nodes to reduce the amount of monitoring data to be conveyed to the control and management plane, while improving detection times. Following such idea, the authors in [15] and [7] proposed a distributed Monitoring and Data Analytics (MDA) framework [10] that includes MDA agents running close to the observation points in the network nodes, as well as a centralized MDA controller running in the control and management plane besides the Software Defined Network (SDN) controller. Such MDA framework is the base to build autonomic optical networks, especially in the case of utilizing white-boxes, which might include specific optical monitoring devices [16].

Recently, the authors in [2] proposed several solutions to monitor the performance of lightpaths at the transponder (Tp) side to verify their proper operation, as well as to detect BER degradations, thus, anticipating connection disruptions. The authors studied several soft-failure causes affecting signal QoT and proposed algorithms to detect and identify the most probable failure. Some of these failures happen in the optical switching intermediate nodes, so monitoring the signal solely at the end nodes does not allow their localization. Therefore, monitoring techniques to analyze and evaluate QoT in-line are required.

As FS and FT failures noticeably affect the optical spectrum of the lightpaths in Spectrally-Switched Optical Network (SSON), OSAs can be used to monitor the spectrum along the transmission line aiming at detecting and localizing those type of failures. Practically speaking, the realization of such solutions becomes possible with the emergence of a new generation of compact, cost-effective OSAs with sub-GHz resolution in the form of optical components [5] allowing real-time monitoring of the optical spectrum of the lightpaths and their corresponding OSNR.

Regarding Filterless Optical Networks (FON), they have recently attracted significant attention as a cost-effective metro solution to interconnect 100G coherent-based nodes in a drop and waste network architecture [18]. FONs can also perform very well in small size regional and submarine transmission networks. In contrast to filtered optical networks, where the signals are dropped at their destination, they continue to spread over the transmission line in FON. This fact leads to spectrum waste and efficiency penalties. On the other hand, since the operating lightpaths do not pass through filtering nodes, FONs can be considered as a kind of *gridless* network where frequency slots are not rigidly defined.

The inherent *gridless* potential of FON could allow the channels to be placed very close to each other aiming to alleviate spectrum waste. Nonetheless, the bottleneck is that (un)intentional Laser Drift (LD) of a transponder, possibly due to failure or its misconfiguration, might lead to overlapping of neighboring channels. This contrasts with SSON, where LD effects are much more moderate. In addition to the overlapping issue, LD could have a detrimental impact on the lightpath itself due to detuning w.r.t. the receiver optical filter. While Coherent Detection (CoD) receivers can easily track the central frequency of the transmitted signal by evaluating the offset with local oscillator, Direct Detection (DD) systems detect only the intensity of the optical signal and can hardly estimate the LD.

Thus, regardless of the transmission technology (i.e. CoD or DD), *signal overlap* is a critical issue for FON. However, the *misalignment* between the signal and the receiver filter happens only in DD systems.

Therefore, cost-effective approaches to monitor FONs are needed to allow network operators to take prompt actions in case of improper operation of a device in their domain. Most of the current surveillance systems rely on the capabilities of coherent receivers to collect measurements [4]. Regardless of their complexity, these approaches limit the performance of surveillance systems, which are intended to monitor the whole domain in real-time with the minimum extra cost and complexity. Moreover, such measurements are not available if DD systems are exploited.

Similarly as for the detection of soft-failures in SSONs, the availability of OSAs deployable in the optical nodes opens a new horizon for the development of surveillance platforms that can benefit from the analysis of optical spectrum. In this regard, optical spectrum monitoring, with the help of cost-effective OSA, can be considered as a novel solution for monitoring the proper operation of lightpaths in FONs.

One of the key parameters determining the cost of the OSAs is their resolution, which has a significant impact on the accuracy of optical spectrum-based detection and identification solutions. Therefore, a target OSA resolution should be studied. It is worth mentioning that while traffic broadcasting is seen as a drawback of FONs; it allows using one single OSA installed in the last span to acquire all signals in the FON. Therefore, remarkable cost savings can be achieved by properly delimiting their target resolution.

Considering the above-mentioned challenges of FONs and the benefits brought about by monitoring optical spectrum using OSAs, it is of paramount importance to devise robust and reliable monitoring solutions that allow FON to exploit novel CoD and DD based transmission systems while go into operation with lower margins.

8.1.2 Optical spectrum

Monitoring the optical spectrum of lightpaths requires Optical Spectrum Analyzers (OSA) to be controlled in the networks' nodes. Although, it was not financially justifiable to use such devices as monitoring tools in the optical networks [5], with the emergence of compact cost-effective coarse-granular OSA, it is becoming viable to consider this monitoring scheme in the optical networks.

Optical spectrum acquired by OSAs is represented by a set of <*frequency, power*> pairs, in which the frequency interval of this vector represents the monitoring resolution of OSAs. Fig. 8.1 presents an optical spectrum with two neighboring Quadrature Phase-Shift Keying (QPSK) 100 Gb/s signals (s1 and s2) acquired by an OSA. In particular, signal s1 in Fig. 8.1 was generated using an experimental system while signal s2 was generated by a commercial one. Note that a band guard is left between both signals to avoid them to overlap.

In general, QPSK and Quadrature amplitude modulation (QAM) optical signals present, once filtered, a flat spectral region around the central frequency, sharp edges, and a round region between the edges and the central one. Thus, when a signal is properly configured, its central frequency should be around the center of the assigned frequency slot to avoid filtering effects, and it should be symmetrical concerning its central frequency.

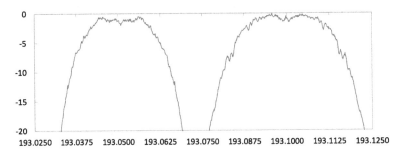

Figure 8.1 Real QPSK optical spectrum acquired by an OSA.

As real measurements are not always available, simulators such as VPI [17] allow to generate and measure the optical spectrum at different points of a testbed. Besides, different OSA granularities can be emulated, as it can be seen in the different plots of Fig. 8.2.

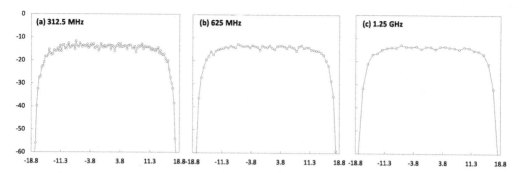

Figure 8.2 VPI Simulation of a QPSK optical spectrum acquired by an OSA for different granularities: (a) 312.5 MHz, (b) 625 MHz, and (c) 1.25 GHz.

8.1.3 Failures affecting the optical spectrum

As this chapter focuses on the detection and the identification of above-mentioned *soft-failures* with the help of monitoring optical spectrums, we review them in more detail below.

When the signal is properly configured, its central frequency should be around the center of the assigned spectrum slot to avoid filtering effects, and it should be symmetrical with respect to its central frequency. However, any of the following failures results in slight replacement of optical signal or its improper filtering.

- *Laser drift* (LD) happens due to the instability in the transmitter, which leads to certain amount of central frequency shift of the lightpath. Such central frequency shift with respect to the assigned slot may result in performance degradation. LD may happen in SSONs or FONs. This phenomenon may result in the *overlapping* of two neighboring signals; a critical problem in FONs. Additionally, in the direct detection systems, LD may result in the *misalignment* of the transmitter and the filter of the receiver.
- *Filter tightening* (FT) may arise due to misconfiguration of the filter, which drives the edges of the optical signal to get noticeably rounded. Tight filtering suppresses both ends of a lightpath and results in QoT degradation.
- *Filter shift* (FS) makes the optical signal to become asymmetrical with respect to its central frequency. Such asymmetric filtering becomes worse when the signal passes through a cascaded of filters and eventually results in QoT degradation.

Owing to the fact that soft failures might eventually evolve to hard failures, it is of paramount importance not only to detect them a priori before connections disruption but also to localize their cause in order to take proper action, e.g., finding a restoration path for the affected connections avoiding the element in failure [14].

Wavelength Selective Switches (WSS) perform optical routing and switching operations at the intermediate nodes in a filtered optical network. When a lightpath passes

Figure 8.3 Optical spectrum for a normal one (a), one experiencing FS (b), and FT (c).

through a chain of WSSs (which is typical in mesh optical networks where a route comprises of several switching nodes), it experiences a phenomenon known as *filter cascading*. Due to filter cascading, the effective bandwidth (bandwidth at –3 dB) of the lightpath reduces. This phenomenon deforms the optical spectra of the lightpaths and has a significant impact on optical spectrum monitoring solutions proposed in this chapter. For illustrative purposes, the effect of three different levels of filter cascading is shown in Fig. 8.3. Additionally, Fig. 8.4 shows how the effective bandwidth of the signal decreases w.r.t. number of cascaded WSSs.

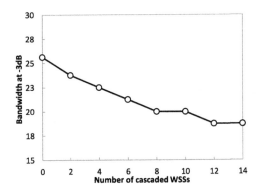

Figure 8.4 The evolution of -3 dB bandwidth of the signal w.r.t. cascading level (b).

8.2. Feature-based spectrum monitoring

In this section, we start exploring the benefits of analyzing the optical spectrum of lightpaths for soft-failure detection and identification in Spectrum Switched Optical Network (SSON). We present a framework exploiting ML-based algorithms that uses descriptive models of the optical spectrum of a lightpath in different points along its route to detect whether the optical signal experiences anomalies reflecting a failure in the intermediate WSSs. Our proposal targets the two most common filter-related soft-failures; Filter Shift (FS) and Filter Tightening (FT), which noticeably deform the expected shape of the optical spectrum. In this regard, filter cascading is a key challenge as it affects the shape of the optical spectrum similarly to FT. The approaches are specifically designed to avoid the misclassification of properly operating signals when normal filter cascading effects are present. Decision Tree (DT) and Support Vector Machine (SVM) algorithms are considered as candidate ML algorithms to perform the classification task. Extensive numerical results are ultimately presented to compare the performance of the proposed approaches in terms of accuracy and robustness.

8.2.1 Motivation and objectives

Failure identification and localization can reduce failure repair times greatly. There have been several works targeting monitoring the performance of lightpaths at the Tp side to verify their proper operation, as well as to detect BER degradations thus, anticipating connection disruptions. However, monitoring the signal at the egress node does not allow localizing failures and therefore, monitoring techniques to analyze and evaluate QoT in-line are required. In this regard, the availability of cost-effective OSAs, integrable in the optical nodes, allows real-time monitoring of the optical spectra of the lightpaths. Therefore, optical spectrum features can be exploited by ML-based algorithms to detect degradations and identify failures.

In this section, we study two different approaches to detect filter related failures. The approaches are based on a set of classifiers that make predictions using meaningful features extracted from the optical spectrum. These approaches can be considered to deal with filter cascading effects: *i*) the *multi-classifier approach*, in which different classifiers are employed for signals experiencing different levels of filter cascading and *ii*) the *single-classifier approach*, in which the lightpaths' features are pre-processed to compensate for the filter cascading effect allowing the use of a single classifier for lightpaths disregarding the level of filter cascading. Ultimately, the optical spectrum analysis can be used by centralized algorithms able to localize failures in the network.

The proposed approaches for failure detection and identification can be deployed in the agents, close to the devices generating measurements, whereas other algorithms, including the one for failure localization, need to be deployed in the MDA controller, so as to provide the global network vision required.

8.2.2 OSA for soft-failure detection and identification

Real-time optical spectrum monitoring provides opportunities for soft-failure detection and identification; particularly, those failures significantly deforming the optical spectrum of a lightpath. For precise detection and identification, algorithms need to be capable of classifying a properly operating lightpath from a failed one, which entails that a set of descriptive features should be identified for classification purposes. Building upon such features, ML-based classifiers can be trained to perform the classification task.

8.2.2.1 Soft-failure detection, identification, and localization

The failure detection, identification, and localization process involves modules running in the MDA agents and modules running in the MDA controller, as shown in Fig. 8.5. In the MDA controller, the FailurE causE Localization for optIcal NetworkinG (FEELING) algorithm is primarily responsible for supervising the failure detection and identification modules running in the MDA agents. Ultimately, it performs the failure localization task.

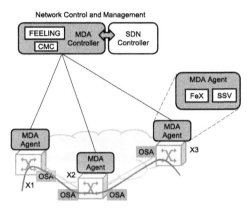

Figure 8.5 Architecture for failure identification and localization.

Optical spectra acquired by OSAs are collected and become available in MDA agents, where it is used to feed the Feature Extraction (FeX) module. An example of a 30-GBaud QPSK modulated optical spectrum acquired by an OSA of 312.5 MHz resolution is shown in Fig. 8.6.

The FeX module processes the acquired optical spectrum for a given lightpath, which consists of an ordered list of frequency-power ($<f, p>$) pairs in the allocated frequency slot. After equalizing power, so the maximum power is set to be 0 dBm, the derivative of the power with respect to the frequency is computed; Fig. 8.6(b) illustrates the derivative of the example optical signal, where sharp convexity can be observed close to the edges. Next, the FeX module characterizes the mean (μ) and the standard deviation (σ) of the power around the central frequency ($fc \pm \Delta f$), as well as a set of primary features computed as cut-off points of the signal with the following power levels:

- edges of the signal computed using the derivative, denoted as ∂,
- a family of power levels computed w.r.t. μ-$k\sigma$, denoted as $k\sigma$, and
- a family of power levels computed with respect to μ-mdB, denoted as $-mdB$. Each of these power levels generates a couple of cut-off points denoted as $f1_{(\cdot)}$ and $f2_{(\cdot)}$.

In addition, the assigned frequency slot is denoted as $f1_{slot}$, $f2_{slot}$. Other features, which are computed as linear combinations of the relevant points, focus on characterizing a given optical signal (see embedded equations in Fig. 8.6(a)); they include: *i*) bandwidth (*bw*) computed as $bw_{(\cdot)}=f2_{(\cdot)}-f1_{(\cdot)}$; *ii*) central frequency (*fc*) computed as $fc_{(\cdot)}=f1_{(\cdot)}+0.5*bw_{(\cdot)}$, as well as the shifting of the central frequency $\Delta fc_{(\cdot)}=fc_{(\cdot)}-fc_{(slot)}$; and *iii*) symmetry with respect to a reference (frequency slot or derivatives), computed as $sym_{(\cdot)\text{-ref}}=(f1_{(\cdot)}-f1_{ref})-(f2_{ref}-f2_{(\cdot)})$.

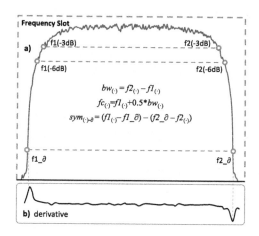

Figure 8.6 Relevant points of a QPSK modulated signal.

Some features are more appropriate for filter-related failure detection and identification, such as bandwidth and symmetry, whereas other features, such as the central frequency, are more appropriate for laser drift identification. For illustrative purposes, Fig. 8.7 shows how some of the identified features evolve with the severity levels of FS and FT. Fig. 8.7(a, c) show the evolution of bw_{-3dB}, bw_{-6dB}, $sym_{-3dB-\partial}$, and $sym_{-6dB-\partial}$ w.r.t. the magnitude of FS; while Fig. 8.7(b, d) show the evolution of those features w.r.t. magnitude of FT. It can be interpreted that $sym_{-3dB-\partial}$ and $sym_{-6dB-\partial}$ are meaningful features to identify FS, while bw_{-3dB} and bw_{-6dB} are more sensitive to the effects of FT.

When the extracted features from the measured signal are available, a classification module, named Signal Spectrum Verification (SSV), running also in the MDA agents analyzes them to detect a soft-failure. The SSV module is implemented as a multiclass classifier that produces a diagnosis, which consists of: *i*) a predicted class among '*Normal*', and '*FilterFailure*'; and *ii*) a subset of relevant signal points for the predicted class. In the case that a filter failure is detected, another classifier is used to predict whether the failure is due to FS or to FT.

As discussed before, one of the key challenges in the identification of filter related failures is the misclassification of a normal signal that has passed through several filters, i.e., affected by filter cascading, as a signal that has suffered from filter failure. Therefore, to improve failure identification accuracy, the FEELING algorithm must be able to

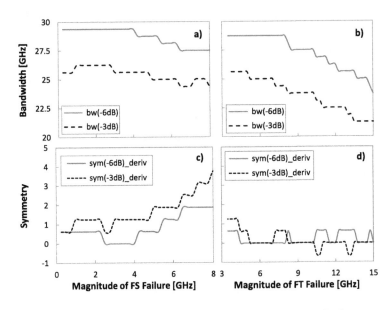

Figure 8.7 The evolution of the features w.r.t. magnitude of FS (a, c) and FT (b, d).

distinguish between actual failures and normal effects arising from filter cascading. In the next section, we propose and study two approaches to prevent such misclassification.

8.2.2.2 Options for classification using FeX

Let us first explain how the FeX relevant features can be used to classify different types of spectra. Fig. 8.8 shows how even a pair of such features can discriminate different types of spectra. Assuming a set of measurements after 2 WSSs and belonging to normal operation, FS, and FT, Fig. 8.8(a–b) show $sym_{-3dB-\partial}$ w.r.t. bw_{-3dB} and $sym_{-6dB-\partial}$, respectively. As represented, observations belonging to different classes can be easily discriminated with even just these two features in place. Now, let's take the same set of observations after 12 WSSs; the results are plotted in Fig. 8.8(c–d). As shown, it becomes very challenging to distinguish the observations belonging to different classes. It is worth mentioning that, as a result of filter cascading, signal features change in a similar way that it happens when a FT failure takes place.

Let's take a closer look at the figures to understand this issue. As shown in Fig. 8.8(a–b), just the observations belonging to FT are gathered in the bottom left corner of the figures; however, in Fig. 8.8(c–d), almost all the observations are gathered in the bottom left corner of the figures. In other words, filter cascading effect pushes the identified features in the direction that all observations look like FT observations. This increases the likelihood of misclassifying a properly operating lightpath as a failed one. In the following, we propose two different strategies preventing such misclassification. The

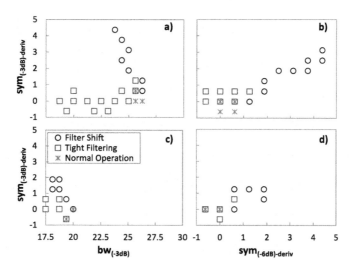

Figure 8.8 The efficiency of the identified features for classification after 2 WSSs (a, b) and after 4 WSSs (c, d).

strategies, built in the SSV module, are based on processing the features extracted by the FeX module. Selected features for classification are: bw_∂, $bw_{5\sigma}$, bw_{-3dB}, bw_{-6dB}, $sym_{5\sigma-\partial}$, $sym_{-3dB-\partial}$, and $sym_{-6dB-\partial}$.

Multi-classifier approach

The most straightforward solution is to use different classifiers as a function of the number of WSSs that a given lightpath has passed through. As shown in Fig. 8.9(a), a set of classifiers are required in every intermediate node and the appropriate one is used when an optical spectrum is acquired. This approach can be considered as the baseline, as the selected classifier decides based on the features extracted directly from the acquired spectrum and do not need any kind of feature pre-processing. However, a very large dataset of optical spectra with different levels of filter cascading is required for training all the classifiers, which is the main drawback of this approach.

To avoid using multiple classifiers, some pre-processing needs to be done so one single classifier can be used despite the level of filter cascading. The second approach proposes a strategy to pre-process the extracted features.

Single-classifier approach

Filter cascading strongly affect some of the features that a classifier uses for prediction. Therefore, if the alteration of those features due to filter cascading is compensated, a single classifier could be considered regardless of the number of filters a signal passes through. The features of a signal acquired after passing N filters can be compensated by

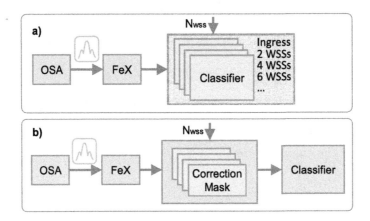

Figure 8.9 Feature-based approaches: a) multi-classifier, b) single-classifier.

adding/subtracting the differences between the values of a properly configured signal at that node w.r.t. those just after the Tp. These differences are stored in a vector called *correction mask*; note that, different levels of filter cascading require different correction masks to be used.

Correction masks can be computed a priori, assuming the effects that the spectrum of a normal signal experiences while passing through different number of filters. It is worth mentioning that the calculation of the correction masks requires just the spectrum of a single properly configured lightpath passing through the desired number of filters, from zero to the maximum allowed cascaded filters in a network; this is in contrast to the previous approach, where the training phase requires that spectral data with different failures and with various magnitudes to be captured after every filter up to the maximum allowed number of filters. The Correction Mask Calculator (CMC) module placed in the MDA controller (see Fig. 8.5) is responsible for generating the correction masks to be sent to the MDA agents. Note that all the correction masks need to be available in the MDA agents, so the proper one can be selected. Following this approach, the classifier can be trained based on the observations of a passing through just a single filter, making the training phase less data-hungry by far compared to the previous approach.

Fig. 8.10 summarizes the failure detection and identification workflow followed in both single-classifier and multi-classifier approaches. Once the optical spectrum of a signal has been acquired, the features are extracted. In case of single classifier approach, the features are corrected applying the specific correction mask that corrects filter cascading effects for the number of filters that the signal has traversed. Next, failure analysis can be carried out.

The first alternative classifier is based on DTs, whereas the second one selects SVMs. Both classifiers aim at identifying whether a filter failure is affecting a connection and if so, which is the type of failure: FS or FT. In the case that a failure has been detected, its

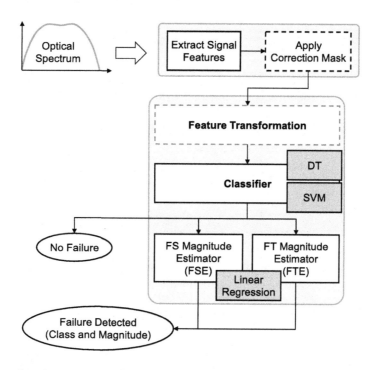

Figure 8.10 Failure detection and identification workflow.

magnitude needs to be estimated. Two filter failure magnitude estimators can be called depending on the detected failure; both are based on linear regression.

Feature transformation for single-classifier approach

The single classifier approach can be further improved if a feature transformation step is applied as shown in Fig. 8.10. For the sake of simplicity, we consider the magnitude of the failures as additional features for training the classifiers, so we use the magnitude estimator before a failure has been detected. In this way, original features are linearly combined to create new ones that might aggregate information in the hope of improving classification accuracy.

Let us now get insight about the training process of the classifiers (see pseudocode in Table 8.1). The algorithm receives a dataset of labeled examples that is firstly balanced by adding copies of instances from the class with less number points in the dataset to have the considered classes (normal, FS, FT) equally represented (line 1 in Table 8.1). A set of configurations that contain specific parameters for the classification algorithm selected will be used during the training process.

The parameters considered to fit DTs are the number of observations per leaf; for every n a DT model is obtained. As for SVM fitting, the parameters are the degree of the

Table 8.1 General classification training algorithm pseudocode.

INPUT *dataset, Configs, maxSplits*
OUTPUT *model*

1:	dataset←balanceClassesByReplication(dataset)
2:	**for each** config in Configs **do** initialize GoC[config]
3:	**for** *i*=1..*maxSplits* **do**
4:	<trainingSet,testingSet>←randomSplit(dataset)
5:	initialize configParams
6:	**for each** config in Configs **do**
7:	model ← fit(trainingSet, config)
8:	errorTraining ←predict(model, trainingSet)
9:	errorTesting ←predict(model, testingSet)
10:	*gocConfig* ← *GoC*[config].addNew()
11:	gocConfig.configParams ← config.params
12:	gocConfig.errorTraining ← errorTraining
13:	gocConfig.errorTesting ← errorTesting
14:	*bestConfig*←computeBestConfig(*GoC*)
15:	**return** fit(dataset, bestConfig)

polynomial kernel (*kernelDegree*) for complexity control and the cost of misclassifying (*misClassCost*) for the size of the SVM. For every configuration, a number of randomly-generated splits of the data set for training and testing will be performed. To store the goodness of each configuration, the *GoC* array will be used in the rest of the algorithm and it is firstly initialized (line 2). Next, a new dataset split is generated, where the training set is used for fitting a model for the classifier with the specific selected configuration (lines 3–7). Once a model is computed, predictions using the training and testing data set are carried out (lines 8–9); the training and testing errors between the model prediction and the actual values are stored in the *GoC* array together with the current configuration parameters (lines 10–13). Finally, the results obtained for the different configurations and training/testing data splits are evaluated to select the configuration with minimum error (line 14). Such configuration is eventually used to fit a model using the whole dataset to improve the algorithm performance (line 15).

8.2.3 Soft-failure localization

The FEELING algorithm running in the MDA controller is detailed in Table 8.2; FEELING is called upon the detection of excessive BER at the reception side of an optical signal. The algorithm first calls feature extraction module, in the ingress and last intermediate nodes to perform signal verification and obtain a diagnosis (lines 1–5 in Table 8.2). It is worth noting that *diagIngress* is a tuple <*class, X, features*>, where *X* is the captured optical spectrum. In the case that the diagnosis of both nodes is normal, FEELING ends with no failure detected (lines 6–7).

Table 8.2 FEELING Algorithm.

INPUT *lightpath*
OUTPUT {*<node, class, magnitude>*}

1:	*ingress* ← *lightpath*.getNodeFromRoute (1)
2:	*lastInterm* ← *lightpath*.getNodeFromRoute (-2)
3:	*FM* ← getFilterMasks(*lightpath*)
4:	*diagIngress* ← getFailureDiagnosis (*ingress, FM* (1))
5:	*diagLast* ← getFailureDiagnosis (*lastInterm, FM* (-2))
6:	**if** *diagIngress.class* = *diagLast.class* AND
	diagIngress.class = *Normal* **then**
7:	**return** {<1, Normal, ->}
8:	*XNodeChange* ← *diagIngress.X*
9:	*diagChange* = *<class, magn>* ←Classifier (*diagIngress.X*)
10:	**if** diagChange.*class*<> *normal* **then**
11:	*FailureSet*←<1, *diagChange*>
12:	**else** *FailureSet* ← Ø
13:	**for** *i*=2..*lightpath*.RouteLength()-1 **do**
14:	*node_i* ← *lightpath*.getNodeFromRoute (*i*)
15:	*Xi*← getSignalPoints (*node_i*)
16:	*diagNode_i* ← Classifier (*Xi*)
17:	**if** *diagNode_i.class* <> *diagNodeChange.class* OR
	diagNode_i.magn - diagNodeChange.magn > α **then**
18:	*XNodeChange* ← *Xi*
19:	*FailureSet* ← *FailureSet* U {<*i, diagNode*>}
20:	**return** *FailureSet*

In the case of a different diagnosis, FEELING starts a procedure to detect filter related problems at intermediate nodes using the classifiers module to compare diagnosis and magnitudes between nodes in the route of the lightpath. This process starts with the diagnosis at the ingress node that it is used as the initial reference node (lines 8–11). Then, the diagnosis of every intermediate node is compared against the one of the reference changing node and failure set is updated if either a new filter failure is detected or the magnitude increased above a certain threshold (lines 12–19). After processing all intermediate nodes, the list of failures detected is eventually returned (line 20).

8.2.4 Illustrative results

In this section, we numerically compare the performance of different approaches described in the previous sections. Firstly, the transmission set-up modeled in VPI Photonics is described; the set-up is used to generate the optical spectrum database required for training and testing the proposed algorithms. Next, the two feature-based approaches implemented as DT and SVM classifiers are evaluated thus, revealing their benefits/drawbacks for filter failure identification tasks. Finally, the performance of the proposed approaches for failure detection and localization is compared.

8.2.4.1 VPI set-up for data collection

The VPI set-up is shown in Fig. 8.11. In the transmitter side (N1), a 30 GBd DP-QPSK signal is generated, passes through 7 intermediate nodes performing optical switching and ends in a coherent receiver (N9) that compensates for the impairments introduced throughout the transmission. Nodes are interconnected by single mode fiber spans; after each span, optical amplifiers compensate for the accumulated attenuation of the fiber. The transmitter and receiver are assumed to be installed in optical nodes, which are modeled with two 2nd order Gaussian filters emulating optical switching functionality for Add/Drop and pass-through performed by WSSs [11], [6]; filters bandwidth is set to 37.5 GHz, leaving 7.5 GHz as a guard band.

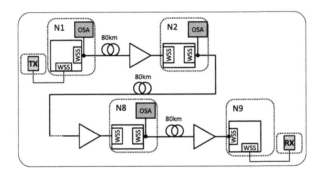

Figure 8.11 VPI setup.

One OSA per outgoing link, configured with 312.5 MHz resolution, is considered in every node to monitor the optical spectrum. As previously discussed, a correction mask should be considered for the features affected by filter cascading, as features get modified while passing through WSSs. Fig. 8.12 shows an example of the amount of reduction in the bw_{-3dB} feature of a lightpath in the set-up and the corresponding correction mask, obtained by fitting a 2nd order polynomial.

Aiming at emulating failure scenarios, we modify the characteristics of the 2nd WSS of each node (from N1 to N8) in the set-up; its bandwidth and central frequency are modified to model FT and FS failures, respectively. A large dataset of failures was collected by inducing failures of magnitude in the range [1-8] GHz for FS and in the range [1-15] GHz for FT, both with 0.25 GHz step-size, where the magnitude of FT is

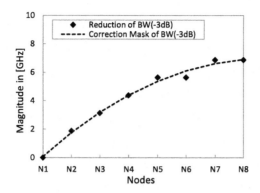

Figure 8.12 Correction mask of bw_{-3dB} of the setup shown in Fig. 8.11.

defined as the difference between the ideal bandwidth of the filter (37.5 GHz) and its actual bandwidth during the failure.

8.2.4.2 ML-based classification comparison

We compare the performance of the proposed approaches in terms of its *accuracy*, defined as the number of correctly detected failures over the total failures. Fig. 8.13 shows the accuracy of detecting FS and FT, respectively at node N1 in terms of the magnitude of the failure. Note that in N1 both multi-classifier and single-classifier are the same as no filter mask is required.

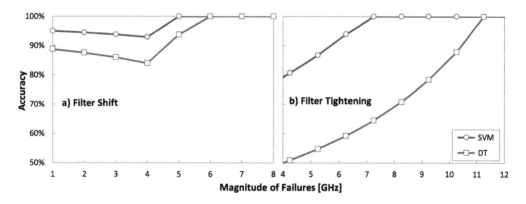

Figure 8.13 Accuracy at N1 of DT and SVM for FS (a) and FT (b).

Every point in Fig. 8.13(a–b) aggregates failure scenarios with different magnitude by considering all the observations belonging to a particular failure magnitude and above. As shown, the accuracy of detecting FS larger than 1 GHz is around ~96% when classifiers are based on SVMs, while it hardly approaches 89% when they are based on DTs. On the other hand, the accuracy of SVMs reaches 100% for failures larger than 5 GHz, while this level of accuracy for DTs is achieved for failures larger than 6 GHz. Regarding FT detection, the best accuracy of the proposed classifiers for low magnitudes (below 6 GHz) is around 80% (achieved for SVMs), which is due to the fact that the shape of the optical spectrum is quite similar to the normal scenario, making it very challenging for the classifier to distinguish. This is in contrast to the case of FS, whose effect is more evident even for low magnitudes due to its asymmetric impact on the optical spectrum. For the magnitudes above 7 GHz, the SVM-based classifier perfectly detects the failure. Note that DT-based classifiers achieve perfect accuracy for magnitudes above 10.5 GHz.

Let us now compare both approaches implemented with DT and SVM-based classifiers for detecting failures in all 8 nodes of the set-up. Recall that multiple classifiers are needed for the first approach and several filter masks are required for the second approach. The results are shown in Fig. 8.14(a-b) for FS and FT, respectively, where every point aggregates the results for all the nodes. As observed, SVM-based classifiers significantly outperform DT-based ones in both approaches and failures. As a result, SVM-based classifiers can be selected as the preferred option for feature-based approaches.

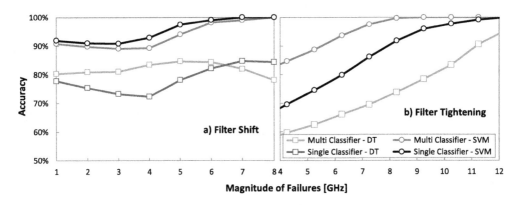

Figure 8.14 Average accuracy over N1-N8 of DT and SVM for FS (a) and FT (b).

Comparing the different SVM-based approaches, the single-classifier performs slightly better in the case of FS detection while the performance of the multi-classifier approach is much better than the single-classifier one in the case of FT failure. Therefore, we can conclude that training multiple classifiers with the data collected at nodes experiencing different levels of filter cascading performs better than correcting the features with the purpose of using a single classifier, as the impact of filter cascading is similar to the effect of FT on the shape of optical spectrum.

8.2.4.3 Benefits of using a single OSA

In this part, considering the single classifier-based approach, we focus on detecting the failures in some nodes after the point where the failure happens, showcasing the *efficiency* of the proposed methods with respect to the evolution of the optical signal along the transmission path. In addition, by following this approach, the number of utilized OSAs in the network can be reduced. Fig. 8.15 shows the minimum magnitude after which the accuracy of classifiers remains 100% in terms of the location of OSA compared to the point that failures happen; 0 on the x-axis means that OSA is placed at the node where the failure happens (N1 in Fig. 8.11), while 7 means that is placed 7 nodes away from the location of the failure (N8 in Fig. 8.11). It can be understood that the SVM-based classifier is more effective regardless of the location of the OSA and it perfectly detects the failures above a magnitude threshold. Even though the DT-based classifier shows an acceptable performance for FS failures up to 3 nodes distance from the location of the failure, it fails when considering FT failures.

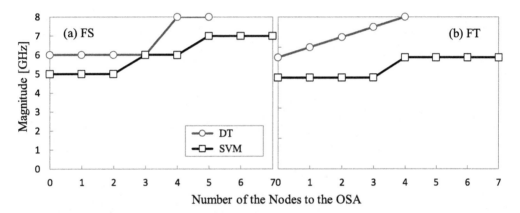

Figure 8.15 Robustness of the ML algorithms while one OSA is considered.

8.2.4.4 Benefits of feature transformation for classification

Once the failures are detected, Filter Shift Estimator (FSE) and Filter Tightening Estimator (FTE) can be launched to return the magnitude of the failures (see Fig. 8.10); estimators are based on linear regression. Estimated values of FS and FT with respect to their expected values are illustrated in Fig. 8.16a and Fig. 8.16b, respectively. As shown, the estimators can predict the magnitude of failures with very high accuracy, with Mean Squared Error (MSE) equal to 0.09091 and 0.00583 for FSE and FTE, respectively.

In addition to the use of these estimators to explore the magnitude of the failures, they can be used in the feature transformation step as anticipated in Section 8.2.2.2. In fact, the output of FSE and FTE can be considered as two principal components of an imaginary two-dimensional vector space as shown in Fig. 8.17. In such space,

FS and FT failures evolve in different directions of the vector space. As illustrated, the observations belonging to normal operation and the small magnitudes of the failures coincide. However, they become perfectly distinguishable as the magnitude of failures increases.

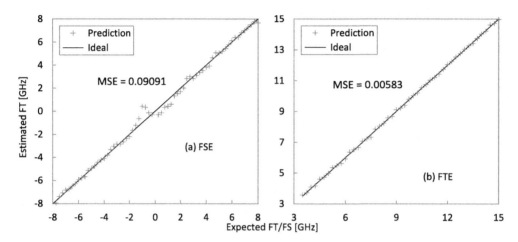

Figure 8.16 Prediction accuracy of FS (a) and FT (b) estimators.

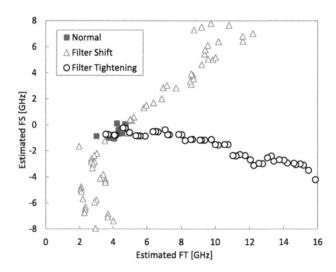

Figure 8.17 Estimated FS vs estimated FT as a two-dimensional vector space.

Let us evaluate the benefits of exploiting the outputs of FSE and FTE estimators as additional features for training the classifiers; Fig. 8.18 presents the obtained results. For the sake of conciseness, we group the magnitudes into three groups of low (L), medium

(M), and high (H) magnitudes, instead of reporting all of them independently. Regarding the location of the OSAs, we report just three locations. Analyzing Fig. 8.18(a), one can realize that the accuracy of DTs can be substantially improved, notably for low and medium magnitude, when using the estimations of FSE and FTE as new features. Additionally, it makes the classifier based on DT more robust while using OSAs far away from the location of the failures. However, it yet cannot outperform the classifier based on SVMs, even with these additional features. We also see that adding these new features does not enhance the performance of SVMs, revealing that such classifier can internally exploit the primary features to the maximum extent. Therefore, the classifier based on SVMs does not require an additional preprocessing step to generate more features; note that the magnitude of the failures are just linear combinations of primary features. Conversely, the substantial improvement seen in the DT classifier reveals that DTs cannot maximally exploit the information carried by the primary features and requires some pre-processing to grasp more information, which is a weakness of DTs compared to SVMs.

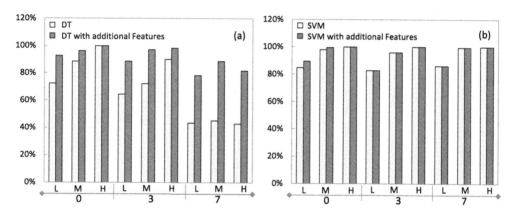

Figure 8.18 Comparisons of classifiers with and without additional features.

8.2.4.5 Failure localization

Let us evaluate the performance of FEELING. Fig. 8.19(a) and Fig. 8.20(a) illustrate localization accuracy for FS and FT, respectively when considered filter mask correction. Accuracy in terms of the proportion of correct localizations is provided as a function of the magnitude of the failure, the conclusion is that, as soon as the failure magnitude increases, localization accuracy also increases. For FS higher than 5 GHz (Fig. 8.19(a)) and for FT smaller than 28 GHz overall accuracy reaches 100%.

Finally, it is important to recall that FEELING is triggered upon excessive BER is detected in the reception of a lightpath. Assuming a BER increase due to a gradual degradation of a filter, Fig. 8.19(b) and Fig. 8.20(b) are provided to illustrate the relation between BER change detection thresholds and failure localization. These figures depict the simulated BER as a function of failure magnitude. For illustrative purposes, let us imagine that, due to two different configurations, excessive BER is detected at $8 \cdot 10^{-5}$ and $5 \cdot 10^{-4}$. Without entering into details, the former could correspond to a BER threshold violation anticipation while the latter could represent an actual threshold violation. A BER equal to $8 \cdot 10^{-5}$ could correspond to a degraded filter shifted around 4 GHz or narrowed until 32 GHz, a failure that is localized with an accuracy around 90%. On the other hand, BER equal to $5 \cdot 10^{-4}$ is obtained for failures whose magnitude is large enough to perfectly localize them. Hence, BER degradation and failure identification and localization must be configured with a global perspective to achieve optimal overall performance.

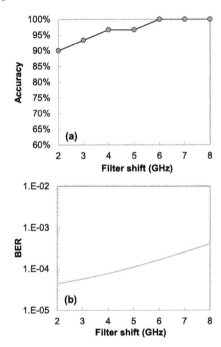

Figure 8.19 FEELING performance for *FS*.

8.2.5 Conclusions

In this section, we have studied the benefits of exploiting OSAs for identification and localization of filter related failures. Two different approaches for filter-related soft-failures detection and identification have been proposed and their performance based on two different ML algorithms was compared in terms of accuracy and robustness.

Multi-classifier and single-classifier, even though their performance is comparable, have notable differences in their implementation complexity. While the multi-classifier

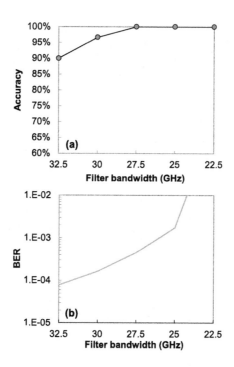

Figure 8.20 FEELING performance for *FT*.

approach requires a huge dataset for the training phase, single–classifier approach re-quires N times less data, being N the maximum number of nodes an optical connection might pass-through. However, this is at the cost of pre-processing optical spectrum fea-tures, which requires the calculation of the correction masks. Table 8.3 summarizes the difference of the two proposed approaches.

For failure localization, a ML based algorithm (called FEELING) was proposed, taking advantage of continuous monitoring of the optical spectrum using cost-effective OSAs. The results showed that FEELING identifies/localizes filter related failures with accuracy above 90%.

In view of that the accuracy of feature-based approaches presented in this section when the cascading is present, the next section studies a different approach and proposes to circumvent the filter cascading issue in which the optical spectral themselves are preprocessed prior to interpreting them to find failures.

8.3. Residual-based spectrum monitoring

In this section, we continue with the exploration of the benefits of analyzing the optical spectrum of lightpaths for soft-failure detection and identification. We present a novel approach called *residual-based* that uses the residual signal computed from sub-

Table 8.3 Key characteristics and results for the considered approaches for failure detection and identification.

	Pre-processing	Classif. method	Training phase	Number of classifiers	Availability at the node level	Accuracy	Robustness w.r.t. # of nodes
Multi Classifier	not required	SVM	requires observations of every level of filter cascading	# of nodes to support	all classifiers	good	good
Single Classifier	pre-processing of the features	SVM	requires observations of just a single level of filter cascading	1	one classifier + correction masks	good	good

tracting the signal acquired by OSAs from an expected signal synthetically generated. Similar to the single classifier based approach presented in the previous section, the residual-based approach requires less data for training the classifier; instead it requires two new modules to be located in the MDA agents to compute the expected signal and the corresponding residuals.

We further extend this approach to facilitate ML algorithm deployment in real network. A two-phase strategy is proposed: *Out-of-field training* is expected to use data from simulation and testbed experiments with generic equipment whereas *In-the-field* adaptation is applied to support heterogeneous equipment. Extensive numerical results are ultimately presented to compare the performance of the residual-based approach with the feature-based approaches proposed in the previous section in terms of accuracy and robustness. We also demonstrate the efficiency of residual adaptation mechanism for a particular use-case.

8.3.1 Residual-based approach for optical spectrum analysis

An alternative approach to resolve the filter cascading issue, compared to the ones in the previous section, is to pre-process the acquired optical spectrum by comparing it to the one that would be expected after passing the same number of filters (see Fig. 8.21). This comparison/computation produces a *residual signal* representing the differential deformation in its shape that might be due to a failure. Note that this approach does not use the FeX module proposed in the previous section. In order to compute the residual signal, we consider two new modules, as shown in Fig. 8.21: *i*) the Expected Signal Calculation (ESC) and *ii*) the residual computation module.

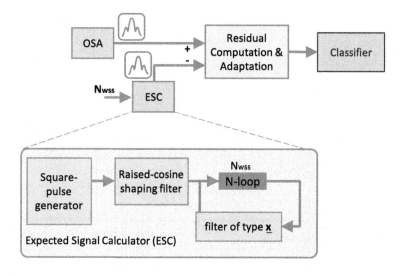

Figure 8.21 Soft-failure detection based on residual analysis.

The ESC module generates a theoretically-calculated optical spectrum emulating a properly operating lightpath. The aim of the ESC module is to synthetically reproduce an averaged noise-free version of the optical signal. In order to do so, the signal is modeled as an ideal square pulse, with bw_{-3dB} equal to the baud rate of the optically modulated signal, shaped by a raised-cosine shaping filter with 0.15 roll-off factor. Then, in order to model different levels of filter cascading, a 2nd order Gaussian filter, emulating a WSS, is used (Fig. 8.21). This, results in an emulated noise-free spectrum, similar to a noise-free 100G DP-QPSK (or 200G DP-16QSPK) modulated signal (Fig. 8.22). Every time a new optical spectrum is acquired for a given lightpath, the residual computation module subtracts it from the expected signal's spectrum produced by the ESC module.

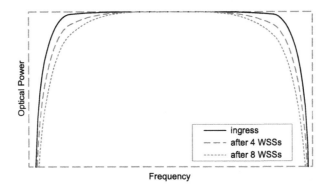

Figure 8.22 Synthetically generated expected signals' spectra.

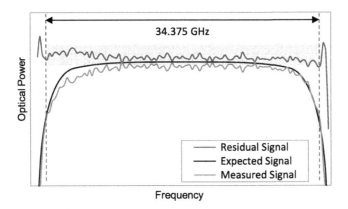

Figure 8.23 Residual signal calculation.

Fig. 8.23 presents an illustrative case of the computed residual signal from the expected and the measured signals' spectra. As the residual signal experiences undesired

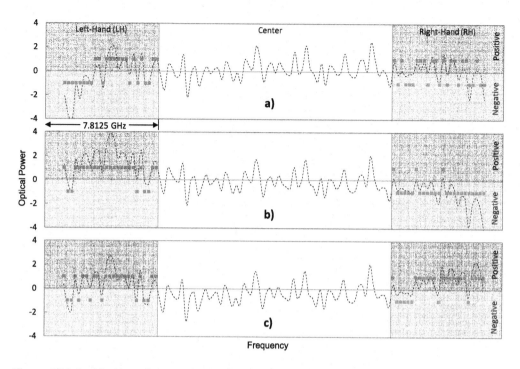

Figure 8.24 Residual based approach workflow considering a normal signal (a), FS (b), and FT (c).

changes at the two ends of the acquisition window of the OSA, we just consider a central spectral window of size 34.375 GHz out of 37.5 GHz. Note that in case of a normal signal, the residual value fluctuates around a mean value along the whole signal spectrum range. To analyze the residual signal, we normalize the values, so as the mean equals 0; in such case, the most likely situation is to have as many positive as negative values. However, in the event of a filter failure, that similar proportion between positive and negative residuals will be altered.

Fig. 8.24 shows an example of how residuals behave in all considered cases. Out of the whole range of the residuals, the left and the right hand-sides capture the effects of soft-failures and are the operational regions for the analysis presented next. The spectral window of the left (LH) and the right (RH) hand-sides are set to 7.8125 GHz, as they contain the sufficient number of points to capture the effects taking place in the edges. In the normal case (Fig. 8.24a), the residuals oscillate uniformly between positive and negative values (dots in Fig. 8.24 are computed from the normalized residuals applying the *sign()* function). In the case of filter shift (Fig. 8.24b), the residuals show a clear distortion toward positive and negative parts in the LH and RH side, respectively. Finally, in the case of filter tightening (Fig. 8.24c), the residuals move toward the positive part in both the LH and RH sides.

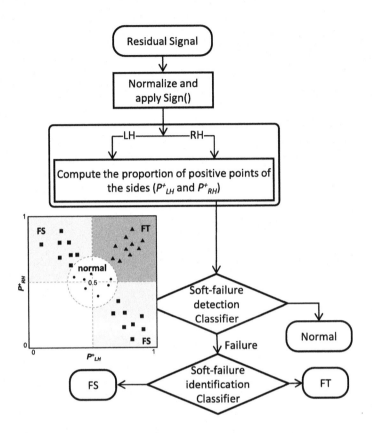

Figure 8.25 Residual-based classification procedure.

In conclusion, comparing LH and RH sides, one can predict whether the signal is normal (symmetric sides and unbiased distribution) or either it is affected by filter shift (asymmetric sides) or a filter tightening (symmetric and biased distribution).

In light of this analysis, we propose the residual-based procedure presented in Fig. 8.25, where the residuals are first normalized with respect to their mean value and centered in zero; then, the *sign*() function is applied to convert normalized residuals into points of amplitude +1 or -1. After selecting the sides, the proportion of positive points in the LH and the RH sides (P_{LH}^+ and P_{RH}^+, respectively) is computed. For illustrative purposes, Fig. 8.25 plots possible observations in the semi-plane P_{LH}^+, P_{RH}^+. According to the rationale previously presented, normal observations should be kept within an area centered around $(0.5, 0.5)$, FS observations should be in the quadrants $(>0.5, <0.5)$ or $(<0.5, >0.5)$, while FT observations should be in the quadrant $(>0.5, >0.5)$, both FS and FT outside the normal area. Therefore, the coordinates of the observations in the semi-plane P_{LH}^+, P_{RH}^+ can be used as features for DT and SVM-based classifiers. In the proposed procedure, two classifiers are trained: the first one for soft-failure detection and the second one for its identification.

8.3.2 Facilitating ML algorithm deployment using residual signals

The premises described in the previous section consider one single filter type, which limits the deployment of ML approaches to real operator networks that usually consist of equipment from different vendors. The most straightforward solution to overcome this limitation is to have different models being trained upon various types of filters that might be available in the network. Nonetheless, it makes the training phase very complex and data-hungry. Yet, it will not be easy to comprehend the sequence of filters a priori and the responses of a slightly non-identical filter in the network might not be very well detected, necessitating even more combination of models to have an appropriate generic model.

Here, we propose to extend the premises described in the previous section to facilitate ML algorithm deployment in real networks; it consists in:

- training one single accurate and robust ML model based on simulations and/or experiments carried out in laboratory or testbed facilities and
- devising a proper adaptation mechanism that makes adjustments on the data for the specific signal being analyzed, which might have traversed different filter types along its route from the transmitter.

Note that this strategy also facilitates the introduction of new filter types, as current vendors deploy new equipment releases in the network. The *residual-based* approach, due to its dependency on the synthetic behavior of the filter responses, has potential characteristics to get adapted to different types of filter.

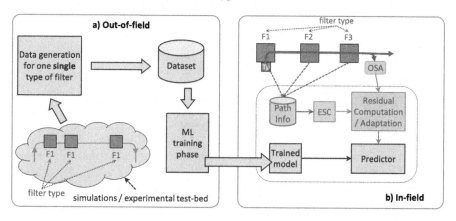

Figure 8.26 Out-of-field ML training and in-field model adaptation.

Fig. 8.26 illustrates the proposed out-of-field ML training and in-field model adaptation strategy. Scenarios with one single type of filters (labeled F1 in Fig. 8.26a) are considered in simulation and/or lab experiments to produce a large dataset that is used for ML training purposes. When the ML model is deployed in the field, an adaptation procedure takes into account the specific types of filter that a given signal has passed through (Fig. 8.26b). The in-field adaptation is performed in *i*) the ESC module by considering the specific filters that the signal has passed through; see three filter transfer functions in Fig. 8.22, and *ii*) the residual computation module that normalizes and adapts the residuals for the signal under analysis.

In the procedure described in Section 8.3.1, the calculated residual is normalized with respect to the mean value of the central part of the residual, so the mean becomes 0. This normalization approach is operational when the same type of filter exits in both out-of-field training and in-the-field operation of the ML algorithm. However, when applied to other filter types it does not work well. In view of this, we propose an adaptation procedure (Fig. 8.27) that consists in dividing the residual signal in three segments (Fig. 8.28) and apply different normalization methods to every segment, reflecting the filter characteristics; the normalization reference of every segment is obtained by applying linear regression to the un-normalized version of the residual signal obtained for that segment.

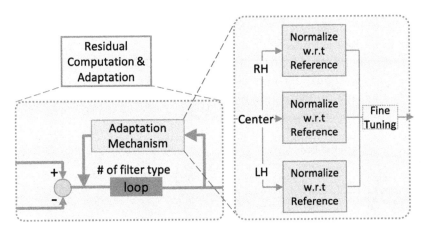

Figure 8.27 Residual computation and adaptation mechanism.

When the signal arrives to the residual computation and adaptation module, the module also receives three different normalization references considering the filter characteristics in the form of polynomials of order 1. For this stage, the number of adaptation mechanism loop equals to the number of filter types that the lightpath has passed through. By subtracting every segment of the un-normalized residuals from the corresponding normalization reference, a filter-agnostic residual signal is obtained.

Note that, as the amount of filter cascading effect depends on the transfer function of the filter, there might be an undesirable power shift in the residual signals when the lightpath traverses different filter types. This power shift is compensated in the fine-tuning step. The amount of power shift can be computed locally assuming that the mean value of the residual remains zero when the signal is in proper operation mode. Ultimately, a single classifier trained with the measurements collected in the lab based on a reference filter type, can be used for optical spectra experiencing filtering effects from different types of filter.

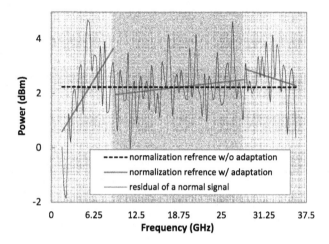

Figure 8.28 Normalization references for the 4th order Gaussian.

The result of this adaptation method is illustrated in Fig. 8.29. The residual signals of a lightpath passing through three different types of filters with Gaussian transfer function of order 2, 3, and 4 are illustrated in Fig. 8.29a. Normalization shifts the residuals so its mean to be 0 (Fig. 8.29b). Note that the differences among residuals are clearly seen at the edges, whereas they are virtually identical in the central part before and after normalization. Adaptation focus on compensating the effects of the different filters and the results are clearly visible at the edges (Fig. 8.29c); note that the most relevant parts of the residuals to detect filter-related soft-failures are that of the edges. As shown, even though the signals pass through different types of filters, they result in an identical residual signal, removing the filter-dependent characteristics of the residual signal.

8.3.3 Illustrative results

In this section, we first numerically compare the performance of the residual-based approach described in Section 8.3 with the feature-based approaches presented in Section 8.2. The same transmission set-up described in Section 8.2.4.1 is used here to generate the optical spectrum database required for training and testing the proposed algorithms.

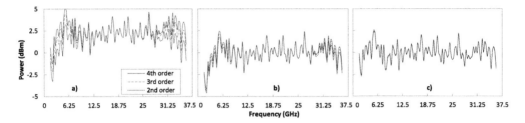

Figure 8.29 Un-normalized residual (a), normalized w/o adaptation (b), and normalized with adaptation (c).

8.3.3.1 Comparison of residual-based and feature-based approaches

Let us first focus on the residual-based approach and analyze the potential of using P_{LH}^+ and P_{RH}^+ as discriminatory features for training the classifiers. Fig. 8.30 plots the samples of the dataset used for training the residual-based classifiers at N1. It is clear, in view of Fig. 8.30, that the selected features allow classifying the different cases easily. Although not shown in detail, we first compared DT and SVMs classifiers and concluded that SVMs perform better also for this approach. Therefore, the results presented next compare the performance of all three approaches using SVM-based classifiers.

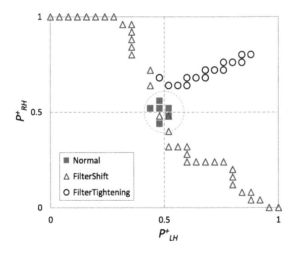

Figure 8.30 P_{LH}^+ and P_{RH}^+ features for failure detection and identification.

Fig. 8.31a-b show the accuracy of detecting FS and FT solely at node N1. As illustrated, multi-classifier and single-classifier approaches show identical performance as they are essentially the same at node N1, where no filter mask is necessary to be used. However, the residual-based approach outperforms feature-based approaches for the detection of FS and FT. The performance of the residual-based approach is noticeable for detecting soft-failures, as it reaches 100% accuracy for 3 GHz FS magnitude and 5 GHz FT magnitude, 2 GHz smaller than the other two approaches.

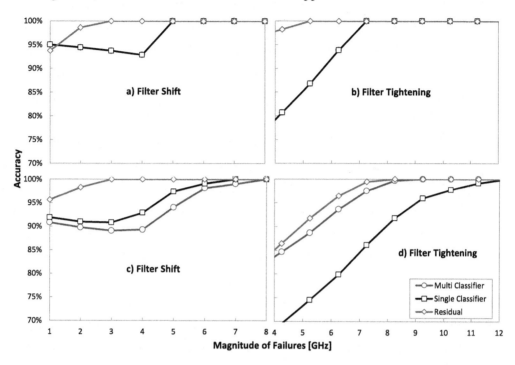

Figure 8.31 Accuracy of the different approaches.

Let us now look at the average accuracy of the approaches over all the nodes; the results are shown in Fig. 8.31c-d for FS and FT failures, respectively. The residual-based approach remains the best solution by far for detecting and identifying both, FS and FT failures. In contrast, the multi-classifier-based approach shows the worst overall accuracy for FS failures, as the features selected for classification get very close to each other and it becomes very difficult for the support vectors to distinguish them perfectly. Interestingly, the situation is different for FT failures, where the multi-classifier approach performs better than the single-classifier.

In addition to classification accuracy, the *robustness* of the approaches with respect to the number of traversed nodes is of paramount importance for practical implementations. Hence, let us compare the robustness of different approaches in terms of the smallest failure magnitude after which the classification accuracy reaches 100%. Plots in Fig. 8.32a-b represent such robustness in terms of the location of FS and FT failures, respectively. Note that, points for N1 in Fig. 8.32a-b correspond to the first points in Fig. 8.31a-b where 100% accuracy is achieved. As observed, the residual-based approach shows the highest level of robustness compared to the feature-based ones for both FS and

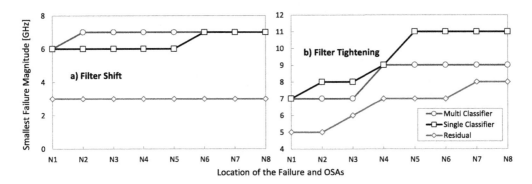

Figure 8.32 Robustness of the different approaches.

FT failures. It can be understood that the residual-based approach is robust regardless of the location of the failure as it perfectly detects and identifies failures with magnitude above the values in Fig. 8.32.

8.3.3.2 The efficiency of residual adaptation mechanism

Let us now to discuss numerical results to demonstrate how the proposed adaptation mechanism enables the residual-based approach to be applied to optical spectrum of a signal after passing through different types of filters in the network. For the experiments, we used the same VPI Photonics setup used before. Aiming at emulating failure scenarios, we modified the characteristics of the 2nd WSS of every node in the setup; its bandwidth and central frequency were modified to model FT and FS failures, respectively. A large dataset of failures was collected by inducing failures of magnitude in the range [1–8] GHz for FS and in the range [1–15] GHz for FT. We configured optical filters to be 2nd order Gaussian for training and re-configured them to become 3rd and 4th order Gaussian for testing, where the same failure scenarios were simulated.

We look firstly at the benefits of applying the adaptation mechanism for identifying the normal cases. We found that accuracy (number of correctly detected cases over the total number of cases) is poor when no adaptation is applied and becomes perfect with residual adaptation. Next, we looked at the benefits of applying residual adaptation for detecting failures. Three cases were studied:

- 2nd *order* for both out-of-field training and in-field testing; note that no adaptation is needed (the case in the previous section),
- 3rd *order*, in which 2nd order filters were used for training and 3rd order filters w/ adaptation were used for testing.
- 4th *order*, in which 2nd order filters were used for training and 4th order filters w/ adaptation were used for testing.

The results are reported in Fig. 8.33, where Fig. 8.33a and Fig. 8.33b show the average node accuracy of identifying FS and FT, respectively, for failures in all 8 nodes and

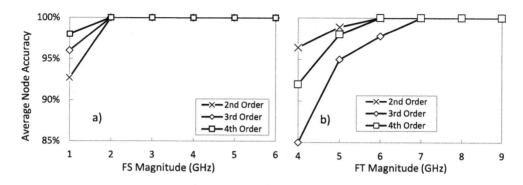

Figure 8.33 Average node accuracy w.r.t. failure magnitude for a) FS and b) FT.

varying levels of failure magnitudes. The accuracy is promising for all the cases under study, even though it degrades for very small magnitudes in which the spectrum looks like normal cases. In fact, failure detection is 100% in all cases. However, the failure identification step should be improved to reduce even more the minimum magnitude for failure identification (Table 8.4).

Table 8.4   Results Comparison.

Scenario	Failure Detection	Failure Type Identification	
		Min FS Magnitude	Min FT Magnitude
only 2nd or 4th order	100%	2 GHz	6 GHz
only 3rd order	100%	2 GHz	7 GHz
mix of 2nd and 4th order	100%	5 GHz	7 GHz

To highlight the impact of cascaded nodes, Fig. 8.34 presents the average accuracy for FS and FT with respect to the node where the failure occurs; failure magnitudes of range [1-4] GHz for FS and [4-7] GHz for FT were considered. As shown, the accuracy drops at the very last nodes as accumulated filter cascading effects makes it very challenging to distinguish between different cases.

Ultimately, the efficiency of the algorithm for transmission system with two different filter types was evaluated. To this end, we modified the above-described setup to have 2nd order Gaussian filters in the first 4 nodes and 4th order Gaussian filters in the last 4 ones. As reported in Table 8.4, failure detection accuracy is 100% while performance degradation happens when failure identification is executed. As a result, the minimum failure magnitude to be detected with 100% accuracy is 5 and 7 GHz, for FS and FT, respectively, just a bit higher than in the case of one single filter type, which validates the performance of the proposed residual adaptation method.

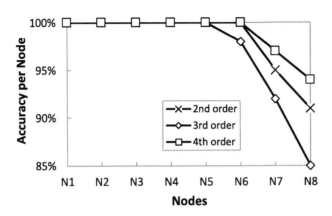

Figure 8.34 Accuracy per node w.r.t. the sequence of cascaded nodes.

8.3.4 Conclusions

In this section, we have continued exploiting the benefits of OSAs for identification and localization of filter related failures. A new approach for filter-related soft-failures detection and identification, called residual-based, has been proposed and its performance was compared in terms of accuracy and robustness against the two approaches proposed in the previous section.

The residual-based approach, which is based on a single classifier strategy, significantly outperforms the feature-based ones and brings down the complexity of training phase compared to multi-classifier approach. However, it requires two additional modules to be available in the MDA agent. Table 8.5 summarizes the key differences of the three approaches.

To conclude, it is beneficial to bring data analytics as close as possible to the MDA agents as the complexity of the proposed modules is low enough to be integrated in programmable units that are expected to be available in the future whitebox-based technologies for switching and Tp optical nodes.

We have also shown that ML algorithm can be trained out–of–field with measurements from testbeds and/or simulations using one single reference filtering solution (possibly belonging to a single vendor). A reliable in–field adaptation mechanism is demonstrated to enable heterogeneous filtering solutions (belonging to different vendors).

While filter related failures may happen in SSON, they are not of any concern in FON. However, other different failures might happen in FONs that will require the adaptation of the approaches proposed in this and the previous sections to the characteristics of FONs

Table 8.5 Key characteristics and results for the considered approaches for failure detection and identification.

	Pre-processing	Classification method	Training phase	Number of classifiers	Availability at the node level	Accuracy	Robustness w.r.t. # nodes
Multi Classifier	not required	SVM	requires observations of every level of filter cascading	# of nodes to support	all classifiers	good	good
Single Classifier	pre-processing of the features	SVM	requires observations of just a single level of filter cascading	1	one classifier + correction masks	good	good
Residual Based	pre-processing of the optical spectrum	SVM	requires observations of just a single level of filter cascading	1	one classifier + ESC and residual signal computations modules	very good	very good

8.4. Monitoring of filterless optical networks

The previous section targeted the use of optical spectrum monitoring to detect and identify filter-related problems in SSON. In this section, we are going to present a real-time surveillance system in FON, which exploits data analytics and cost-effective OSAs. Specifically, we present a set of novel signal identification, classification, and tracking algorithms for FONs and study the most proper OSA resolutions for their efficient deployment in the networks. In this regard, we propose to use one OSA in FONs to continuously scan the whole C-band and utilize the acquired spectrum to perform signal identification and classification tasks.

Moreover, we propose two different optical signal tracking solutions, which are launched upon a notification from the node controller:
- *feature-based tracking*, which uses features extracted from the optical spectrum, and
- *residual-based tracking*, which uses a set of residual signals calculated by subtracting the measured signal by the OSA from an expected version of the signal.

We propose two different flavors of the feature-based approach; one of them uses the features directly extracted from the optical signal, while the other one manipulates the features to generate super features resolving some of the issues of the former approach. In the case of the residual-based approach, in addition to its excellent *individual* signal tracking, it is enriched with the capability of using tracking information from one signal to enhance the tracking accuracy of neighboring signals (we call this *contextual*).

In order to demonstrate the performance of our proposals, we challenge them in two different scenarios of a FON:
- we simulate a direct detection PAM4 transmission system and apply the proposed algorithms to the PAM4 optical spectra, and
- we apply the algorithms to a set of experimental measurements of a QPSK transmission system.

8.4.1 Motivation of optical monitoring in FONs

In contrast to filtered optical networks, FONs eliminate or minimize the number of active switching elements in the optical line systems. An example of FON connecting packet nodes is presented in Fig. 8.35. In FONs, traffic is broadcasted throughout the network; for instance, in Fig. 8.35 four lightpaths are created: R1->R4 (labeled 1), R2->R3 (2), R2->R5 (3), and R3->R5 (4).

Let us imagine that Tps for lightpaths 2 in R2 and 4 in R3 experience a Laser Drift (LD) failure shifting the signals to the right in the spectrum. In this case, signal 3 might be affected as the spectrum of signal 2 overlaps it, whereas signals 2 and 4 may be affected if the receiver is not capable of tracking their Central Frequency (CF) and misalignment between signal and the optical filter at the receiver occurs.

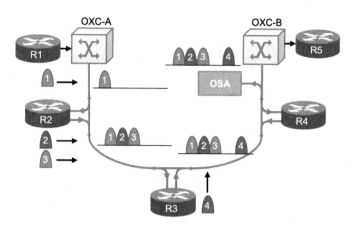

Figure 8.35 Example of a FON.

In contrast to coherent transmission systems, in PAM4, a small detuning of the transmitter leads to a noticeable performance penalty in the receiver. DD receivers are not capable of tracking the laser wavelength of the transmitter and any amount of LD results in misalignment between the signal launched by the transmitter and the optical filter in the receiver (see Fig. 8.36a). LD can potentially produce signal overlap when channel spacing between two neighboring signals is reduced (see Fig. 8.36b). Misalignments and overlaps ultimately introduce significant performance penalty.

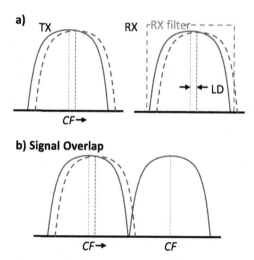

Figure 8.36 Signal misalignment and overlap.

In order to resolve these issues, one conservative solution could be assigning larger channel spacing thus, deteriorating even more the spectral efficiency of FON. A more effective solution is to consider a reliable and robust spectrum surveillance and optical signal tracking procedure to detect LDs accurately, so Tps can be adequately retuned. Consequently, margins (e.g., channel spacing) can be reduced while assuring proper Tp operation. Note that from Fig. 8.42 and Fig. 8.46, we could reduce channel spacing provided that very stable systems are deployed in the network.

The proposed spectrum monitoring approach has an advantage while exploited for FONs, which is the small number of OSAs required for real-time network spectrum monitoring; note that in a similar SSON, one OSA per link (five in total) would be required. In this section, we propose to use one single OSA installed in the last span, where all signals in the FON can be acquired. Captured spectrum needs to be analyzed real-time so active lightpaths in the FON are monitored and prompt actions are taken before a properly operating lightpath becomes affected by a failed Tp.

Note that the frequency range of a signal might not be exactly determined and slightly change along lightpaths' lifetime. Therefore, a capture of the optical spectrum in a defined range might contain only part spectrum of a target lightpath or might include part of the spectrum of other neighboring lightpaths. In consequence, we propose algorithms that periodically scan the whole C-band and rely on an ordered list of lightpaths, including relaxed frequency ranges for each one, obtained from the SDN controller; the scan process is intended to ensure that signals in the network and lightpaths in the list match in terms of frequency ranges. Any found difference (i.e., signals not in the list and lightpaths not in the FON), as well as detected anomalous signal CF shifts that might end in impacting neighboring lightpaths are reported to the SDN controller. Analyzing the current signals' spectrum allocation and lightpaths information from the controller, we thus aim at checking whether each signal is confined within the frequency range allocated to a lightpath (*normal* signals); conversely, three anomalies can be identified (illustrated in Fig. 8.37a), namely:

- a signal is partially out of the spectrum allocated to a lightpath (*outOfRange*);
- a signal is in a spectrum range not allocated to any lightpath (*unknown*); and
- no signal has been detected in the spectrum allocated to a lightpath (*missing*).

The detection of any of these anomalies triggers a notification with *critical* severity level to the controller, whereas *normal* signals need to be tracked afterwards to predict a potential anomaly.

8.4.2 Signal identification and classification

The proposed procedure starts when a new C-band scan is acquired by the OSA. The first step is to detect the allocated spectrum to each signal; by using the derivative of the power w.r.t. the frequency, as we introduced in Section 8.2.2. The sharp power rising at the left frequency edge followed by the power falling at the right frequency edge of each signal in the spectrum can be detected. Next, the algorithm in Table 8.6 is used to classify the set of identified signals S w.r.t. to the list of lightpaths P.

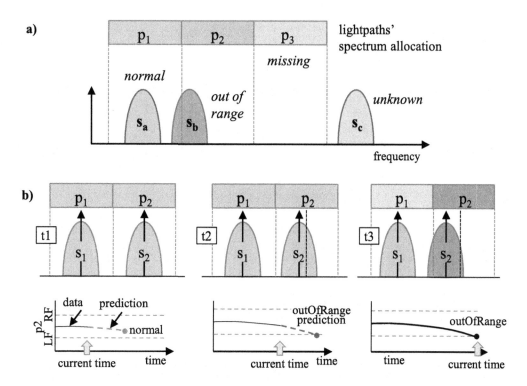

Figure 8.37 Signal classification example (a) and anomaly detection (b).

After some initializations (lines 1–3 in Table 8.6), the algorithm iterates on the signals to find the lightpaths where the allocated spectrum includes part of their range (lines 4–5); if no lightpath is found, the signal is classified as *unknown* (line 6), whereas it is classified as normal if the allocated spectrum of just one lightpath totally overlaps the signal (lines 7–9). Otherwise, signals are classified as *outOfRange* and assigned to the first overlapping lightpath if more than one exists (lines 10–16). Finally, the set of *missing* lightpaths (if any) are obtained and the classification results eventually returned (lines 17–18).

Non-*normal* signals trigger notifications to the controller and they can be discarded for further analysis. The next step focuses on tracking *normal* signals to predict any possible violation of their spectrum allocation that could impact on neighboring signals. The proposed optical signal tracking approaches are detailed in the next section.

8.4.3 Optical signal tracking

In this section, we present two different approaches to track the movement of an optical spectrum.

Table 8.6 Signal Classification Algorithm.

INPUT *S, P*
OUTPUT *normal, outOfRange, missing, unknown*

1:	*normal=outOfRange=missing=unknown←∅*
2:	*ID* ←getAllIds(*P*)
3:	*S'*←∅
4:	**for each** s∈*S* **do**
5:	*P'* ←findOverlappingLightpaths(*P,s*)
6:	**if** *P'* ≠ ∅ **then** *unknown←unknown* ∪ {s}
7:	**else if** |*P'*|==1 **and** totalOverlap(*s,P'*) **then**
8:	*normal←normal* ∪ {<*P'*.getId(), s>}
9:	*ID ← ID* \ {*P'*.getId()}
10:	**else** *s.I←* *P'*.getId()
11:	*S' ← S'* ∪ {s}
12:	**if** *S'*≠ ∅ **then**
13:	**for each** s∈*S* **do**
14:	*I* ←{*s.I*} ∩ *ID*
15:	*outOfRange ← outOfRange* ∪ {<*I*.first,*s*>}
16:	*ID ← ID* \ *I*.first
17:	**if** *ID*≠ ∅ **then** *missing←missing* ∪ *ID*
18:	**return** *normal, outOfRange, missing, unknown*

8.4.3.1 Feature-based tracking

This approach relies on extracting some meaningful features describing the key characteristics of the optical spectrum. In this approach, we consider the FeX module introduced in Section 8.2.2.2.

Considering the value of the features obtained by FeX module, we can follow two different approaches to track the movement of an optical signal, which is essentially equivalent to the amount of LD experienced by that signal. The first one (named as *Individual Feature*) relies on using a few meaningful features calculated directly from FeX module, while the second one (named as *Super Feature*) does not rely solely on the extracted features, but uses a multiple linear regression model to transform the extracted features from FeX to a more informative one, which turns out to be a more accurate estimation of LD using a combination of several features. Both approaches are detailed below.

Individual feature

In this approach, we use the relevant features extracted by FeX module to estimate the LD and track the evolution of the signals. LD is defined as the difference between the expected *CF* of the signal and the estimated one ($LD = CF_{exp} - CF_{est}$). Using the outputs of FeX, CF can be calculated by subtracting/adding half of the $bw()$ from/to the cut-off points of different power levels at the right/left side; for instance CF_{-3dB} can be calculated in one of these ways $CF_{-3dB} = f_{2(-3dB)} - bw_{(-3dB)}/2$ or $CF_{-3dB} = f_{1(-3dB)} + bw_{(-3dB)}/2$. While CF_{est} is useful for estimating the LD, power levels at the left and right frequencies (i.e. P_{f1} and P_{f2}) are more relevant to detect signal overlaps. Therefore, these relevant points can be used to track the evolution of the signal with time and eventually to predict whether it is likely to exceed the spectrum allocation within a given future time window.

An example of this procedure is illustrated in Fig. 8.37b, where signal $s2$ is gradually approaching neighboring signal $s1$. In this case, the prediction of $s2$ P_{f1} at time $t2$ states that it will exceed its spectrum allocation and thus, a notification with *warning* severity level is triggered towards the controller before an *outOfRange* anomaly is detected (which actually happens at time $t3$). This approach has low complexity as it just requires the output of FeX module and does not require modeling any estimator a priori. However, it has a drawback, which is the dependency of its results on the resolution of the OSAs.

Super features

While in the previous approach the LD can be estimated by performing simple addition/subtraction of the outputs of FeX module, in this approach, we need to manipulate the outputs of the FeX module to obtain an accurate model for estimating the amount of LD. This approach requires a small database of optical spectra including various levels of LDs and the corresponding relative points of each spectrum returned by FeX module to model a multiple linear regression that can be later used to estimate the LD. Even though this approach becomes a bit more complex as compared to the previous one, due to the need for a database and large number of required inputs from FeX, it performs much better and resolves the OSA resolution dependency of the previous approach.

8.4.3.2 Residual-based tracking

In the residual based tracking, we face the problem from another perspective. Our proposal relies on computing an expected version of the signal and uses that as baseline to track the movement of the signal. Therefore, the LD estimation module requires the expected shape and the location of every signal in the spectrum. As expected signals need to be generated considering the characteristics of the lightpath under study, for the sake of clarity; we describe the procedure using an example of a PAM4 optical signal. The measured spectra of PAM4 signals and their corresponding expected ones are shown in Fig. 8.38, where a 1.2 GHz-resolution OSA was emulated by averaging power values on windows of such width. Note that the optical carrier vanished so it cannot be used for LD estimation.

Fig. 8.38a shows a properly configured PAM4 signal, whereas Fig. 8.38b shows the case when the signal experiences 600 MHz of LD. The next step is to subtract the expected signal from the measured one, which produces a third signal called *residual* signal. If the measured and expected signal are on the same CF, the residual signal oscillates around 0 (Fig. 8.38a). On the contrary, a positive or negative slope is observed in the residual signal, depending on the direction of the drift (Fig. 8.38b). For illustrative purposes, Fig. 8.39a plots the residual signals for LD from 0 to +4.8 GHz with 600 MHz step size.

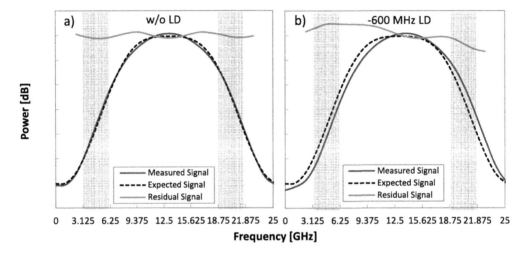

Figure 8.38 PAM4 optical spectrum a) w/o LD, b) w/ 600 MHz LD.

For modeling the LD estimator, the 3.125 GHz portions of left and right edges highlighted in blue (light gray in print version) in Fig. 8.39 were considered as features. These portions are essentially the edges of the spectrum that are expected to be affected due to the LD. Note that in the particular case of Fig. 8.39, where the traces are based on OSAs of 1.2 GHz resolution, these portions contain only 3 different power-frequency points. The evolution of these points w.r.t. the magnitude of LD is illustrated in Fig. 8.39b. The final step is to consider these features and apply multiple linear regression to obtain a model for LD estimation.

When the maximum expected LD is small compared to the spectral occupation of the signal (in Fig. 8.38, we consider the maximum 5 GHz of LD compared to 25 GHz spectral occupation of PAM4 small), a single multiple linear regression model is sufficient to perform the LD estimation task accurately. However, if the expected shift in the signal becomes comparable with the spectral occupation of the channel (as in the case shown in Fig. 8.40), more than one model is required to have an accurate estimation of the shift. Considering such large shift makes sense, for instance, when

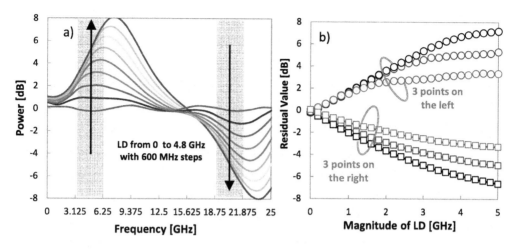

Figure 8.39 Residuals for different LD magnitudes (a) and the evolution of residual features wr.t. the LD (b).

the target task is channel spacing reduction. Fig. 8.40 shows the optical spectrum of a 100G QPSK-modulated signal, as well as its corresponding expected version and the calculated residuals.

Fig. 8.40a shows the signal when LD is assumed to be zero, while Fig. 8.40b shows the case when 2 GHz of LD is considered. In Fig. 8.40c, the residual traces are illustrated for LD magnitude of 0 to 24 GHz with step size of 2 GHz. As shown, the residual signal at lower frequencies seems to be saturated and does not contain useful information for the LD estimation of high magnitude. Therefore, in such scenarios, different multiple linear regression models can be obtained and called depending on the situation. In the case of Fig. 8.40, two models are enough; one for LD magnitude below 10 GHz and one for magnitudes higher than that and less than 25 GHz. The spectral regions considered as features for both models are highlighted and the evolution of features of both portions are plotted in Fig. 8.41.

The above procedure works well when the signals are *individually* considered. However, in the case of signal overlap like between signals 2 and 3 in Fig. 8.35, that procedure needs to be improved, as it finds LD in signal 3, which is actually properly configured. To avoid the effects of spectrum overlap, the LD estimation procedure analyzes the spectrum forward from left to right and considers *contextual* information; once it detects LD in one signal (e.g., in signal 2), it analyzes the following one (e.g., signal 3) considering the actual position of the previous one.

This approach is a bit more complex than the previous ones, as it requires an additional module to calculate the expected signal for every lightpath. Additionally, like the *Super Features* based approach, it requires a database of optical spectra for obtaining the multiple linear regression model.

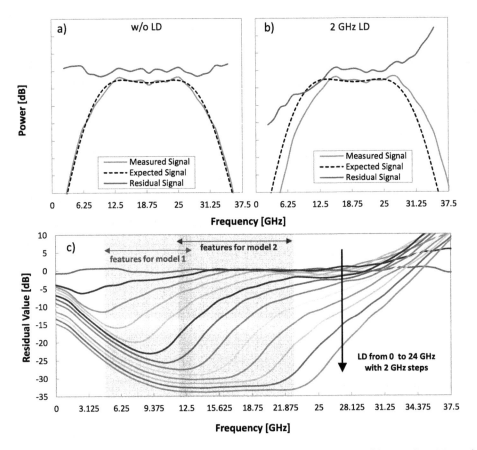

Figure 8.40 QPSK optical spectrum captured by a 1.2 GHz OSA a) w/o LD, b) w/ 2 GHz LD, and c) residuals for different LD magnitudes.

8.4.4 Illustrative results

In this section, we present results demonstrating the efficiency of the proposed algorithm. We apply the developed algorithms to two different set of observations:

- obtained by performing a set of simulations for a PAM4 transmission system,
- obtained by setting up a lab experiment for a QPSK transmission systems.

Note that, both of the above approaches were followed to generate enough data to model and then evaluate the proposed approaches.

8.4.4.1 PAM4 scenario

In the first scenario, we simulate a PAM4 system. For our experiments, we considered a 12.5 GBaud (25 Gb/s) PAM4 for cost-effective metro networks, where the electrical shaping filter of the transmitter is modeled as a 12.5 GHz Root-Raised-Cosine

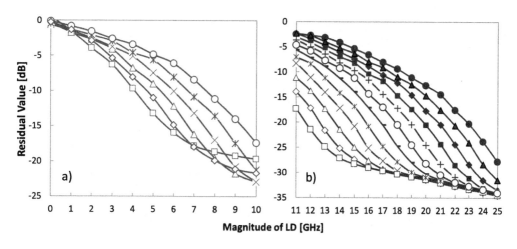

Figure 8.41 The evolution of residual features w.r.t. the magnitude of LD.

(RRC) shaping with mild roll-off factor of 1, driving a Mach-Zehnder Intensity Modulator with 30 dB extinction ratio. Propagation along 50 km of uncompensated standard G.652 SMF is considered, including EDFAs to compensate for losses; OSNR is set equal to 28 dB. At the receiver, the desired channel is extracted through a 25 GHz optical filter modeled as in [11]. Optical power at the receiver is –10 dBm and a post-detection RRC filter with roll-off 1 and 10 GHz bandwidth is applied on the photo-detected signal; thermal noise from photodetectors is also included. BER measurements are performed by means of symbol-by-symbol hard threshold detection; no DSP is performed on the acquired waveforms prior to BER measures, except timing recovery to find the optimum decision point within the symbol time, as well as the optimum decision thresholds.

In the first set of experiments, we focus on analyzing one single PAM4 signal (labeled 4 in Fig. 8.35) that is simulated to drift right with steps of 200 MHz up to 5 GHz; the BER degradation was plotted in Fig. 8.42, where after 3 GHz of LD, the BER starts to increase sharply due to detuning w.r.t. the receiver optical filter. Selecting the spectra of PAM4 of the simulation, the signal identification and classification algorithm perfectly identifies both signals and matches them to two existing lightpaths; the algorithm classifies as *normal* signals when both are still in their expected spectral window, whereas it identifies the signal that has moved out of its expected spectral window as *outOfRange*.

Additionally, it is important to investigate whether our proposed LD estimation algorithms work well with acquisitions of OSAs of different resolutions. Therefore, we carried out LD estimation considering OSAs of different granularities; the results are plotted in Fig. 8.43. The results are illustrated for the three approaches described in the previous section. As shown, the estimation accuracy of the approaches relying on the

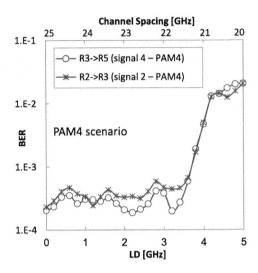

Figure 8.42 BER degradation vs LD and/or channel spacing for PAM4 systems.

residuals and super features is almost perfect regardless of the OSA resolution (residual based approach performs even slightly better), while the accuracy of the approach relying on the individual feature degrades as OSA of coarser granularity is considered.

In the second set of experiments, three PAM4 signals spaced 25 GHz apart, are modeled. Similar to Fig. 8.35, the CF of the central one (signal 2) is detuned towards the channel on the right (signal 3) with steps of 200 MHz up to the point their spacing becomes 20 GHz. Now, we are interested in evaluating whether the proposed approach enables reducing the channel spacing to reduce the margins.

In this case, LD estimation for signal 2 works as accurate as in the previous experiment for signal 4. However, when the individual analysis procedure is applied to analyze signal 3 we observed that the accuracy was poor; the results are presented in Fig. 8.44a, where the estimation error is plotted as a function of the spacing between the signals. These results are as a consequence of part of signal 2 overlapping signal 3, which for coarse-granular OSA produces a significant loss of accuracy. Fig. 8.44b presents the results obtained when contextual analysis is applied. After finding that signal 2 presents some amount of LD, the contextual approach generates an estimated signal that takes into account such fact. This contextual expected signal generates residuals that counteract the effects of signal overlapping and allows preserving the accuracy observed in the first set of experiments. In fact, is that extraordinary accuracy what enables reducing channel spacing, as any slight LD can be detected and procedures for laser retuning triggered.

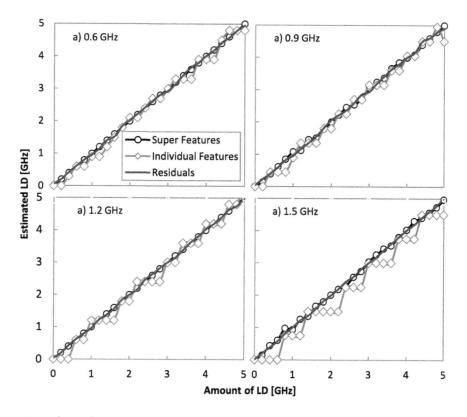

Figure 8.43 The tracking accuracy of the PAM4 optical spectrum.

8.4.4.2 QPSK scenario

In this second scenario, we apply the algorithms to a set of experimental measurements of QPSK signals. We setup an experimental testbed where two neighboring 100 Gb/s signals (labeled $s1$ and $s2$) were launched. Signal $s1$ was generated using an experimental system, while signal $s2$ was generated by a commercial system. Signal $s2$ was considered to operate properly, while $s1$ is forced to move toward the neighboring one at 1 GHz steps from an initial 50 GHz spacing between signals (Fig. 8.45), simulating a laser drift failure.

Due to the filterless characteristics of the network, signal $s2$ is affected after a certain the amount of laser drift. Fig. 8.46 shows the pre-FEC BER of signal $s2$ as a function of channel spacing between the two signals. A sequence of C-band spectra was acquired using a commercial OSA with 100 MHz resolution; a number of captures with coarser resolution, from 300 MHz to 3 GHz, were subsequently generated from the original capture to analyze the impact of the resolution on the accuracy of the proposed algorithms.

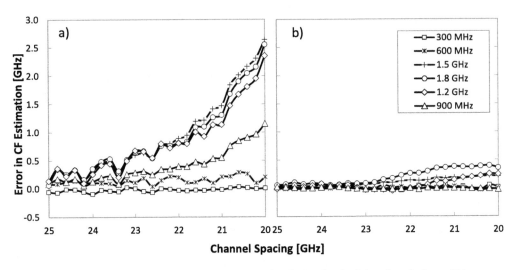

Figure 8.44 The tracking accuracy of the PAM4 signal with a) individual signal analysis, and b) contextual analysis.

Similar to the PAM4 scenario, selecting the spectra in Fig. 8.45, the signal classification algorithm perfectly identifies both signals and matches them to two existing lightpaths; the algorithm classifies as *normal* signals when it analyzes the spectrum in Fig. 8.45a, whereas it identifies $s1$ as *outOfRange* when analyzes that in Fig. 8.45b. Nonetheless, it is worth studying the accuracy of signals' detection vs. OSA resolution. To that end, we emulated 5000 different lightpath frequency ranges for every spectrum capture with no overlap.

The accuracy results of $f_{1(-3dB)}$ / $f_{2(-3dB)}$ computation vs. OSA resolution are reported in Fig. 8.47a. The finer the OSA resolution the lower the error in points' computation, which impacts on signal identification.

Let us now study the accuracy of signal overlapping detection as a function of the OSA resolution. In this case, we track $f_{1(-3dB)}$ and $f_{2(-3dB)}$ to infer when both signals begin to overlap. Fig. 8.47b shows the results of the detection; with 300 MHz and 600 MHZ OSA resolution, the overlap is perfectly detected; the inner graph inside Fig. 8.47b shows how the sudden change in $f_{1(-3dB)}$ of signal $s2$ allows detecting the overlap. When OSA resolution is up to 2.1 GHz, 1 GHz of effective overlap is needed to detect it, whereas the overlap is not detected for coarser OSA resolutions. Note that, we defined effective overlap as the spectral distance between the optical spectrum traces of two signals at power − 20 dB.

Ultimately, we report the accuracy of the three LD estimation approaches to the case of QPSK-modulated signals. The results are illustrated in Fig. 8.48. Similar to the PAM4 case, the tracking accuracy of the approaches that are based on residuals and super

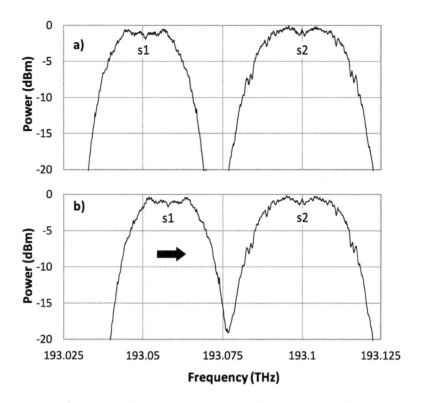

Figure 8.45 Optical spectrum under normal conditions (a) and when LD causes overlapping (b).

features is almost perfect. However, the accuracy of the approach based on individual features becomes worse as the coarser OSA resolution is considered.

8.4.5 Conclusions

In this section, a data analytics–based surveillance system exploiting the optical spectra of lightpaths, to detect anomalies and failures, has been proposed. The proposed surveillance system comprises of signal identification, classification, and tracking modules. The main purpose of signal identification and classification modules is to check whether all the lightpaths are placed in their expected spectral windows and if not, to notify the controller. The optical signal tracking module, on the other hand, is in charge of monitoring the movement of properly operating lightpath, due to improper operation or misconfiguration of a transponder. Optical signal tracking module can be used: i) to avoid the misalignment of a transmitter with the optical filter of its corresponding receiver in a direct detection system, and ii) to predict whether two neighboring lightpaths may overlap due to a failure. It can be also used to realize system operating with low channel spacing contributing to low-margin operation of the network. The

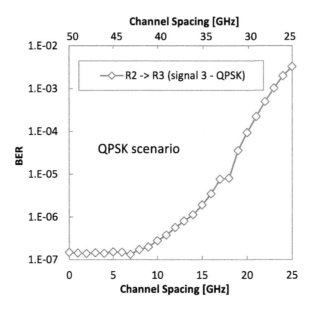

Figure 8.46 BER degradation vs LD and/or channel spacing for a QPSK system.

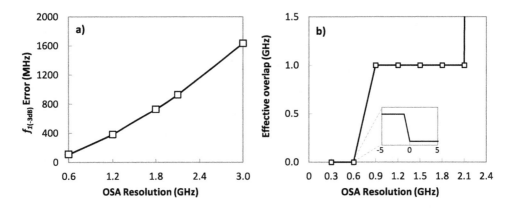

Figure 8.47 $f_{1(-3dB)}$ computation vs OSA resolution(a) and overlap detection vs OSA resolution (b).

performance of all the approaches has been demonstrated by applying to a set of measurement collected through simulation in a PAM4 direct detection system and a set of measurements acquired in an experimental testbed of 100G QPSK signals.

8.5. Concluding remarks and future work

Optical spectrum monitoring is one of the promising solutions for detecting soft-failures. Several data analytics have been developed in this chapter exploiting the optical

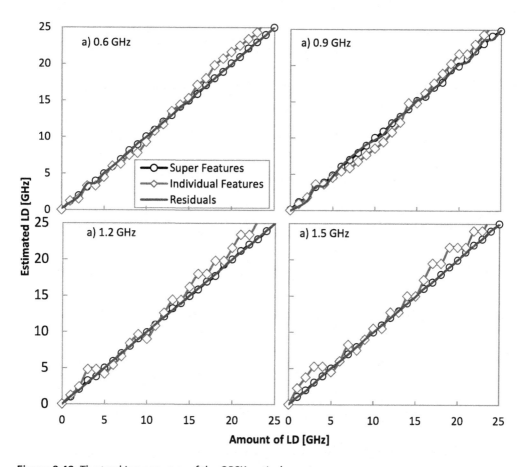

Figure 8.48 The tracking accuracy of the QPSK optical spectrum.

spectra of lightpaths to detect and identify soft-failures in filtered (SSON) and filterless (FON) optical networks. In SSONs, the proposed solutions have been demonstrated to perfectly detect and identify the most common filter related failures: asymmetric filtering and tight filtering. In FONs a real-time spectrum surveillance and signal tracking mechanism has been proposed and demonstrated on simulated and experimental data showcasing its outstanding performance to detect and identify failure, as well as to reduce operational margins.

In addition to optical spectra-related soft-failures, aging and external causes like, for example, temperature variations might degrade optical devices. In this regard, it would be useful using a tool to estimate the QoT of a lightpath and compare to that measured in the transponders. Found deviations can be explained by changes in the value of input parameters of the QoT model representing the optical devices, like noise figure in optical amplifiers and reduced OSNR in the WSSs. By applying reverse engineering, the value of those modeling parameters can be estimated as a function of the observed QoT of the lightpaths, which would be useful to anticipate soft-failure detection and localization, as well as accurately identify degradations before they have a major impact on the network. Overall idea and preliminary results are available in [3].

List of acronyms

QAM	Quadrature Amplitude Modulation
BER	Bit Error Rate
CF	Central Frequency
CMC	Correction Mask Calculator
CoD	Coherent Detection
DD	Direction Detection
DP	Dual Polarization
DSP	Digital Signal Processing
DT	Decision Tree
EDFA	Erbium Doped Fiber Amplifier
EON	Elastic Optical Network
ESC	Expected Signal Calculator
FEC	Forward Error Correction
FEELING	FailurE causE Localization for optIcal NetworkinG
FeX	Feature Extractor
FON	Filterless Optical Network
FS	Filter Shift
FSE	Filter Shift Estimator
FT	Filter Tightening
FTE	Filter Tightening Estimator
LD	Laser Drift
MDA	Monitoring and Data Analytics
ML	Machine Learning
MSE	Mean Squared Error
OPM	Optical Performance Monitoring
OSA	Optical Spectrum Analyzer
OSNR	Optical Signal to Noise Ratio
PAM4	4-level Pulse Amplitude Modulation
QoT	Quality of Transmission
QPSK	Quadrature Phase-Shift Keying
RRC	Root-Raised Cosine
SDN	Software Defined Network
SNR	Signal-to-Noise Ratio
SSON	Spectrally-Switched Optical Network
SSV	Signal Spectrum Verification
SVM	Support Vector Machine
Tp	Transponder
VPI	VPIphotonics: Simulation Software and Design Services
WSS	Wavelength-Selective-Switches

References

[1] A.P. Vela, M. Ruiz, L. Velasco, Distributing data analytics for efficient multiple traffic anomalies detection, Elsevier Computer Communications 107 (2017) 1–12.

[2] A.P. Vela, M. Ruiz, F. Fresi, N. Sambo, F. Cugini, G. Meloni, L. Potì, L. Velasco, P. Castoldi, BER degradation detection and failure identification in elastic optical networks, IEEE/OSA Journal of Lightwave Technology (JLT) 35 (2017) 4595–4604.

[3] S. Barzegar, E. Virgillito, M. Ruiz, A. Ferrari, A. Napoli, V. Curri, L. Velasco, Soft-failure localization and device working parameters estimation in disaggregated scenarios, in: Proc. IEEE/OSA Optical Fiber Communication Conference (OFC), 2020.

[4] Z. Dong, F.N. Khan, Q. Sui, K. Zhong, C. Lu, A.P.T. Lao, Optical performance monitoring: a review of current and future technologies, IEEE/OSA Journal of Lightwave Technology (JLT) 34 (2015) 525–543.

[5] Flexgrid high resolution optical channel monitor (OCM) [on-line], www.finisar.com. (Accessed June 2018).

[6] Finisar Whitepaper, Filter bandwidth definition of the waveshaper S-series programmable optical processor [on-line], www.finisar.com. (Accessed June 2018).

[7] Ll. Gifre, J.-L. Izquierdo-Zaragoza, M. Ruiz, L. Velasco, Autonomic disaggregated multilayer networking, IEEE/OSA Journal of Optical Communications and Networking (JOCN) 10 (2018) 482–492.

[8] Victor Lopez, Luis Velasco, Elastic Optical Networks: Architectures, Technologies, and Control, The Springer Book Series on Optical Networks, 2016.

[9] C. Mas, I. Tomkos, O. Tonguz, Failure location algorithm for transparent optical networks, IEEE Journal on Selected Areas in Communications 23 (2005) 1508–1519.

[10] L. Velasco, A. Chiadò Piat, O. González, A. Lord, A. Napoli, P. Layec, D. Rafique, A. D'Errico, D. King, M. Ruiz, F. Cugini, R. Casellas, Monitoring and data analytics for optical networking: benefits, architectures, and use cases, IEEE Network Magazine 33 (2019) 100–108.

[11] C. Pulikkaseril, L. Stewart, M.A.F. Roelens, G.W. Baxter, S. Poole, S. Frisken, Spectral modeling of channel band shapes in wavelength selective switches, OSA Optics Express 19 (2011) 8458–8470.

[12] D. Rafique, L. Velasco, Machine learning for optical network automation: overview, architecture and applications (invited tutorial), IEEE/OSA Journal of Optical Communications and Networking (JOCN) (2018).

[13] M. Behringer, M. Pritikin, S. Bjarnason, A. Clemm, B. Carpenter, S. Jiang, L. Ciavaglia, Autonomic networking: definitions and design goals, in: IETF RFC 7575, 2015.

[14] L. Velasco, A.P. Vela, F. Morales, M. Ruiz, Designing, operating and re-optimizing elastic optical networks (invited tutorial), IEEE/OSA Journal of Lightwave Technology (JLT) 35 (2017) 513–526.

[15] L. Velasco, Ll. Gifre, J.-L. Izquierdo-Zaragoza, F. Paolucci, A.P. Vela, A. Sgambelluri, M. Ruiz, F. Cugini, An architecture to support autonomic slice networking [invited], IEEE/OSA Journal of Lightwave Technology (JLT) 36 (2018) 135–141.

[16] L. Velasco, A. Sgambelluri, R. Casellas, Ll. Gifre, J.-L. Izquierdo-Zaragoza, F. Fresi, F. Paolucci, R. Martínez, E. Riccardi, Building autonomic optical whitebox-based networks, IEEE/OSA Journal of Lightwave Technology (JLT) (2018).

[17] www.vpiphotonics.com/.

[18] B. Zaluski, B. Rajtar, H. Habjanix, M. Baranek, N. Slibar, R. Petracic, T. Sukser, Terastream implementation of all IP new architecture, in: Proc. Int. Convention on Information and Communication Technology, Electronics and Microelectronics (MIPRO), 2013.

## CHAPTER NINE

Machine learning and data science for low-margin optical networks
The ins and outs of margin optimization

Camille Delezoide, Petros Ramantanis, and Patricia Layec
Smart Optical Fabric and Devices Lab, Nokia Bell Labs, Nozay, France

9.1. The shape of networks to come

Optical networks are at a turning point. The spectral efficiencies achieved with state-of-the-art commercial equipment are within an order of magnitude of their natural upper bound: the Shannon limit. In other words, we are irremediably running out of ways to accommodate the continuously increasing demand while reducing the cost-per-bit. In this context, optical network margins constitute an almost untapped reservoir of capacity. However, contrary to the potential understanding of margins as something unnecessary, they ultimately contribute to the reliability of optical networks. Consequently, optimizing margins is about striking the best trade-off between costs, capacity and reliability, and network monitoring data is a key resource to achieve it.

All modern network elements, e.g. transponders, amplifiers and routing nodes have the built-in capacity to provide a myriad of indicators describing the physical state of the optical layer in real time. Such indicators could be leveraged in many ways to refine margins. This is now made technically possible by powerful commercially-available telemetry tools able to collect and process large quantities of monitoring data. Optical networks are also becoming increasingly programmable, which means that all network elements will eventually be fully controllable remotely through software. Combined with monitoring, this paves the way for the automation of maintenance and optimization tasks. With the upcoming ability of the network to adapt to changing conditions, the need for margins will also be reduced. In such cases however, knowledge of the possibility to trade margins off for more capacity only comes after deployment. This means that in order to benefit from margin reductions, networks will need to be very adaptive, i.e. with many degrees of freedom. Very high level of adaptability will be achieved with three key ingredients: rate flexibility, resource sharing (e.g. multipoint and multiflow), and virtualization. With such ingredients, the optimal trade-off between reliability and network capacity can be achieved, with seamless optical service enablement and extra network capacity treated as just another commodity to be sold on a global market. Yet, the projected evolution of optical networks towards more heterogeneity, in both

Machine Learning for Future Fiber-Optic Communication Systems
https://doi.org/10.1016/B978-0-32-385227-2.00016-4

technology and vendor, represents significant challenges for optimal network operation due to the inherent increase in complexity. The discussed properties of future optical networks and their dependencies are represented in Fig. 9.1 below.

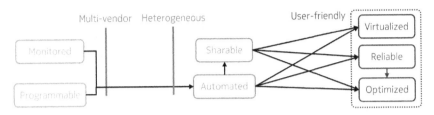

Figure 9.1 Developing network properties and their dependencies.

In this context, machine learning algorithms recently appeared as new and powerful tools to solve complex problems, and now represent a dominant trend in optical network research. Although optical network margins could be largely optimized through conventional engineering techniques, machine learning tools suitably combined with expert knowledge of the problem at hand are liable to achieve further optimization. This is particularly true when essential information is missing, and/or when known analytical models fail to explain accurate observations from the field. The main objective of this chapter is not to directly apply machine learning to margin optimization, but to sufficiently well explore the ins and outs of margin optimization so that machine learning could be effectively employed to solve it.

We first provide in Section 9.2 a quick overview of what is commonly termed margin in the context of worst-case optical network design. Then, upon the observation that such margin is only one of many design aspects that contribute to both capacity and reliability, we propose in Section 9.3 a generalization of the concept of margin based on the Shannon limit, with the aim to clarify what are the true levers and constraints of margin-related network optimization. Following this theoretical exploration of the concept of margins, the rest of the chapter is dedicated to the analysis of experimental data from a large backbone network. In Section 9.4, we start by assessing how margins are statistically distributed across the studied network, and how they vary in time. On the one hand, long term observations lead us to discuss common assumptions related to aging. On the other hand, large and recurrent fluctuations of the monitored quality of transmission (QoT) lead us to the numerical evaluation in Section 9.5 of the trade-off between capacity and availability as a measure of the reliability. With the understanding that any margin reduction will lower the availability, we leverage in Section 9.6 the dataset to explore how simple rate adaptation mechanisms would perform in jointly optimizing capacity and availability, i.e. how to maximize the capacity gain from a margin reduction while minimizing the impact on network reliability. Leveraging the conclusions of all previous sections, we finally discuss in Section 9.7 the various pitfalls when evaluating solutions based on machine learning, but also the great benefits they can bring to low-margin optical networks when properly combined with existing tools and expert knowledge.

9.2. Current QoT margin taxonomy and design

In the context of optical networks, the term *margin* is typically associated with the *Quality of Transmission (QoT)*. The QoT of a lightpath specifically describes the integrity of the transmitted signal. To avoid any confusion with the generalized margin from the next section, we will replace the literature term *margin* by *QoT margin*. Based on [2], setting margins is a complex empirical process involving both network vendors and operators. The overall goal of this process is to guarantee a minimal QoT during the entire lifetime of the network so that a failure from insufficient QoT, e.g. below the FEC limit, should not occur. However, in case of catastrophic events such as fiber cuts, power outages or equipment breakdown, no amount of margins can prevent lightpath failures and only optical protection schemes can efficiently mitigate the impact of such events. Thus, the margin setting process we review here is designed to make QoT considerations completely independent from the handling of catastrophic events.

Let us consider the process for a single lightpath. The first step is to estimate QoT_{design} as the worst-case QoT at BOL. This estimation relies on a QoT model, on the model's parameters values and on their uncertainties. Typically, the model is used to calculate the QoT when we assume that all parameters are at their worst-case values. Then, the maximal model error is subtracted to the result to obtain QoT_{design}. As a result of the worst-case assumption, the QoT measured at BOL should theoretically always exceed QoT_{design}. Next, all devices and physical phenomenons potentially responsible for a QoT decay over the years are listed. Based on the projected network lifetime, each list entry of index k is attributed a maximal expected impact, typically as a positive penalty $P_k[dB]$. Note that $X[dB]$ means that X is expressed in dB scale; quantities are expressed in linear scale otherwise. Once again, worst-case performance degradation is assumed. Such a penalty can for instance be of 0.1 dB/year per traversed degree of routing node, the assumed worst-case impact of the aging of these specific network elements on QoT. Empirically, the linear sum of all contributions is considered to be the maximal difference expected between the QoT at BOL and the QoT at the end-of-life (EOL):

$$QoT_{EOL}[dB] = QoT_{BOL}[dB] - \sum_k P_k[dB] \tag{9.1}$$

We show in [19] that this is equivalent to assuming that penalties are small enough for the linear approximation to hold, and that all phenomenons leading to a QoT decay are correlated, statistically-speaking. In parallel to system margins, the operator defines

QoT_{min} as the minimal QoT allowed at any time. This value may be larger or equal to the forward error-correction (FEC) limit QoT_{FEC}. This defines the operator margin:

$$M_{ope}[dB] = QoT_{min}[dB] - QoT_{FEC}[dB] \qquad (9.2)$$

The sum of the design constraints result in the definition of QoT_{accept} as the minimal QoT that must be measured at BOL for the operator to validate the connection delivered by the vendor:

$$QoT_{accept}[dB] = QoT_{FEC}[dB] + M_S[dB] \qquad (9.3)$$

where the system margin M_S includes the operator margin M_{ope}:

$$M_S[dB] = \sum_k P_k[dB] + M_{ope}[dB] \qquad (9.4)$$

At this point, we derive the unallocated margin M_U from:

$$M_U[dB] = QoT_{design}[dB] - QoT_{accept}[dB] \qquad (9.5)$$

During design, a connection is judged feasible only if $M_U[dB]$ is positive. At commissioning, using either transponder's QoT monitoring capacity or third-party test equipment, the operator determines $QoT_{meas}(BOL)$, the measured QoT at BOL. This value finally allows to quantify the design margins M_D, as:

$$M_D[dB] = QoT_{meas}[dB](BOL) - QoT_{design}[dB] \qquad (9.6)$$

This definition of the design margin matches that in [17] but differs from that in [2] which is most likely the result of a typo in [2]. We finally define the total QoT margin $M_{tot}(t)$ as can be deduced at any time from a QoT measurement:

$$M_{tot}[dB](t) = QoT_{meas}[dB](t) - QoT_{FEC}[dB] \qquad (9.7)$$

The various margin groups and the key performance indicators associated are illustrated in Fig. 9.2.

9.3. Generalization of optical network margins

During network greenfield or brownfield planning, QoT margins are designed so that the QoT of any lightpath remains above some level at any given time during its lifetime. However, QoT margins cannot protect the lightpath in case of catastrophic events known to occur frequently and with significant impact on availability, such as equipment breakdown, power failures and fiber cuts. The impact of such events can

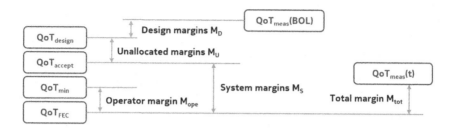

Figure 9.2 Summary of QoT margin groups appearing in literature.

only be mitigated through the introduction of redundancies, taking many forms from duplicated power converters to optical protection paths. Similarly to QoT margins, redundancies represent a significant extra cost for the network, and their only point is to guarantee a targeted level of reliability. Therefore, there is no fundamental reason to exclude redundancies from optical network margins. The issue raised here is the addition of very different forms of network margins to form a generalized margin indicator. This can seem as comparing apples and oranges. In the case of optical protection, margins can be naturally expressed in terms of capacity instead of QoT. For instance, the capacity margin for a connection with 1+1 protection would be 100%. QoT margins as described in the previous section could then be converted into a capacity equivalent so that a more general margin combining protection and QoT margins can be derived. Yet, thinking network margins in terms of capacity and not QoT inherently interrogates on what are the fundamental and practical limitations of the capacity supported by an optical connection. The Shannon limit handily provides a theoretical limit for the capacity that can be carried by a communication channel at a given signal to noise ratio (SNR) and symbol rate R. For fiber optical communications, this schematically translates into a limit capacity achievable over a given lightpath which depends on the number of channels that can be multiplexed – by any mean available – in the fiber. This capacity limit thus appears as a product of the spectral efficiency and the overall bandwidth of the fiber, the spectral efficiency being the ratio between the channel capacity C and the spectral bandwidth B allocated to this channel. Independent from the number of spatially multiplexed channels – over a single multimode or over multiple fibers – the spectral efficiency appears as a convenient metric to describe how efficiently each propagation medium is used.

9.3.1 Optimal spectral efficiency

By definition, a margin needs an arbitrary reference to compare the performance – here the spectral efficiency – actually achieved with what could have been achieved under different circumstances. In our context, the reference value needs to be the optimal spectral efficiency, i.e. the highest achievable value considering that some system parameters can be changed while others cannot. In one of its simplest forms, the modulation format of a flexible transponder can be adjusted to the QoT as studied in Section 9.6. In more complex scenarios, several parameters could be considered to be changeable in order to evaluate both costs and gains from any substantial modification.

Under white noise, the Shannon limit for any single-mode, polarization multiplexed optical channel can be expressed as [10]:

$$SE_{lim} = 2\frac{R}{B}\log_2(1 + SNR) \tag{9.8}$$

where SE_{lim} is the maximum spectral efficiency (SE), R is the symbol rate, B is the allocated bandwidth and SNR is the signal-to-noise ratio. The relevant definition of SNR is more thoroughly discussed in Section 9.4.1. The Shannon limit defines the maximum spectral efficiency that can be achieved for a given set of parameters (R, B, SNR). However, since the SNR itself depends on many parameters including R and B, the conversion of the Shannon limit into a maximal spectral efficiency achievable for a given connection is not straightforward. Let us assume that the SNR can be written as a holistic function of a finite amount N of parameters. Those can be numeric values corresponding to typical parameters impacting the SNR, such as fiber attenuation, channel power, amplifier noise figure, transponder noise, etc. They can also describe some aspects of the existing or future network relevant to SNR such as empty slots loaded with ASE noise, dispersion-managed fiber links or active optical nonlinearity mitigation. For any given network optimization problem, we can consider a subset of M<N parameters - the optimization parameters - allowed to change while the others are fixed. We write these parameters A_k with k ranging from 1 to M. From (9.8), SE_{lim} can also be written as a function of the same parameters A_k. Based on that, we postulate the existence of a set of M optimization parameters $\mathbf{A_{opt}}$ that maximizes SE_{lim}, and define the optimal spectral efficiency for a given optical connection SE_{opt} accordingly:

$$SE_{opt} = SE_{lim}\left(\mathbf{A_{opt}}\right) \tag{9.9}$$

9.3.2 Field margins

On the field, the set of optimization parameters can naturally differ from $\mathbf{A_{opt}}$ and takes the value $\mathbf{A}_{field}$, resulting in a suboptimal Shannon limit SE_{field} and associated field margin M_{field}:

$$SE_{field} = SE_{lim}\left(\mathbf{A}_{field}\right) < SE_{opt} \quad \| \quad M_{field} = \frac{SE_{opt}}{SE_{field}} \tag{9.10}$$

In our definition, field margins are directly interpreted as the potential spectral efficiency gains achievable through the modification of the defined optimization parameters. Let us for instance consider the case of a network where legacy dispersion compensation modules (DCMs) have been left on the field because of OPEX limitations whereas all channels are coherent, i.e. where DCMs are obsolete. Here, due to enhanced optical nonlinearities, DCMs induce SNR penalties to all coherent channels. Considering that the presence or absence of DCMs is an optimization parameter, the potential gains in maximal spectral efficiency obtained through the removal of DCMs contribute to the field margin. This may appear odd considering that the concept of margins generally relates to system reliability, which is not directly impacted in the DCM example. However, the SNR gains from the removal of DCMs - if not used to upgrade the capacity - would increase the QoT margin, thus improve the reliability. Therefore, any potential SNR gain can be understood as a virtual margin. This is similar to the concept of potential energy in physics, e.g. that of an apple in a tree, only revealed once its tie to the tree is severed.

The example of guard bands described below also justifies the use of the term margin in our context, and the use of the spectral efficiency to express it. Guard bands (GB) consist in extra bandwidth allocated to a channel to mitigate both linear and nonlinear crosstalk, and filtering impairments leading to QoT penalties and/or clock recovery failures. Despite the fact that guard bands contribute to reliability and consume network resources, they are not considered as part of the classical margins, leading to an incomplete understanding of the concept of margin optimization. Guard bands however can be easily included in the field margins. We can assume the existence of $B = B_{opt}$ maximizing the spectral efficiency, i.e. achieving $\text{SE}_{opt,GB} = \text{SE}_{lim}(B_{opt})$. Regardless of B_{opt}, setting B obeys to standardized engineering rules: B must be a multiple of 12.5 GHz, and is generally uniformly defined at the network level although this tends to evolve with the flexgrid standard. These rules result in $B = B_{GB} = B_{opt} + GB$, corresponding to $\text{SE}_{field,GB} = \text{SE}_{lim}(B_{GB})$. We thus express the impact of the guard bands on the field margin through the guard band margin M_{GB} defined as:

$$M_{GB} = \frac{SE_{opt,GB}}{SE_{field,GB}} = \frac{B_{GB}}{B_{opt}} \cdot \frac{\ln\left[1 + SNR\left(B_{opt}\right)\right]}{\ln\left[1 + SNR\left(B_{GB}\right)\right]} \qquad (9.11)$$

This specific submargin of M_{field} is quantified in [5].

Compared to the classical design view treating the SNR as a quantity simply resulting from deployment, i.e. over which one has very little or no control, field margins are essential to model the fact that some actions can significantly improve the SNR and/or the spectral efficiency network-wide, with consequences on the capacity and/or the reliability of the network.

9.3.3 Uncertainty margins

Up to now, we considered SE_{field} as deriving from known design choices and deterministic parameter values. Although some parameters such as modulation format or coding rate are exactly known, most parameters impacting the connection's QoT are uncertain, although the uncertainty can be reduced after deployment as depicted in [17]. Under the uncertainties related to deployment and/or parameter evolution in time, network planning requires to define SNR_{WC} as the assumed worst SNR value for a given time period. This implies that the range of possible SNR values after deployment is finite and can be known at design, so that there is 100% probability that the field SNR will be over SNR_{WC}. In practice however, since the field SNR has a strong stochastic component stemming from field uncertainties, the range of possible SNR values is infinite. Thus, there is no ideal way to define SNR_{WC}. At best, the statistical distribution of the SNR can be evaluated so that SNR_{WC} is chosen based on the probability that the SNR will be above the FEC limit. Thus, the choice of SNR_{WC} is necessarily subjective and fundamentally impacts the design, understanding that the lower the SNR_{WC}, the higher the reliability, but the higher the cost per transported bit.

To introduce the uncertainty margin, we arbitrarily define SNR_{field} as the statistical SNR expectation, i.e. with first order approximation, the SNR obtained when all random parameters are equal to their expected values. The statistical aspects of SNR estimation are more thoroughly discussed in [19]. We further define the uncertainty margin as:

$$M_{unc} = \frac{SE_{field}}{SE_{WC}} = \frac{\ln\left(1 + SNR_{field}\right)}{\ln\left(1 + SNR_{WC}\right)} \tag{9.12}$$

9.3.4 Unallocated and implementation margins

Accounting for uncertainty margins, SE_{WC} is the highest spectral efficiency that a network designer can rely on when assuming an ideal coding scheme. An actual implementation is bound to achieve less. Also, the net rates hence spectral efficiencies are necessarily quantized. For a given allocated slot-size B, we define $SE_{max,ope}$ as the maximal spectral efficiency that can be delivered in practice by a given transponder assuming $SNR = SNR_{WC}$. The implementation margin is defined as:

$$M_{imp} = \frac{SE_{WC}}{SE_{max,ope}} \tag{9.13}$$

The implementation margin solely depends on the transponder and integrates its granularity i.e. its capacity to adjust its net rate to an arbitrary demand in capacity. Finally, the spectral efficiency at which the transponder actually operates, SE_{ope}, does not have to be equal to $SE_{max,ope}$. This is typically the case when the length of a route is well below the maximum reach corresponding to the current state of the art for the capacity demand, or when operator margins are imposed. We define the unallocated margins accordingly:

$$M_{una} = \frac{SE_{max,ope}}{SE_{ope}} \tag{9.14}$$

As defined, the unallocated margin slightly differs from the definition in [2] and integrates the operator margin. The idea of this new definition is to quantify by how much the spectral efficiency could be increased simply by adjusting the rate of the deployed transponder.

9.3.5 Protection margins

Let us consider a protected connection which can be treated – as far as margins are concerned – as two independent lightpaths, a working lightpath (w) and a protection lightpath (p). The working (resp. protection) lightpath carries a capacity C_w (resp. C_p) over an allocated bandwidth B_w (resp. B_p) resulting in an operational spectral efficiency $SE_{ope,w}$ (resp. $SE_{ope,p}$). Since the spectral efficiency of the working path $SE_{ope,w}$ does not accurately reflect all network resources allocated to the connection as margins, we define the effective spectral efficiency (ESE) accounting for the extra bandwidth B_p allocated for protection:

$$ESE = \frac{C_w}{B_p + B_w} = \frac{B_w}{B_p + B_w} SE_{ope,w} \tag{9.15}$$

Note that the ESE can easily be generalized for the situations where there is more than one protection path. We introduce η as the protection ratio, that is the ratio between the working capacity C_w and the protection capacity C_p. In 1+1 protection, we have $C_w = C_p$ leading to $\eta = 1$. In some situations, the protection may be partial with $\eta < 1$, or extra redundant with $\eta > 1$. Using η, the ESE and protection margin M_{prot} are written as:

$$ESE = \frac{SE_{ope,p} \, SE_{ope,w}}{SE_{ope,p} + \eta SE_{ope,w}} \quad \| \quad M_{prot} = \frac{SE_{ope,w}}{ESE} = 1 + \eta \frac{SE_{ope,w}}{SE_{ope,p}} \tag{9.16}$$

9.3.6 Total spectral efficiency margin and QoT margin equivalency

The total margin for a protected optical connection can simply be written as a product (or the sum in dB scale) of all the margins defined above:

$$M_{tot} = \frac{SE_{opt}}{ESE} = M_{field} \, M_{unc} \, M_{imp} \, M_{una} \, M_{prot} \tag{9.17}$$

For comparative purposes, we can convert the total margin in spectral efficiency M_{tot} into a SNR margin $M_{SNR,tot}$. To do this, we write $SNR_{opt}=SNR(\mathbf{A_{opt}})$ as the SNR corresponding to the optimal spectral efficiency SE_{opt}, and SNR_{ESE} as the SNR achieving $SE_{lim} = ESE$ with the same rate and slot-size. We can show that SNR_{ESE} verifies:

$$SNR_{ESE} = \left(1 + SNR_{opt}\right)^{1/M_{tot}} - 1 \tag{9.18}$$

For equivalency with the QoT margin typically expressed in dB, we thus express the total SNR margin as:

$$M_{SNR,tot} = \frac{SNR_{opt}}{SNR_{ESE}} = \frac{SNR_{opt}}{\left[1 + SNR_{opt}\right]^{1/M_{tot}} - 1} \qquad (9.19)$$

The generalized margin proposed addresses the incompleteness of QoT margins to satisfactorily describe all capacity reservoirs stemming from design choices. It provides a single metric gathering all margins relevant to overall network reliability, which allows to evaluate the impact of a wide variety of design choices on the network margin level and reliability. By further decomposition, it allows to distinguish among the general margin which part could be employed to increase the capacity with already deployed equipment, which part would require an upgrade of the network, and finally which part requires technology that is not yet available. Machine learning tools are good candidates to efficiently solve complex generalized margin optimization problems with multiple optimization parameters.

9.4. Large scale assessment of margins and their time variations in a deployed network

9.4.1 Assessing the quality of transmission

The *signal-to-noise ratio (SNR)* [10] is a universal metric in science and engineering to assess the quality of any transmission anywhere in the transmission system. In the context of coherent optical communications, the SNR is generally evaluated within the digital signal processing (DSP) of the receiver right before forward error correction (FEC). This SNR is often called *electrical* SNR (ESNR) or *generalized* SNR (GSNR) due to the fact that it is evaluated in the electrical - more exactly the digital - domain and integrates all noises and distortions, i.e. any deviation from the intended or ideal signal. This is in opposition to the *optical* SNR (OSNR) redefined in a recent IEC standard [13], that explicitly compares signal and ASE noise power in the optical domain, typically right before the receiver. Note that in OSNR measurements, the "signal" measured necessarily integrates noises from the transmitter and optical nonlinearities, leading to many interpretation issues. In contrast, the (E-G) SNR can be directly measured as it is defined without ambiguity, it can also be estimated from closed-form expressions or numerical simulations and is directly connected to the Shannon limit, making it a very convenient metric to characterize the QoT in an optical network.

The SNR of coherent transmissions can be evaluated without traffic interruptions from two independent methods. Firstly, a measure of the SNR is obtained by analyzing the coherent constellation right before FEC. The same method is used to measure the *error vector magnitude (EVM)*. In fact, we verify that

$$SNR_{CA} \simeq EVM^{-2} \tag{9.20}$$

where SNR_{CA} is the SNR measured from constellation analysis. Secondly, we can measure the SNR through FEC decoding. The FEC module of the receiver first provides the parity error count from which the pre-FEC bit error ratio (BER) is deduced. Then, under the Gaussian Noise (GN) model assumptions [4], there are simple and accurate closed-form relations between the SNR, the Q^2, and the pre-FEC bit error ratio (BER) [15]. In particular, the SNR is approximately equal to the Q^2 in PDM-QPSK. For further reference, we write SNR_{EC} the SNR measured from this method. When the noise level is low, SNR_{EC} is a less accurate measurement of the SNR than SNR_{CA} due to the low number of errors counted. In opposition, a high noise level leads to a superposition of symbols in the constellation, so that SNR_{EC} is the better SNR measurement under such circumstances. In [18], we thus proposed a weighted SNR measurement, SNR_w combining both constellation analysis and error count methods to achieve a constant accuracy regardless of the noise level:

$$SNR_w = wSNR_{CA} + (1 - w)SNR_{EC} \tag{9.21}$$

This method is specifically important for the high-speed characterization of connections with a high level of QoT margins, i.e. when the refresh rate is too low to count enough errors for the measurement to be accurate. For low-speed QoT monitoring however, e.g. a refresh rate of 15 minutes as in the dataset described below, the SNR measurement error from the error count remains negligible.

9.4.2 Description of the dataset

In 2017, a dataset containing 14 months of performance monitoring of a large North-American optical backbone network was released [16]. While the exact network topology is not available, values of Q^2, received optical power, chromatic dispersion and differential group delay from February 2015 to April 2016 are reported and organized in 4000 data tables, one for each monitored connection. Each connection is composed of a single, 100 Gb/s, PDM-QPSK optical channel propagating over a single lightpath, and all lightpaths are shared by multiple channels. Reported Q^2 values are averages over 15-minute-periods, and are most likely measured from the error count method. Since all monitored lightpaths are PDM-QPSK, we directly convert the Q^2 values reported in the dataset into SNR values. After processing Q^2 values as SNR values, we convert each SNR into a time-dependent SNR margin $M_{SNR,tot}$ using:

$$M_{SNR,tot}[dB](t) = SNR[dB](t) - SNR_{FEC}[dB] \tag{9.22}$$

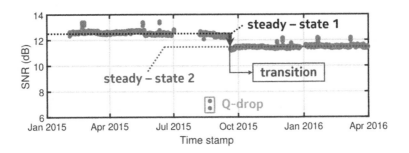

Figure 9.3 Evolution of SNR for "channel 112".

In absence of deeper knowledge on the monitored connections, we consider the SNR forward-error correction (FEC) limit for the default 100 Gb/s rate to be 5 dB, as provided in [12]. This value, corresponding to a pre-FEC bit-error ratio (BER) of $3.77e^{-2}$, is consistent with the performances of state-of-the-art commercial transponders.

9.4.3 Example of SNR variations in time

To illustrate the time variations of the monitored SNRs, we plot in Fig. 9.3 the SNR as function of time for a selected connection labeled "channel 112" and representative of the dataset. Let us focus on two important observations. On one occasion on August, 21 2015, the measured SNR temporarily drops by more than 5 dB. Since dataset values are averaged over a 15-minute-period, the actual SNR variation may have been larger. Such sudden drops of performance have been previously reported and labeled as *Q-drops* [11]. Although not explained, authors observed that Q-drops are correlated with subsequent outages. This suggests that Q-drops may be due to major physical layer issues which are either very limited in time or rapidly compensated by some automation, which either leads to or simply correlates with full connection breakdowns. Another important observation is the transition in September 2015 from an SNR stable around 12.5 dB to an SNR around 11.5 dB. Such transition from one steady-state to another can only denote a significant modification of the optical connection. In a deployed network, this can occur for various reasons: fiber cuts and repairs, replaced hardware, removal of dispersion compensation modules (DCMs), defragmentation, removal or addition of channels, etc. Such observations indicate the necessity to statistically assess how the SNR varies in time for each connection.

9.4.4 Distributions of the minimal, maximal and median margins for all connections in the dataset

Here we calculate the minimum, maximum, median SNR values as the well as the SNR standard deviation or each connection, i.e. for each of the 4000 time series. Based on these calculations, we plot in Fig. 9.4 various graphs providing a network-wide overview of how the QoT margin of each connection varies in time, using (9.22) for conversion. In Fig. 9.4(a), we plot the cumulative distribution function (CDF) of the SNR standard deviations. We observe that while 80% of connections have a standard deviation below 0.2 dB, and 2.5% of connections are over 1 dB. As more thoroughly explained in [6], high standard deviations are due to permanent variations of the QoT due to maintenance operations and repairs. Next, we plot in Fig. 9.4(b) the CDFs of the minimum, maximum and median SNR margins. We observe that due to Q-drops, there is a significant difference between the distributions of minimum and median SNR margins. In Fig. 9.4(c), we plot the amplitude of the SNR margin, i.e. the difference between the maximal and minimal margins observed, versus the standard deviation. Because of Q-drops, we observe that for a large cluster of connections, the margin amplitude is high while the standard deviation is low. We also observe another cluster of connections for which the amplitude is correlated with the standard deviation, which occurs because of the steady-state transitions as observed in the case of "channel 112" in Fig. 9.3.

In the context of network design, the worst-case approach would consist here in solely considering the margin minimum, for instance to plan for upgrades. However, since minimal values are generally statistical outliers, this would lead to vastly underestimate the potential of the network for optimization. The observation of Fig. 9.4(b-c) confirms the need to reconsider the worst-case approach, and to evolve towards a statistical approach of network design as developed in [19]. Nevertheless, we observe from Fig. 9.4(c) that around 50% of optical connections have a minimal SNR margin of 6 dB or more, and 10% have a minimal margin exceeding 9 dB. This underlines the great potential of this network and potentially others for significant capacity optimization, even with conservative approaches derived from the worst-case paradigm.

9.4.5 System margins and long term performance variations

In known design practices, system margins are associated with how the QoT varies in time. Depending on the sources [17], authors estimate that system margins are typically anywhere between 3 dB and 6 dB. Since commercial 100G PDM-QPSK transponders are contemporary to the dataset, it is likely that connections have been monitored close to the begin-of-life (BOL). Therefore, the SNR margins observed in the previous section include most of the system margins. System margins thus appear as a dominant contribution to overall margins.

In the rationales for system margins, network designers first account for the progressive addition of channels propagating in the fiber links. This slowly increases the nonlinear noise due to the Kerr effect, thus reduces the SNR of pre-existing connections. Such reductions shall appear after a maintenance period as a downward step, as may be the case in Fig. 9.3. The amplitude of the downward step could easily be estimated using available QoT estimation models [4], provided the model parameters

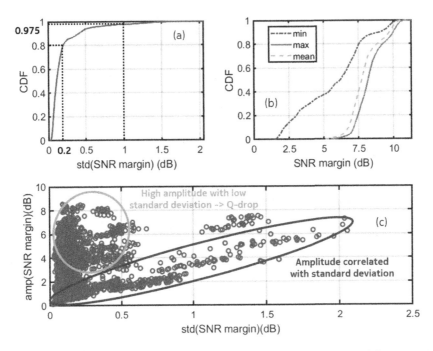

Figure 9.4 SNR margins statistics calculated for the 4000 connections over their full time extents: (a) CDF of the standard deviation of the SNR margin, (b) CDFs of the minimum, maximum and median SNR margin, and (c) margin amplitude (max-min) versus margin standard deviation.

are known. Then, it is often considered that performances of network elements such as fibers and amplifiers degrade over time. This phenomenon is typically called aging. Although aging values circulate in the literature [17], the origin of such values is - to the best of our knowledge - never disclosed. This lack of reliable aging data can lead to inappropriate margin settings. On top of such permanent SNR decreases, positive or negative SNR variations can occur for other reasons. As all optical devices, network elements are sensitive to external factors such as temperature [3]. This sensitivity can be related to birefringence effects [8]. Depending on how network elements are deployed, temperature may have more or less influence, e.g. an aerial cable compared to a buried cable, or depending if the storage room is temperature-controlled. Such effects can typically explain the SNR standard deviations observed in Fig. 9.4(a) for most of the connections, with an order of magnitude of 0.1 dB. Last but not least, we have shown [6] that on some occasions that the average SNR can significantly increase. Such improvements of the QoT can be explained by very specific maintenance operations, such as the removal of legacy 10G channels and associated dispersion compensation units in the optical links used by the monitored connections [12], or due to the reallocation of the connections' wavelengths in the context of a defragmentation.

In the following, we leverage the dataset to quantify the cumulated impact of all possible SNR variations on the long term evolution of the SNR, with potential repercussions on how to optimally reclaim system margins.

9.4.6 Distribution of long term SNR variations

Here, we first calculate for each connection the steady-state values at beginning and end of the time series, respectively SNR_{begin} and SNR_{end}. More precisely, SNR_{begin} and SNR_{end} are the median SNR values over the first month and the last month, respectively. The median is chosen instead of the average for its lower sensitivity to statistical outliers such as Q-drops. For simplicity, we call *aging* the difference for a given connection between SNR_{end} and SNR_{begin}. Since variations, positive or negative, are more likely to be observed over longer monitoring periods, we consider the *aging coefficient* as the *aging* normalized by the duration of the period over which the *aging* is calculated.

$$aging\ coefficient = \frac{aging}{\Delta T_{mon}} = \frac{SNR_{end} - SNR_{begin}}{T_{end} - T_{begin}} \tag{9.23}$$

where $\Delta T_{mon} = T_{end} - T_{begin}$ is total duration of the monitoring. We plot the corresponding histogram in Fig. 9.5.

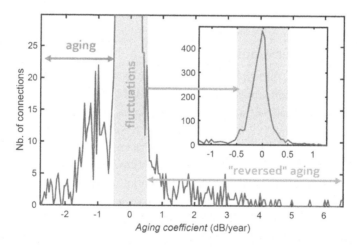

Figure 9.5 Distribution of the *aging coefficient*. Inset: zoom on the central lobe.

We observe that most connections - 85% - have an *aging coefficient* within ±0.5 dB per year. The median value is −0.05 dB/year meaning that roughly 50% of the connections have a positive *aging coefficient*. Left of the histogram, we observe the optical connections for which the cumulative impact of SNR variations results in a negative *aging* value, i.e. *regular aging*. Roughly 10% of monitored connections have an *aging coefficient* below -0.5 dB/year. On the right side, 5% of connections exhibit what can be called *reversed aging*, i.e. an aging coefficient above 0.5 dB/year, i.e. above what can be considered to be due to fluctuations. More, 3% of all connections have an *aging coefficient* above 1 dB/year, 1.6% above 2 dB/year, 0.8% over 3 dB/year and 0.15% over 5 dB/year. Note here that in such instances of *reversed aging*, the measured values should not suggest that a steady increase of the SNR may be observed. Indeed, SNR increases are only in the form of step-like variations after service interruptions, and obtained values result from the indirect averaging in the calculation of *aging coefficients*. The results show that long term SNR variations are clearly not homogeneous across the monitored network. Also, contrary to what is generally assumed, such variations are not solely downwards. Here, this is only true for half of the connections. Therefore, it clearly appears that a permanently efficient use of network resources can only be achieved by dynamically adapting the transported capacity in a per-connection basis. This is studied in the following sections.

9.5. Trade-off between capacity and availability

The trade-off between maximal capacity and reliability/availability has previously been underlined by several authors [1,14], but not quantified. In this section, we leverage the dataset to numerically assess the impact of virtual, static and arbitrary margin reductions on the availability of the connections.

9.5.1 Method and definitions

We first evaluate the SNR thresholds corresponding to a pre-FEC BER of $3.77e^{-2}$ [12] as a function of the net rates achievable with a 32 GBd PDM coherent optical channel, i.e. 100 Gb/s in QPSK, 150 Gb/s in 8QAM, 200 Gb/s in 16QAM etc using formulas in [15]. The correspondences of formats, rates and SNR FEC thresholds are displayed in Table 9.1, where some faulty errors from [6] were corrected. We consider the intermediary rates to be achieved by format adaptation, through hybrid modulation formats or probabilistic shaping, and neglecting implementation penalties. For simplicity, we assume that the SNR thresholds for the intermediary rates can be determined using a spline interpolant on the standard formats. Furthermore, we assume that changing the rate does not modify the SNR. This common assumption of the GN model is largely verified in linear or weakly nonlinear propagation regimes in uncompensated links. In other cases, extra nonlinear penalties would have to be applied as in [12], eventually leading to a reduced capacity gain achievable with elasticity. The extra penalties could easily be calculated with the more recent QoT models. In our case however, this would have required a deeper knowledge of the network than what is publicly available.

For each monitored connection, we evaluate the *availability* [21] as a function of the rate setting. The availability A is the ratio between the available time (AT) and the total time (TT). The unavailable time (UT) is simply the difference between total and

Table 9.1 Correspondences of formats, rates and associated FEC thresholds.

PDM format	QPSK	8-QAM	16-QAM	32-QAM	64-QAM
Rate (Gb/s)	100	150	200	250	300
SNR FEC threshold (dB)	5	8.3	11.3	14	16.8

available times $UT = TT - AT$, so that:

$$A = \frac{AT}{TT} = 1 - \frac{UT}{TT} \qquad (9.24)$$

In our simulations, we use Eq. (9.24) to calculate the availability of a connection for a given rate setting from the estimation of the corresponding unavailable time, for which there are two separate contributions.

The first contribution is the actual outages that took place during the 14 months of the data collection, with a static rate at 100 Gb/s for all connections. Since these outages are independent from our rate adaptation simulations, we call them *external* outages in the following, and associate them to the *external* unavailable time UT_{ext}, i.e. the unavailable time due to *external* outages. In the dataset, outages are not explicitly reported. However, we know from [11] that missing monitoring data indicates a service interruption, i.e. an outage or a maintenance window. Therefore, we estimated UT_{ext} for each connection by summing up all time periods during which data is missing, a single missing entry corresponding to 15 minutes of unavailable time. Note that this definition of what constitutes an unavailable time differs from the standard according to which maintenance windows should be excluded.

The second contribution to the unavailable time UT is the virtual outages caused by our simulated rate upgrades. In opposition to the natural, *external* outages, we call these latter outages *internal*, and associate them to the *internal* unavailable time UT_{int}. For each rate setting, UT_{int} is estimated by summing up all 15-minute periods for which the average SNR is below the FEC limit corresponding to the simulated rate. The same principle is applied in Section 9.6 with a dynamically changing rate. In the end, we can write:

$$A = \frac{AT}{TT} = 1 - \frac{UT_{ext}}{TT} - \frac{UT_{int}}{TT} = A_{ext} + A_{int} - 1$$

$$A_{ext} = 1 - \frac{UT_{ext}}{TT} \quad \text{and} \quad A_{int} = 1 - \frac{UT_{int}}{TT} \qquad (9.25)$$

where we introduce A_{ext} and A_{int} as the *external* and *internal* availabilities, respectively. Note that in (9.25), availabilities are expressed in direct form, i.e. not in percents. By construction, *external* and *internal* outages are independent and cannot occur at the same time. Therefore, the only cause for unavailability at the default 100 Gb/s rate is external, resulting in $UT_{int} = 0$ min and $A_{int} = 100\%$ at 100 Gb/s. Also, contrary to A_{int}, A_{ext} is independent from the simulated rate setting. Finally, we point out that both the 15-minute-refresh rate and the overall monitoring duration of 14 months limit the availability resolution. For instance, a single 15-minute outage over 14 months leads to A=99.9975%, i.e. between four and five-nines. Exploring availability values above 99.9975% with the same monitoring duration would require shorter refresh rates, e.g. 6 minutes for five-nines, and 37 seconds for six-nines resolution.

9.5.2 Case study: impact of a rate upgrade on availability for "channel 112"

In Fig. 9.6(a), we plot the monitored SNR with thresholds corresponding to rates from 100 Gb/s to 250 Gb/s by steps of 25 Gb/s. This plot illustrates the various contributions to the unavailability of the connection. For instance, a significant contribution to UT_{ext} is the missing data spanning from July to August 2015. Regarding UT_{int}, let us briefly consider a simulated rate at 150 Gb/s corresponding to a FEC threshold at 8.3 dB (cf. Table 9.1). Then, the only *internal* outage in such simulation occurs because of the Q-drop on August, 21 2015 (cf. Fig. 9.3), resulting in $UT_{int}(R = 150 \text{ Gb/s}) = 30$ min. For rates higher than 175 Gb/s, the SNR thresholds are much closer to the typical SNR values for this connection, resulting in a fast decay of A_{int}. For a deeper analysis, we estimate the availability for all rates between 100 Gb/s and 300 Gb/s with a 0.1 Gb/s granularity. Results are plotted in Fig. 9.6(b) in log scale, i.e. in number of nines. For comparison, we also represent the external availability A_{ext}, equal to 85.9% for the studied connection.

A wide majority of the collected SNR values are over 11 dB. This explains why, beyond 100 Gb/s, the internal availability A_{int} remains close to 100% until the rates have SNR thresholds exceeding 11 dB. Also observable in Fig. 9.6(a), this occurs close to 200 Gb/s. The step-like variation of A_{int} around 130 Gb/s in Fig. 9.6(b) occurs because of the 15-minute granularity of the virtual unavailable periods. Beyond 200 Gb/s, the absence of observed values in between the two steady-states converts into an availability plateau, i.e. a region where the rate can be increased with little to no impact on the availability. Finally, when the SNR threshold exceeds 13 dB, there is no sufficient SNR value at any time to support the corresponding rates. Consequently, A_{int} tends to zero.

The plot in Fig. 9.6(b) also allows to compare the rate-dependent internal availability A_{int} to the constant external availability A_{ext}. We can determine a remarkable rate value, R_{crit}, for which $A_{int} = A_{ext}$. It is equal to 202 Gb/s for the studied connection. R_{crit} separates the two fundamental limitations of the connection's availability A, as formally established in (9.25). For rates lower than R_{crit}, the availability is limited by *external* outages whereas for rates higher than R_{crit}, it is limited by *internal* outages, i.e. the

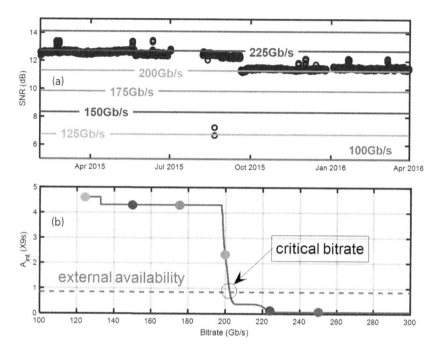

Figure 9.6 (a) SNR in time and SNR thresholds corresponding to rates from 100 Gb/s to 250 Gb/s by steps of 25 Gb/s; (b) Estimated availability A_{int} as a function of the rate for channel 112 and comparison with A_{ext} (dashed line) in logarithmic scale, i.e. in number of nines (X9s).

outages occurring when the monitored QoT is too low compared to the FEC threshold corresponding to the simulated rate setting.

9.5.3 Setting margins based on availability estimations

The observations derived from this case study provide first insights on how monitoring data can be leveraged to reduce margins optimally, i.e. with minimized impact on availability. Two approaches may be considered. The first approach would consist in allowing rate upgrades, i.e. margin reductions, as long as the predicted *internal* availability A_{int} remains – for instance – above five nines. In opposition, the second approach would be to consider that the rate can be upgraded as long as the predicted impact on A is negligible.

Whereas the first approach attempts to minimize the probability of *internal* outages, i.e. outages due to the rate elasticity, the second approach puts the same probability in a broader context, by implicitly comparing it with the probability of an *external* outage. This is especially important for networks where the availability spans over several orders of magnitude depending on the connection, as depicted in [11]. From (9.25), we understand that if for instance A_{ext} is below two nines, the *internal* outages will not significantly lower the overall availability A as long as A_{int} remains above three nines. Indeed, if $A_{ext} = 99\%$ and $A_{int} = 99.999\%$, then $A = 98.999\%$ whereas $A_{int} = 99.9\%$ leads to $A = 98.9\%$ with an unchanged A_{ext} value. In this example, the second approach allows to significantly reduce the constraint on A_{int} compared to the first approach, which can in turn allow much higher rate upgrades.

9.5.4 Margins and network throughput

In our simulations, a connection could be set with an arbitrarily high rate leading to 0% availability. Therefore, the estimation of the network throughput in terms of capacity must account for both rate setting and resulting availability.

In the context of virtual optical networks (VONs), authors in [14] introduced the time average aggregated data rate (TAADR) as optimization metric. The TAADR is defined as the sum over all connections k of the product between the connection's rate R_k and the corresponding availability A_k:

$$\text{TAADR} = \sum_{connections} R_k A_k \tag{9.26}$$

The TAADR is a suitable indicator of the network capacity throughput when the rates of all connections are static. Since we later aim to study the impact of dynamic, data-driven rate adaptation on network throughput, we need to adapt the TAADR metric to be compatible with time-varying rates.

In the static case, the product between rate and availability coincides with the time-average of the capacity delivered by a connection. In the dynamic case, we can consider that the delivered capacity C_k is time-dependent, equal to the rate under normal operation, and zero during outages or maintenance periods. Thus defined, C_k includes the availability and is compatible with a time-dependent rate. Now, we can generalize the TAADR in the form of a network throughput T defined as the sum over all connections of individual connection throughput T_k, i.e. the time-averaged capacity $T_k = <C_k>_t$:

$$T = \sum_{connections} T_k = \sum_{connections} <C_k>_t \tag{9.27}$$

9.5.5 Capacity-availability trade-off and margin sweet spot

Here, we leverage the dataset to evaluate the network throughput as a function of an *internal* availability target enforced network-wide. This means that all connections are virtually upgraded at the maximum capacity that results in an internal availability over the target. Calculation methods are more thoroughly discussed in [6]. Results are plotted in Fig. 9.7.

With all 4000 monitored connections at 100 Gb/s, the throughput would be 400 Tb/s with 100% availability for all connections. Due to outages and maintenance peri-

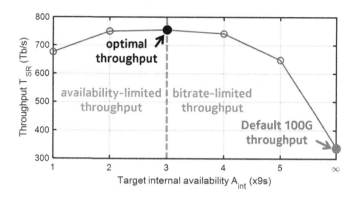

Figure 9.7 Network throughput T as a function of the *internal* availability target.

ods however, the throughput achieved is lower, equal to T_{ref}=340 Tb/s. This through-put is used in the following as reference for performance comparison. With rate settings designed to achieve at least 99.999% internal availability for every connection, the sim-ulated throughput is 648 Tb/s, almost double compared to the reference 100G case. We further observe that the network throughput reaches its optimal value, 755 Tb/s, for an internal availability target at 99.9%. Elsewhere, the throughput is either limited by low availability or by low rates. We finally observe that the throughput variation between the 99% and 99.99% targets does not exceed 2%.

The bell curve obtained in Fig. 9.7 unequivocally demonstrates the trade-off be-tween capacity and availability at the network scale, with several important conse-quences. First, a network should not be operated in the availability-limited domain, left of the throughput optimum, where both availability and throughput are subopti-mal. This corresponds to excessively low margins. Depending on the point of view, the optimal margin setting could be when the throughput optimum is achieved, or when a satisfactory middle ground is found between improved throughput and necessarily de-graded availability, on the right-side of the throughput optimum. For instance, we find a margin sweet-spot around $A_{target} = 99.99\%$ where the availability A is barely degraded compared to the reference 100G operation, and the throughput close to the maximum achievable with rate elasticity.

The results displayed in Fig. 9.7 clearly establish the trade-off between capacity and availability, and allow to identify a margin sweet-spot. However the *a posteriori* analysis we used is impossible to apply to a live network. Moreover, from the data analysis in Section 9.4, there is clearly more capacity to yield when the rates are dynamically adapted than when they are statically set. In the following, we leverage the dataset to explore ready-to-deploy dynamic rate adaptation mechanisms liable to function at the margin sweet-spot identified here.

9.6. Data-driven rate adaptation for automated network upgrades

9.6.1 Memoryless, dynamic rate adaptation (M-D)

The M-D rate adaptation consists in adapting the rate as fast as possible, every 15 minutes here, solely based on the last monitored SNR value, i.e. the SNR averaged over the previous 15 minutes. We assume the rate adaptation to be instantaneous. To avoid potential rate instability should the SNR fluctuate around a threshold value, we introduce an hysteresis cycle as depicted in Fig. 9.8.

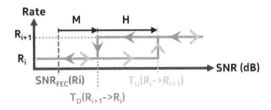

Figure 9.8 Hysteresis cycle to set the rate according to the monitored SNR in M-D mode.

Let us write R_i an arbitrary rate and R_{i+1} the upper rate. The difference between R_{i+1} and R_i is the rate granularity G. As illustrated in Fig. 9.8, the condition to increase the rate from R_i to R_{i+1} is for the SNR to cross $T_U(R_i \to R_{i+1})$ upward. Similarly, the condition to decrease the rate from R_{i+1} to R_i is for the SNR to cross $T_D(R_{i+1} \to R_i)$ downward. We define the residual margin M and hysteresis H, both in dB scale, as:

$$M = T_D(R_{i+1} \to R_i) - SNR_{FEC}(R_i)$$
$$H = T_U(R_i \to R_{i+1}) - T_D(R_{i+1} \to R_i)$$

(9.28)

where $SNR_{FEC}(R_i)$ is the FEC limit for R_i.

In the following, we study the impact of the choice of M-D parameters (H, M, G) on the network throughput and availability. We plot in Fig. 9.9(a) the network throughput increase achieved in M-D mode as a function of the hysteresis H. Note that in our context, a 100% throughput increase means that the throughput is doubled compared to the reference throughput T_{ref}=340 Gb/s, obtained when all connections are at the default 100 Gb/s rate. As expected, the smaller the granularity, the higher the throughput. For a given granularity, the highest throughput is achieved with a 0.01 dB hysteresis value. Although a lower hysteresis ought to increase the number of outages, this phenomenon is mitigated by a small residual margin, i.e. 0.1 dB. Also, the impact of additional outages on throughput for some connections is overbalanced by a throughput gain for the connections - a majority - not requiring hysteresis. In Fig. 9.9(b), we observe that a residual margin improves the network throughput only for the smaller

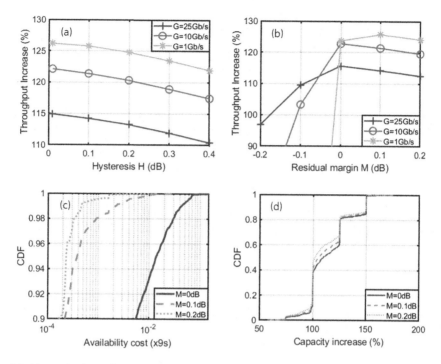

Figure 9.9 Network throughput increase in M-D mode as a function of (a) the hysteresis *H* for *M*=0.1 dB and (b) the residual margin *M* for *H*=0.1 dB. CDFs of the availability cost in number of nines (c) and corresponding throughput capacity increase (d) for *G*=25 Gb/s, *H*=0.1 dB and selected *M* values.

1 Gb/s granularity. Indeed, with coarser granularities, the mismatch between observed performance and rate results in an intrinsic margin, similar to an unallocated margin at design [17]. Like Fig. 9.7 obtained with static rate settings, the bell-shape curves here denote the trade-off between capacity and availability. For each granularity, the throughput is maximized when throughput gains from higher rates balance losses from related outages. Interestingly, the network throughput hardly depends on the residual margin above 0 dB. For deeper insight, we plot in Fig. 9.9(c-d) the CDFs of the availability cost and capacity throughput increase calculated for each connection. The availability cost is the difference between the default availability, at 100 Gb/s, and the availability obtained when simulating rate upgrades, where both availabilities are expressed in number of nines. Beyond the analysis in [6], the availability cost measures how much of the initial availability is traded for capacity relatively to the default value. The CDFs show that a 0.1 dB residual margin is enough to significantly reduce the availability cost of the rate upgrades, with minor impact on capacity increase. We thus identify a sweet-spot with H=M=0.1 dB where with standard G=25 Gb/s, throughput is increased by 115%

compared to T_{ref}, with an availability cost below 10^{-2} nines for more than 99.9% of the connections.

The M-D mode is interesting due to its ability to leverage any temporary or permanent SNR increase to upgrade the capacity. Also, it can quickly adapt to a SNR decrease - e.g. due to a soft failure - by downgrading the capacity, thus maintaining the connection available while the failure is being repaired. However, the applicability of the M-D mode is contingent on the transponder ability to adapt rate in a hitless manner. Although researched [7], such property is not yet a commercial reality. More, high-frequency rate adaptation may be challenging for network operation. For this reason, we explore next an opposite dynamic scheme where the rate may only be adapted after a service interruption.

9.6.2 Causal dynamic rate (C-D) rate adaptation

In C-D rate adaptation, we leverage the data accumulated from the last service interruption to adjust the rate after an observation period of duration T_{obs}, during which the rate is set at 100 Gb/s. Compared to the M-D mode, this could be easily deployed with today's network infrastructure and transponders.

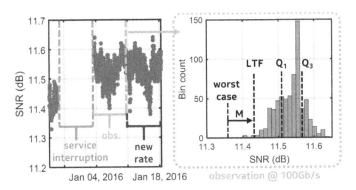

Figure 9.10 Rate adaptation mechanism in C-D mode.

The rate is set so that the FEC threshold is below the lower Tukey fence [20] of the SNR time series during T_{obs}, as depicted in Fig. 9.10. Using Tukey fences is a typical method to identify statistical outliers within a time series. We thus set the rate so that only a statistical outlier can lead to an outage due to insufficient SNR. The lower Tukey fence or *LTF* is estimated by applying the following formula:

$$LTF = Q_1 - 1.5(Q_3 - Q_1) \tag{9.29}$$

where Q_1 and Q_3 are respectively the first and third quartiles of the SNR distribution during T_{obs}. To cover for potentially misleading observations during T_{obs}, we add a residual margin M on top of the Tukey fence criteria. In the end, the selected rate R_{sel}

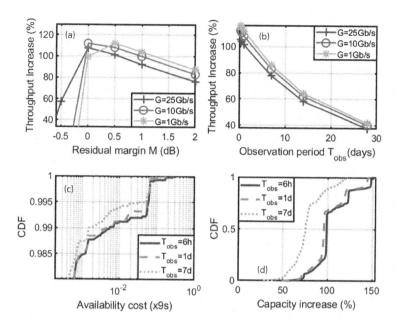

Figure 9.11 Network throughput in C-D mode as a function of (a) the residual margin for T_{obs}=1 day and (b) the observation period for M=0.5 dB. CDFs of the availability cost in number of nines (c) and corresponding capacity throughput increase (d) for G=25 Gb/s, M=0.5 dB and selected T_{obs} values.

verifies:

$$R_{sel} = \max\left(R : SNR_{FEC}(R) < LTF - M\right) \qquad (9.30)$$

We study the impact of C-D parameters (M, T_{obs}, G) on the performances in terms of network throughput increase and corresponding availability cost. Those are calculated similarly to the previous section. We first plot in Fig. 9.11(a) the network throughput increase as a function of the residual margin M for T_{obs}=1 day. The impact of M is similar to what is observed with the M-D mode, i.e. a residual margin optimizes the throughput only for 1 Gb/s granularity; for larger granularities, a residual margin slightly decreases the throughput but improves the availability of all connections. The influence of the observation period T_{obs} is investigated in Fig. 9.11(b-d). In Fig. 9.11(b), we observe that for a given granularity, the highest throughputs are achieved with the smaller observation periods. For deeper analysis, we plot in Fig. 9.11(c-d) the CDFs of availability cost and throughput capacity gain independently calculated for each connection. We observe that increasing the observation period from 6 hours to 7 days drastically reduces the throughput capacity increase, but has almost no impact on availability. With SNR stationarity and Tukey fence criteria, a mere 6-hour observation period achieves a throughput increase of 106% with an availability cost below 10^{-2} nines for more than 99% of the connections. The performances of the C-D mode are thus slightly lower than those observed with the M-D mode, which is largely compensated by the practicality of its implementation.

9.7. Machine learning for low-margin optical networks

The previous sections of this chapter made it clear that optimizing optical network margins is a very general problem that can benefit from multiple research subjects treated throughout this book. The end-to-end learning of complete fiber-optic communication system discussed in Chapter 5, with the prospect to optimize the QoT of all optical transmissions, can contribute to reducing the field and implementation margins discussed in Section 9.3. The methods for quality of transmission (QoT) estimation laid out in Chapter 7 also contribute to the low-margin effort. Indeed, more accurate QoT estimations directly result in a reduced need for uncertainty margins, as also discussed in Section 9.3. In parallel, since lower margins ultimately trade off with availability – as demonstrated and quantified in Sections 9.5 and 9.6 – all technologies likely to improve the availability in optical networks can positively contribute to reducing margins. This is the case for cognitive network fault protection and management treated in Chapter 8, and the network security management and attack/intrusion detection treated in Chapter 10. More specifically, the idea is that gains in availability can be left to improve network reliability beyond what is currently achieved, or re-invested into capacity upgrades while maintaining the current reliability. Finally, optical performance monitoring as discussed in Chapter 6 is a strategic element in all low-margin schemes. For example, as illustrated in Section 9.4, the analysis of optical performance monitoring data e.g. with regards to network aging can allow to further reduce the part of uncertainty margins that encompasses aging.

However, let us keep in mind that except end-to-end learning, all of the directions mentioned above have been exhaustively researched well before the advent of machine learning. This research resulted in many reliable ML-free solutions that are now key to the design and operation of current optical networks. Good examples are the state-of-the-art analytical models deriving from physics to estimate the QoT of lightpaths in a wide variety of configurations. The fact that such solutions are fast, accurate and routinely used in modern networks tends to be overlooked in publications focused on promoting ML-based solutions. More, quantitative comparisons between ML-free and ML-based solutions are seldom proposed. In the rare studies actually comparing performances of both types of solutions, the confrontation of results is typically unfair since ML-based techniques are provided with extra information, e.g. monitoring data. Such studies merely prove the benefits of using experimental data over not using experimental data when building a predictive model, which is a well established fact in empirical sciences. Thus, the current confusion when it comes to properly assess the real benefits of ML in optical networks may be due to the inherited separation between network and machine learning research on the one hand - mostly influenced by formal sciences such as computer science - and optical transmission research on the other hand, mainly deriving from empirical sciences, physics especially. Therefore, applying machine learning to optical networks should be understood as an interdisciplinary exercise, requiring strong knowledge in very different areas of science to reach effective solutions.

In this context, what appears to be the most suitable going forward is to properly merge ML-free and ML-based solutions to get the best from the two worlds, and devise coherent and unbiased strategies to evaluate the comparative performances of all proposed solutions. To that end, the optical network community could get inspiration from the open competitions organized in many data-centric disciplines - e.g. Kaggle competitions - where competing solutions are trialed over the same dataset and ranked according to well-defined performance indicators. An important aspect of such competitions is that the test data, i.e. the data used to rank the solutions, is statically independent from the initially-provided training and validation data, and is only made available at the final stage of the competition where no modification of the algorithm is allowed. This test methodology is particularly suited to remove all subjectivity when it comes to comparing solutions. For instance, QoT estimation solutions could be ranked according to their accuracy, versatility and speed. Ultimately, our community is in dire need of large public datasets enabling such competitions. Such datasets would ideally provide access to all topological and design information as well as extensive monitoring data for a variety of deployed optical networks.

In the following, we further examine how machine learning (ML) could be best employed to solve open network problems related to margins. After summarizing the various tools to emulate optical networks, we discuss the various issues of interpretation that can arise when evaluating the benefits of ML-based solutions depending on the emulation tools used to generate the data. We also propose some general guidelines for an unbiased assessment of the benefits from ML, which must be distinguished from the benefits from leveraging monitoring data. Finally, we discuss various use cases where the fusion of machine learning with existing tools and expert knowledge will most likely lead to significant improvements in network efficiency.

9.7.1 Solutions to emulate optical networks

The ultimate trial of any technical solution is its field deployment, i.e. here in an actual, live optical network. The challenge for any solution that researchers can propose is to convince that the solution will perform as expected, and eventually to prove that it will not create more problems than it solves. To do so, the research community currently leverages four types of tools aimed at emulating live networks: analytical models, numerical simulations, lab experiments and field trials.

The field trial is generally thought to be the best emulation of a live network. This is because many traffic-impacting phenomenons such as environmental factors are almost impossible to accurately emulate otherwise. This said, it is very important to point out that depending on its exact configuration, a field trial can be a very poor emulation of an actual network in comparison to well-designed lab experiments, numerical simulations, or even studies based on analytical models. For instance, a field trial without proper mesh routing of the channels fails to emulate well-known and significant filtering penalties stemming from Wavelength Selective Switches (WSS). The launch power can also be set low to minimize the impact of nonlinear noise and the inherent complexity it involves on QoT prediction whereas actual networks tend to be operated at higher powers, close to the nonlinear threshold. Finally, because field trial lightpaths are generally much shorter than those found in actual networks, the high QoT levels reached do not always represent the diversity of lightpaths found in existing networks, and have a tendency to minimize the relative inaccuracy of QoT prediction tools. What is true for field trial is also true for all methods of emulation, that is expert knowledge is paramount to correctly emulate what occurs in an actual network, and most importantly to understand what simplifying assumptions are made, and their potential consequences on operations should the solution be deployed.

If properly designed, lab experiments can be good emulations of actual networks, with a clear limitation however when it comes to evaluating the impact of environmental factors. Yet, the major limitation of lab experiments is often the cost and complexity involved in reproducing meshed lightpaths. For this reason, lab experiments are dominantly used for producing reference results in simple cases such as submarine propagation - emulated with re-circulation loops - to test the accuracy of new analytical models or numerical simulations tools.

Numerical simulations are in between analytical models and lab experiments when it comes to the inherent quality of the emulation. Recent advances in GPU-based calculations now allow to emulate fully loaded optical lightpaths with reasonable hardware needs and computation times. However, simulating all lightpaths of a fully-meshed network with numerical simulations remain impractical to date. There are also concerns that some of the assumptions made in numerical solutions methods will progressively affect accuracy as the symbol rates increase.

Finally, analytical models are the most cost-efficient and fastest ways to emulate a full network, especially closed-forms expressions where outputs are direct functions of known parameters, i.e. model inputs. Another advantage of an analytical model is its generality, i.e. the ability of the model to predict the outcome of a wide-variety of scenarios simply by adjusting the values of the model's parameters, this without the need for any form of training. For these reasons, analytical models are the only tools that are currently used to design networks from scratch, or even to design network upgrades. The accuracy of the outputs of an analytical model derives from two contributions. First, even a perfect model cannot accurately predict the outcome of an experiment if the parameters of the experiment, the model inputs, are not perfectly known. In measurement theory, the inaccuracy of a physical quantity i.e. the offset between the assumed or estimated value and the real value is quantified by the uncertainty. Thus, uncertainty on the model's inputs result in an uncertainty on the model's outputs. This is true for any analytical model, and also for any machine learning algorithm. Indeed, faulty training data can only lead to faulty predictions, a fact that is typically overlooked since training and validation data are usually generated in the same fashion. In such cases, a bias in the training data – i.e. a systematic error – does not lead to any detectable inaccuracy of the prediction since the same bias exists in the validation data. A problem would only declare if the training and validation data were generated by independent methods. A strategic property of analytical models is the ability to easily evaluate the impact of input parameters uncertainties on the uncertainty of the output through the derivative of analytical expressions [19]. This aspect is especially important in the context of QoT estimation when it comes to designing network margins. The accuracy of the outputs of an analytical model also depends on the model itself. Improving the accuracy of existing models is a very interesting application of machine learning.

9.7.2 Pitfalls when assessing the benefits of machine learning in optical network applications

A common shortcut in studies applying machine learning to QoT estimation is when evaluating the accuracy of the ML-based solution through the average absolute estimation error, or worst, the average error. The average error is a largely biased metric to evaluate the accuracy of a QoT estimation method since a small average error can very well be the result of large but evenly-distributed positive and negative errors. The average absolute error is a less biased performance indicator. However, what currently matters in optical network design is the worst case and not the average case. Indeed, the infrastructure is always defined through the worst case to make sure the network will work in all situations. Thus, a more suitable performance indicator in the context of current optical networks is the maximal estimation error. In the specific context of QoT estimation, the ultimate performance indicator is the maximum positive error, which corresponds to an overestimation of the QoT. Indeed, positive errors potentially lead to faults whereas negative errors are somewhat harmless, only resulting in more margins which can then be leveraged through rate automation as previously discussed in Section 9.6.

In a different context, i.e. end-to-end communication learning, it has been demonstrated [9] that the comparative gains from ML can be partially or fully due to the ML algorithm learning the test sequence used in the simulations or experiments. A fair comparison with a ML-free solution would thus be to grant the use of the transmitted sequence at the reception, which would naturally result in errorless transmissions. Naturally, this would not be reproducible in an actual network where sequences of transmitted symbols are arbitrary. This underlines the need to carefully examine all potential biases that can arise when evaluating ML gains, and take appropriate steps to mitigate such biases. For end-to-end communication learning, this entails for instance to test the algorithms with a wide-variety of arbitrary sequences to properly emulate field conditions.

Another issue when assessing ML-based solutions is the unavailability of proper test data, i.e. data independent from the training and validation data, used respectively to establish a predictive model and then to tune its hyperparameters. A large number of studies proposing the application of ML in the context of optical networks use analytical models to generate training, validation and test data. Thus, the three datasets are essentially the same, which is typically considered as bad practice by ML experts. These studies often extrapolate that since their ML-based solutions are able to generate outputs that match the test data, the same solutions would perform well on the field.

The paradox behind such extrapolation is that, considering the reference for the evaluation of the ML algorithm is an analytical model, stating that the ML algorithm will be accurate on the field implies that provided with the same information and analytical model, a ML-free algorithm can only be more accurate and efficient. Thus, comparative benefits of such ML solutions appear low, especially since the analytical models used are well-known in the scientific literature, hence accessible to everyone with minimal required knowledge.

A property deriving from the *analytically-generated dataset paradox* discussed above is that ML-based solutions should preferably be evaluated through the accuracy and efficiency at predicting independent test data, and systematically compared to ML-free techniques leveraging state-of-the-art analytical models. This naturally implies that none of the training, validation and test data should be generated through analytical models, since until proven otherwise, directly leveraging the same analytical models would always lead to better performance. It then logically appears that the most efficient use of ML would not be to replace analytic models, but to assist them in the various cases where current analytical models are limited. For instance, ML would improve performances if it uncovers input-output correlations that are either unknown in the literature, or for which a robust or computationally efficient analytical model does not currently exist. The convergence of such ML-based solutions may then be used to understand how current analytical models may be extended to make faster and more accurate predictions, potentially mitigating the need for ML-based solutions in the long run.

In other scenarios, the problems at hand can simply be too complex to be solved analytically. Well known examples of such problems in optical communications occur at the reception of coherent signals, and are typically solved by blind digital signal processing (DSP) blocks such as the constant modulus algorithm (CMA) and carrier phase estimation (CPE). These well-known algorithms are actually good examples of how analytical models and numerical convergence methods such as stochastic gradient descent – now used in machine learning – can be combined to efficiently solve complex problems. In the following, we discuss various cases were machine learning is very likely to bring significant improvements over existing solutions.

9.7.3 Optimally using ML to solve open optical network problems
9.7.3.1 The quest for the best rate adaptation mechanism

In the previous section, we studied basic examples of automated rate adaptation mechanisms with two extremes in terms of reactivity. On the one hand, the memoryless adaptation reacts in real-time to all QoT variations. This highly dynamic mechanism has the advantage to quickly adapt to an increase of QoT for maximizing the capacity, and to a decrease of the QoT by lowering the capacity thus protecting the connection from failing. However, the highly-dynamic behavior is inherently limited by the response time of the hardware, and also the infrastructure (associated to the demands) which will probably have trouble reacting synchronously. On the other hand, the quasi-static causal adaptation, setting the rate after an observation period following a service interruption up to the next one, would not react to temporary QoT variations occurring between two service interruptions. Depending on how the QoT varies, this adaptation scheme may be less robust, and yield lesser capacity gains. Thus, the comparative advantage of the causal adaptation essentially lies in its ease of implementation.

To obtain an optimal rate adaptation mechanism, it seems logical that both mechanisms could be leveraged simultaneously to obtain the juxtaposition of semi-static rate adaptation adapting to major changes of the network infrastructure, and a dynamic component that for instance pre-emptively adjusts the bitrate when the QoT decreases. Here, from an algorithm design point of view, there is an infinity of ways to define how such mechanism should react, especially when it comes to setting the optimal values of the various parameters, such as the hysteresis, the observation period and the residual margins. It is also clear that what is optimal for a given network - or even for a single connection - may not be optimal for another connection in the same network, and even less so for connections of other networks. This is because QoT fluctuations can be very different from one connection to another or from network to another, for instance due to the presence of aerial fibers or various line defects.

In the end, because the optimal rate adaptation mechanism strongly depends on QoT variability, which currently seems impossible to accurately model, ML appears as the most effective technical solution to systematically achieving an optimal rate adaptation without the need for further engineering.

9.7.3.2 The quest for the best QoT estimation method

The ultimate QoT estimation tool has to be as fast and accurate as possible when predicting the QoT of field lightpaths, the most challenging being to predict how QoT will vary in time. The accuracy of any QoT estimation method will depend both on the accuracy of the method itself, and the accuracy of its inputs. Improving QoT estimation accuracy can thus be achieved by improving the accuracy of the inputs, and/or improving the method itself.

Several ways to improve inputs accuracy have been proposed in the literature, using model-based regression or machine learning to determine the most likely value of a parameter initially known with an arbitrary uncertainty. This has been demonstrated for filter cascade, fiber type and length, amplifier noise factor (NF) and tilt, launch power, etc. Accordingly, many ML-based solutions can be used in parallel to reduce the uncertainty of a wide variety of parameters that are of significant importance when estimating the QoT. For instance, all proposed ML-free or ML-based solutions could be used to directly reduce the uncertainties for the inputs of state-of-the-art analytical models, thus improving their ability to make accurate predictions. Alternatively, the same solutions may be used to reduce the uncertainties of the data used to train ML-based solutions, and also on the field to improve the quality of the inputs used by the ML-based solution to make predictions.

The other way to improve QoT estimation is to improve the estimation method itself, starting with assuming perfect accuracy for all the inputs. As discussed above, since very accurate analytical models currently exist to predict the QoT of any lightpath, ML-based solutions should ideally focus on completing existing analytical models. This implies to train ML-based solutions to minimize the error between a set of observations and analytical model predictions.

9.7.3.3 The quest for the best QoT optimization

As pointed out in Section 9.3, the Shannon limit is only limiting if we assume there is no technical way to further improve the QoT, more specifically the SNR. This may be close to the truth for transoceanic cables, but much less the case in backbone optical networks. One way to optimize the QoT and thus the ultimate capacity that a lightpath can support is by operating the lightpath at a launch power which maximizes the QoT, i.e. the nonlinear threshold. The major challenge of systematically operating at the nonlinear threshold, i.e. the QoT optimum with regards to launch power, stems from the fact that fiber nonlinearities inherently correlate the power of a given channel to the QoT of neighbor channels. Thus, automating the launch power for hundreds of channels susceptible to interact between each other seem as an impossible engineering task, with a high risk of creating instabilities. This is also considering that upon network reconfigurations such as defragmentations or upgrades, the number and configuration of channels change. Machine learning can be an efficient technical solution to this complex and time-varying optimization problem.

Another way to optimize the QoT would be to allow a high level of flexibility with regards to the frequencies of the channels in an optical network. Fine frequency adjustments in the 1 GHz order of magnitude can allow to significantly reduce filter cascade penalties by efficiently aligning the signal with the filter cascade in the spectral domain. In a broader sense, the way channel frequencies are distributed in a network could be regularly optimized with the goal to minimize nonlinear interferences, thus generally optimize the QoT for all existing lightpaths. The underlying complexity of such an optimization makes this problem another good candidate for the use of machine learning.

9.7.3.4 The quest for an holistic network optimization of margins

In current optical networks, an important reason for the suboptimal use of deployed resources can be that for simplicity, a given technical problem is converted into two or more separate problems, with distinct solutions. The outcome is generally that multiple resources contribute to solving the same initial problem, and can thus be vastly redundant, hence wasted. From Section 9.3, we understand that the general problem of optical network reliability has been solved over the years by at least three independently designed solutions: the guard bands between channels to prevent potential crosstalk issues, the QoT margins set to make sure that all network connections will be available under normal circumstances, and finally the protection strategies designed to resume service under exceptional situations such as fiber cuts. The point of the generalized margin we proposed in Section 9.3 is to quantify most of the network resources that are directly or indirectly contributing to the overall network reliability, and which could be leveraged to increase overall capacity. In the simple case of modulation format adaptation, we demonstrated in Section 9.5 and 9.6 that with a suitable mechanism, capacity could be largely optimized with an almost negligible cost on availability, i.e. on network reliability. Even for this simple use case, we saw that determining the best rate adaptation mechanism and parameters would ultimately require machine learning. Now, let us consider that in addition to rate elasticity, we allow that any increase of capacity can either be used to carry more traffic, or used as partial or full protection for one or more lightpaths, thus allowing to convert QoT margins into network redundancies. More, guard bands could be arbitrarily set based on the margins and robustness of each transmitted channels so as to maximize the spectral efficiency on each active fiber, hence the use of deployed resources to create value. The optimal configuration to find would be one that minimizes the generalized margin, while also minimizing the unavoidable cost of the margin reduction on the availability. The underlying optimization problem implies to determine an optimal set of values for hundreds, if not thousands of parameters for large networks. Clearly, latest machine learning techniques would be of great help in solving such problems.

9.8. Conclusion

In this chapter, we strove to provide the reader with enough context on optical networks and their margins to further pave the way for an optimal use of machine learning, that is in combination of existing tools, in the resolution of various open network problems, from local rate optimization through automated rate adaptation to cross-layer network optimization through the concept of generalized margin.

We first provided a quick overview of what is classically termed margin - i.e. QoT margins - in the context of worst-case optical network design. We pointed out their conceptual limitations with regards to margin optimization, especially when it comes to understanding what are the available resources, and how leveraging unused resources impacts network reliability. To solve this, we proposed a generalization of optical network margins with renewed nomenclature to provide a broader - though not yet extensive - picture of design aspects that impact both network capacity and reliability, which we demonstrate to be intricate in the context of margin optimization. To provide an experimental view of how margins and reliability may be jointly optimized, we leveraged public optical performance monitoring data from a large backbone network. With this dataset, we demonstrated the existence of a margin sweet-spot where the capacity may be largely increased, with a minimal counterpart in terms of reliability. This sweet-spot reliably defines what the just-enough in just-enough margins means. We also demonstrated that this best margin trade-off could be achieved with easy-to-deploy automated rate adaptation.

Finally, we leveraged all the previous sections to discuss the application of machine learning to low-margin optical networks. We pointed out various evaluation pitfalls found in the literature, most likely stemming from the inherited conceptual gaps between machine learning and optical network physics. To bridge these gaps, we proposed unbiased evaluation protocols with the prospect of comparing the performances of the solutions proposed in the literature through open data competitions. In preparation for such competitions, we discussed several use cases where machine learning combined with existing tools and knowledge could reach unprecedented results. However, a major limiting factor is the current absence of public dataset combining long-term performance monitoring of an actual optical network with all the related metadata, i.e. the network topology and all parameters involved the network's design. The availability of such datasets will be an essential step to demonstrate the power of data science tools - including machine learning - to improve the efficiency of optical networks.

References

[1] K.K. Aggarwal, Integration of reliability and capacity in performance measure of a telecommunication network, IEEE Transactions on Reliability R-34 (2) (June 1985) 184–186, https://doi.org/10.1109/TR.1985.5221990.

[2] J. Augé, Can we use flexible transponders to reduce margins?, in: 2013 OFC Proceedings, 2013, pp. 1–3.

[3] C. Berkdemir, S. Ozsoy, On the temperature-dependent gain and noise figure analysis of c-band high-concentration edfas with the effect of cooperative upconversion, Journal of Lightwave Technology 27 (9) (2009) 1122–1127, https://doi.org/10.1109/JLT.2008.929410.

[4] A. Carena, V. Curri, G. Bosco, P. Poggiolini, F. Forghieri, Modeling of the impact of nonlinear propagation effects in uncompensated optical coherent transmission links, Journal of Lightwave Technology

30 (10) (May 2012) 1524–1539, https://doi.org/10.1109/JLT.2012.2189198, ISSN 0733-8724, 1558-2213.

[5] C. Delezoide, Method for a comprehensive evaluation of margins in optical networks, in: 45th European Conference on Optical Communication (ECOC 2019), 2019, pp. 1–4.

[6] C. Delezoide, P. Ramantanis, P. Layec, Leveraging field data for the joint optimization of capacity and availability in low-margin optical networks, Journal of Lightwave Technology 38 (24) (2020) 6709–6718, https://doi.org/10.1109/JLT.2020.3022107.

[7] Arnaud Dupas, Patricia Layec, Dominique Verchere, Quan Pham Van, Sébastien Bigo, Ultra-fast hitless 100Gbit/s real-time bandwidth variable transmitter with SDN optical control, in: Optical Fiber Communication Conference, San Diego, California, OSA, ISBN 978-1-943580-38-5, 2018, p. Th2A.46, https://www.osapublishing.org/abstract.cfm?URI=OFC-2018-Th2A.46.

[8] T. Duthel, C.R.S. Fludger, J. Geyer, C. Schulien, Impact of polarisation dependent loss on coherent POLMUX-NRZ-DQPSK, in: OFC/NFOEC 2008 - 2008 Conference on Optical Fiber Communication/National Fiber Optic Engineers Conference, San Diego, CA, USA, IEEE, February 2008, pp. 1–3.

[9] Tobias A. Eriksson, Henning Bülow, Andreas Leven, Applying neural networks in optical communication systems: possible pitfalls, IEEE Photonics Technology Letters 29 (23) (2017) 2091–2094, https://doi.org/10.1109/LPT.2017.2755663.

[10] R.J. Essiambre, G. Kramer, P.J. Winzer, G.J. Foschini, B. Goebel, Capacity limits of optical fiber networks, Journal of Lightwave Technology 28 (4) (2010) 662–701, https://doi.org/10.1109/JLT.2009.2039464.

[11] Monia Ghobadi, Ratul Mahajan, Optical layer failures in a large backbone, in: Proceedings of the 2016 ACM on Internet Measurement Conference, IMC'16, Santa Monica, California, USA, ACM Press, ISBN 978-1-4503-4526-2, 2016, pp. 461–467.

[12] Monia Ghobadi, Jamie Gaudette, Ratul Mahajan, Amar Phanishayee, Buddy Klinkers, Daniel Kilper, Evaluation of elastic modulation gains in Microsoft's optical backbone in North America, in: Optical Fiber Communication Conference, Anaheim, California, OSA, ISBN 978-1-943580-07-1, 2016, p. M2J.2.

[13] International Electrotechnical Commission, International Electrotechnical Commission, Technical Committee 86, International Electrotechnical Commission, and Subcommittee 86C, Fibre optic communication system design guides. Part 12, ISBN 978-2-8322-3171-5, 2016, OCLC: 961192923.

[14] Inwoong Kim, Xi Wang, Martin Bouda, Olga Vassilieva, Qiong Zhang, Paparao Palacharla, Tadashi Ikeuchi, Q-availability based virtual optical network provisioning, in: Optical Fiber Communication Conference, San Diego, California, OSA, ISBN 978-1-943580-38-5, 2018, p. W1D.3, https://www.osapublishing.org/abstract.cfm?URI=OFC-2018-W1D.3.

[15] Kyongkuk Cho, Dongweon Yoon, On the general BER expression of one- and two-dimensional amplitude modulations, IEEE Transactions on Communications 50 (7) (2002) 1074–1080, https://doi.org/10.1109/TCOMM.2002.800818.

[16] Microsoft, Wide-area optical backbone performance, https://www.microsoft.com/en-us/research/project/microsofts-wide-area-optical-backbone/, 2017. (Accessed 9 December 2020).

[17] Yvan Pointurier, Design of low-margin optical networks, Journal of Optical Communications and Networking 9 (1) (January 2017) A9, https://doi.org/10.1364/JOCN.9.0000A9, ISSN 1943-0620, 1943-0639.

[18] Petros Ramantanis, Camille Delezoide, Patricia Layec, Sébastien Bigo, Uncertainty aware real-time performance monitoring for elastic optical networks, in: ECOC Proceedings, 2021.

[19] Petros Ramantanis, Camille Delezoide, Patricia Layec, Sébastien Bigo, Revisiting the calculation of performance margins in monitoring-enabled optical networks, Journal of Optical Communications and Networking 11 (10) (October 2019) C67, https://doi.org/10.1364/JOCN.11.000C67, ISSN 1943-0620, 1943-0639.

[20] Songwon Seo, A Review and Comparison of Methods for Detecting Outliers in Univariate Data Sets, 2006, p. 59.

[21] M. Tornatore, G. Maier, A. Pattavina, Availability design of optical transport networks, IEEE Journal on Selected Areas in Communications 23 (8) (August 2005) 1520–1532, https://doi.org/10.1109/JSAC.2005.851774, ISSN 0733-8716.

CHAPTER TEN

Machine learning for network security management, attacks, and intrusions detection

Marija Furdek and Carlos Natalino
Department of Electrical Engineering, Chalmers University of Technology, Gothenburg, Sweden

10.1. Physical layer security management

Due to the vital importance of communication networks in supporting the modern society, the digital evolution is widely recognized as the greatest single enabler of sustainable development. However, the lack of trust and security in cyberspace jeopardizes the digital evolution and requires global efforts to improve the levels of trustworthiness supported by 5G-and-beyond networks. Indeed, cyber security risks are seen as the most significant threat to businesses by a large margin, potentially incurring financial, legal, reputational and operational losses [1]. What is more, security of the *network infrastructure* in particular is considered extremely or very challenging by more than half of respondents to a recent survey by Cisco [2].

The resiliency and security of the physical fiber optic infrastructure underpinning global communications has important repercussions on the Internet and the entire ICT sector. Optical fibers carry massive amounts of data and serve aggregate traffic flows between various end users using various protocols. Thus, security breaches at the physical layer can escalate across upper network layers and cause network-wide disruptions. Reliable and secure network infrastructure is fundamental for achieving privacy and protection [3]. While the majority of security breach incidents at the optical layer is expected to be covered by confidentiality clauses, over a thousand of them are reported yearly [4]. The proliferating deployment of fiber plant, typically in unsecured environments, its extensive sharing among a multitude of overlay network services organized into distinct domains with convoluted knowledge sharing agreements, along with the ultra-high data rates and ultra-long optical reach all contribute to the high complexity of optical network management.

10.1.1 Physical-layer attacks

Physical-layer attacks are typically divided into two broad categories based on the goal of the attacker: eavesdropping and service denial attacks. The goal of an eavesdropping

Machine Learning for Future Fiber-Optic Communication Systems
https://doi.org/10.1016/B978-0-32-385227-2.00017-6

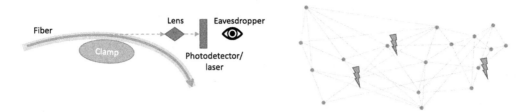

Figure 10.1 Different types of attacks that can be performed against the optical physical infrastructure. Left: eavesdropping performed by creating a temporary optical coupler [7]. Right: service disruption attack performed by coordinated fiber cuts in a national backbone network.

attack is to obtain unauthorized access to the information carried over the fiber. For unencrypted traffic, this can be done via fiber tapping, where a portion of the optical signal is collected by an eavesdropper's optical detector. Access to the signal can be obtained by e.g. accessing monitoring ports of optical devices where a small portion of the signal is tapped for legitimate monitoring purposes, but also by bending the fiber to violate the conditions of total internal reflection and cause the light to leak out of the fiber. In fact, this is the principle behind a well-know monitoring method, illustrated in the left part of Fig. 10.1, that can be abused by adversaries. Some of the tapping methods cause very low loss (e.g., below 1 dB), which might not exceed alarm threshold, causing the attack to remain undetected [5]. An example of fiber tapping attack of the trunks in Frankfurt airport has been reported in 2000 [6].

Service disruption attacks are targeted at degrading or disrupting the service, and can employ different techniques with largely varying properties. Fiber cuts are a relatively straightforward method causing outright service interruption, illustrated in the right part of Fig. 10.1. Optical networks are designed and provisioned with sufficient redundancy to ensure resilience to single or multiple fiber cuts, as this is the most common failure type. Such methods protect from accidental fiber cuts that typically occur when construction equipment accidentally tears through fiber ducts. However, deliberate fiber cut attacks can be designed to inflict tremendous damage by, for example, simultaneously cutting a larger number of links. This may leave parts of the network unconnected and overload other parts with the switched over traffic. The efficiency of the attack can be boosted by cutting critical fiber links. An example of a fiber cut attack has been recorded in the Bay Area where perpetrators dressed as maintenance personnel have been entering manholes and cutting fibers, causing disruptions along the US west coast and triggering a formal investigation by the FBI and a $250.000 reward for information leading to arrest offered by AT&T [8].

More sophisticated methods of service disruption attacks include jamming, where a harmful signal is inserted into the network, e.g. by accessing the patch panel or bending the fiber. The harmful signal can add unfilterable noise to optical channels at the

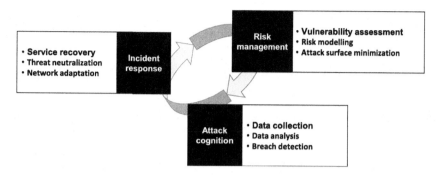

Figure 10.2 Optical network security management framework.

same wavelength (in-band jamming) or can rob the co-propagating channels at other wavelengths of the limited optical amplifier gain, resulting in insufficient amplification (out-of-band jamming). This type of attack has the potential to propagate through the network beyond the insertion point. Other types of attack can be performed without intruding the fiber. In polarization scrambling attack, fiber is squeezed and the fast-varying lateral pressure causes polarization modulation that leads to errors when faster than the coherent receiver's polarization recovery algorithm [9].

10.1.2 Attack management framework

Network security management encompasses aspects related to prevention, detection and remediation of threats, as illustrated in Fig. 10.2. Preventive measures, referred to as security assurance, encompass efforts to reduce the network vulnerability to attacks. These include the assessment of vulnerabilities, identification of attack vectors, modeling of risks and applying the obtained knowledge to design the network such that the attack surface is minimized. Examples of attack-aware network design include the placement of specialized devices in the network to limit the spreading of attacks that can propagate [10], assignment of routes and spectral resources to services such that the potential damage from jamming attacks is limited [11,12], or vulnerability to eavesdropping is reduced [13].

Remediation efforts, referred to as incident response, encompass quick and efficient recovery from security breaches. The response can take place over multiple stages, starting with immediate rerouting of affected connections away from the breach location for quick service recovery, and followed by neutralization of the threat source and long-term network adaptation to immunize against the type of threat if possible.

Detection efforts, referred to as attack cognition, encompass quick and accurate identification of security breaches. They rely on continuous data collection and analysis of various network performance indicators to map the different observed trends to various incident types. Due to the intricate effects of physical-layer breaches to optical signals, for which exact theoretical models currently do not exist, machine learning (ML) techniques have great potential to enable cognitive security management. Coupled with the emerging network telemetry frameworks that feature collection of rich network monitoring datasets, ML models are becoming an important part of optical network security diagnostics.

10.2. Machine learning techniques for security diagnostics

Different physical-layer attacks methods can cause widely varying effects on the Optical Performance Monitoring (OPM) parameters. The effects do not depend only on the attack method, but also on the way the network is designed (e.g., what type of devices are deployed) and the services are provisioned (e.g., on the spectral and spatial properties of the optical channels). As the attack regime may drastically differ from the normal operating conditions, existing physical-layer impairment models do not apply to model the effect of attacks. Thus, defining exact models or applying thresholds for triggering security alarms has not been proven viable for cognitive security management. Instead, ML models have shown tremendous potential to enable the necessary functionalities.

Machine learning approaches that have been shown to contribute to detection and identification of physical-layer attacks can be divided into supervised, unsupervised and semi-supervised learning techniques. In the context of attack diagnostics for physical-layer security, they are primarily used for classification tasks, i.e., deriving decision boundaries between different data classes.

- **Supervised Learning (SL)**

  Supervised learning techniques (e.g. Artificial Neural Networks (ANNs)) rely on having *a priori* knowledge of various attack types, and using a representative, labeled dataset for training. The dataset is labeled by experts so it is clearly mapped which data sample has been obtained under which security condition. Labeling can be performed, for instance, by the experimenters during carefully designed experiments, or by the Network Management System (NMS) upon an occurrence (and correct diagnosis) of an event of interest. Once trained, an SL model can distinguish among all known types of attacks, as illustrated in Fig. 10.3.

  SL is most appropriate for problems where complete information about the learning objective is available, i.e., the set of inputs to the problem is finite and their mapping to the expected outputs is defined. This allows the SL approach to provide fine-granular diagnostic information such as the presence of an attack, its type, intensity, location, or any other property provided during training. However, a representative dataset is not always available due to the evolution of known threats and the emergence of new ones. Moreover, it may not always be possible to collect field data (subjecting the network to attacks for training purposes is typically not

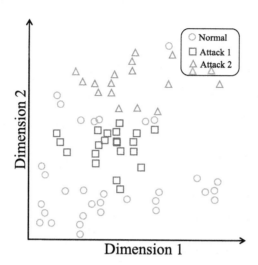

Figure 10.3 An illustration of how supervised learning classification works. The ML model is capable of distinguishing between different types of attacks.

possible), or the mapping between the inputs and the outputs may not be clearly defined, which makes the labeling infeasible or too costly.

- **Unsupervised Learning (UL)**
  Unsupervised learning techniques (e.g. K-means clustering or principal components analysis) do not rely on any prior knowledge about security threats. Instead, the algorithm learns to recognize similarities in the data to cluster them or extract features. UL is, thus, most appropriate when data labels are not available and there is no clear mapping between inputs and outputs, i.e., normal vs. abnormal conditions are not clearly defined. Instead, the data should be grouped by similarity, such that the presumably less common data points representing anomalies (i.e., attacks), appear as outliers substantially separated from the samples representing normal operating conditions, as illustrated in Fig. 10.4.

  Due to the absence of prior information about the conditions, UL can only answer the question on the presence of an attack, without providing further information as SL. To provide this answer, UL traverses the entire dataset and analyses the similarity of all samples to detect an anomaly. However, unlike SL which would require (re)training whenever a new threat emerges, UL can detect previously unseen attack conditions, which makes it an important contender for attack cognition.

- **Semi-Supervised Learning (SSL)**
  Semi-supervised learning techniques (e.g. One-Class Support Vector Machine (OCSVM)) combine the properties of the previous two approaches. Similar to

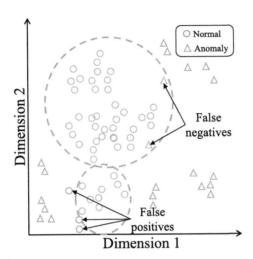

Figure 10.4 An illustration of how unsupervised learning attack detection works. The ML model clusters data points that are close and flags as an anomaly those that are sparse.

UL, SSL does not rely on previous knowledge of attacks and, thus, cannot provide fine-granular diagnostic information but can only detect the presence of an attack. However, similar to SL, it uses knowledge on normal, attack-free operating conditions for training. SSL is applicable when only a small subset of the data is labeled. In the case of anomaly detection, the labeled subset refers to normal operating conditions, which the algorithm then learns to tightly enclose into a spatial region and interpret outliers as anomalies, i.e. attacks, as illustrated in Fig. 10.5.

SSL has similar abilities as UL when it comes to detecting new types of threats and is able to detect only their presence. On the other hand, similar to SL, the inference complexity of SSL is lower than UL as it is trained on data representing normal operating conditions and does not require traversing the dataset again to detect an anomaly.

10.2.1 Experimental setup and data collection

To evaluate the performance of representative SL, UL and SSL techniques and identify their comparable advantages, a series of experiments has been performed by subjecting an optical network testbed to attacks. The testbed comprises six Reconfigurable Optical Add-Drop Multiplexers (ROADMs), one Erbium Doped Fiber Amplifier (EDFA) and ten fiber links, as shown in Fig. 10.6. It is equipped with commercially available coherent transceivers that collect a rich OPM set and expose it to the network management system. In the experiment, two optical channels (OChs) are established across

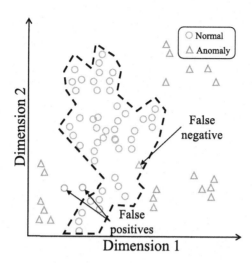

Figure 10.5 An illustration of how semi-supervised learning attack detection works. During training, the ML model builds a region around the normal samples and flags as anomalies those that fall outside of this region.

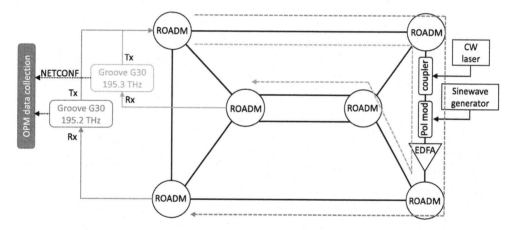

Figure 10.6 The optical testbed subjected to the jamming and polarization modulation attacks in the experiment [9].

the testbed, carrying 200 Gbit/s polarization multiplexed 16QAM optical signals. This forms the baseline (BSL) scenario.

To perform jamming attacks, a continuous-wave (CW) tunable laser is connected via a 3 dB optical coupler as a jamming signal generator. For in-band jamming, the CW laser is tuned to a frequency that overlaps with the OCh under test, slightly detuned from its central frequency to maximize the efficiency of the attack. Two different intensities of the attack are achieved by setting the power level of the jamming signal to 10 dB and 7 dB below the OCh level for a light (INBLGT) and strong (INBSTR) intensity, respectively. For out-of-band jamming, the jamming signal is substantially separated from all OChs, falling out of the OCh plan of the NMS. This creates an unexpected input signal for the optical amplifier downstream of the insertion point, and results in a power reduction on the other OChs when the amplifier works in the constant power mode. Consequently, all OChs traversing the breached link experience a reduction in the OSNR that degrades their performance. The power of the jamming signal is set to 3 dB and 8.7 dB higher than the OChs to emulate a light (OOBLGT) and a strong (OOBSTR) attack intensity, respectively.

To perform polarization modulation attacks, a polarization state modulator was clamped onto the fiber. It is an all-fiber component made of three piezoelectric fiber squeezers (just one was used) that produce stress-induced birefringence and cause polarization modulation. The squeezer is driven by a sinewave signal at 136 kHz frequency, which corresponds to one of the resonant frequencies of the piezoelectric element and therefore produces a deep polarization modulation even with modest driving voltage. The sinewave amplitude was set to 0.4 V and 1.6 V peak-to-peak resulting in light (POLLGT) and strong (POLSTR) attack, respectively.

For each described condition, full OPM dataset was collected in a 1-minute resolution over a period of 24 hours, resulting in 1440 samples per condition. Each sample contains information on OPM values summarized in Table 10.1. Such a rich experimental dataset allows for obtaining a deep insight into performance and challenges related to accuracy, runtime complexity and interpretability of the ML models' outputs.

10.2.2 ML models and their configuration

10.2.2.1 Artificial Neural Network (ANN)

As described in Chapter 1, ANN is a supervised learning technique which imitates the nervous system by forming layers of artificial neurons that exchange information through weighted combinations of input and output values. Before processing the experimental dataset, data was normalized using z-score standardization technique, so that each data point is represented by the multiple of standard deviations from the average value.

The ANN comprised two hidden layers with 50 and 100 neurons using linear and tanh activation functions, respectively. Weights initialization used the Xavier procedure [14]. The output layer was composed of 7 neurons using the softmax activation function, allowing for classification among 7 different security conditions. The training was done over 1000 iterations using Adam optimizer with learning rate of 0.0001, configured to optimize classification accuracy.

Table 10.1   OPM parameters contained in each data sample.

Acronym	Description
CD	Chromatic Dispersion
DGD	Differential Group Delay
OSNR	Optical Signal to Noise Ratio
PDL	Polarization Dependent Loss
Q-factor	Q factor
BE-FEC	Block Errors before FEC
BER-FEC	Bit Error Rate before FEC
UBE-FEC	Uncorrected Block Errors before FEC
BER-POST-FEC	Bit Error Rate after FEC
OPR	Optical Power Received
OPT	Optical Power Transmitted
OFT	Optical Frequency Transmitted
OFR	Optical Frequency Received
LOS	Loss Of Signal

For all parameters except BE-FEC, UBE-FEC, and LOS, the system provides the maximum, minimum and average values in the observation interval.

10.2.2.2 Density-Based Spatial Clustering of Applications with Noise (DBSCAN)

The DBSCAN algorithm belongs to the domain of unsupervised learning and its basic principle is to separate the data samples into clusters and outliers, i.e., anomalies. The design parameter ϵ defines the neighborhood radius of a sample, while *MinPts* defines the number of neighbors necessary for the sample to be considered a core sample. A core sample along with its neighbors creates a cluster, while samples that do not belong to any cluster appear as outliers.

10.2.2.3 One-Class Support Vector Machine (OCSVM)

OCSVM is a semisupervised learning technique which maps input data into multi-dimensional space using a kernel function. During the training phase, the algorithm searches for the hyperplane that encloses the normal working condition data as tightly as possible. During inference, if a new data point falls outside of the boundaries of the learned space, it is considered an anomaly. The three main design parameters of an OCSVM model are the *kernel* type, γ, which specifies the kernel coefficient for some kernels, and ν, which specifies an upper bound on the fraction of training errors.

10.3. Accuracy of ML models in threat detection

To achieve high accuracy in attack detection, the objective is to detect as many attacks as possible with minimal likelihood of false alarms. When an ML model interprets a sample from the normal operating condition as an attack, it generates a *false positive*. When an attack sample is interpreted as attack-free condition, it results in a *false negative*.

A false alarm stemming from a false positive is likely to trigger unnecessary remediation actions, such as connection rerouting as a quick remedy, or shutting off optical amplifiers to isolate a compromised link as a part of long-term recovery. This may incur unnecessary cost to the network operator, such as a waste of spectral resources or an unnecessary dispatch of a field team.

A false negative, on the contrary, results in overlooking an attack and proceeding with business as usual when remediation should be taking place. From a security perspective, the consequences of a false negative may be much more deleterious than those of a false positive. Nonetheless, an ML model should obtain minimal false positive as well as false negative rates to be considered for deployment in production environments.

During normal operating conditions, an ML model can report either true negatives or false positives, with their rates denoted by T_N and F_P, respectively, where $T_N + F_P = 1$. Conversely, during an attack condition, a model can report either true positives or false negatives, whose rates are denoted by T_P and F_N, respectively, where $T_P + F_N = 1$.

A single metric for measuring the accuracy of an ML model is $f1$ score, defined as:

$$f1 = 2 \times \frac{P \times R}{P + R} \tag{10.1}$$

In (10.1), P stands for precision, denoting the sensitivity of a model to false positives, defined as:

$$P = \frac{T_P}{T_P + F_P} \tag{10.2}$$

R stands for recall, denoting the sensitivity of a model to false negatives, defined as:

$$R = \frac{T_P}{T_P + F_N} \tag{10.3}$$

10.3.1 ANN accuracy

The classification accuracy of ANN is analyzed by means of a confusion matrix shown in Fig. 10.7 that plots the mapping between the actual security setting and the ANN output. The performance of ANN is not too far from ideal (all 1s along the main diagonal) as no false positives nor false negatives are observed. However, the algorithm encounters minor issues in correct classification of the attack type and intensity.

A small fraction of strong out-of-band jamming attacks (0.8%) is misclassified as light polarization modulation attacks. In general, incorrect classification of the attack type may result in activating inappropriate remedy measures. An example is when attacks that may propagate through the network are addressed using actions tailored for attacks

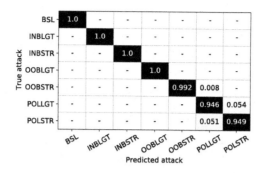

Figure 10.7 Confusion matrix for the ANN-based attack detection and identification.

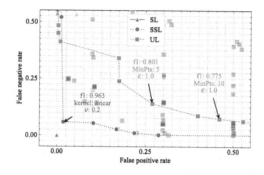

Figure 10.8 Accuracy results for the supervised, semi-supervised and unsupervised learning models.

that are confined to the breached element. In the analyzed case of ANNs, effects of both out-of-band jamming and polarization scrambling should remain confined to the breached link, under the assumption that the first ROADM downstream of the insertion point will filter out the jamming signal. Nonetheless, these two attack methods may require different remediation approaches since the former is caused by the presence of an illegitimate signal, while the latter is caused by fiber squeezing.

10.3.2 DBSCAN and OCSVM accuracy

The accuracy of all three ML techniques is compared in Fig. 10.8. For UL and SSL, each point corresponds to a specific algorithm configuration, while the lines denote the Pareto frontiers of DBSCAN and OCSVM, defined as the lowest value of false positives rate obtained for a given false negative rate.

For DBSCAN, most of the points fall on the Pareto frontier, as the algorithm offers a clear trade-off between the false positives and false negatives. For OCSVM, the different kernel functions generate higher scattering of the results. OCSVM also achieves better performance than DBSCAN, i.e., it obtains 1.7% false positives and 5.3% false negatives for the configuration with the highest $f1$ score. In the best case, DBSCAN achieves

26.9% false positive rate and 13.9% false negative rate. While an operator may choose a different parameter setting to lower the false negative rates at the expense of increasing the false positives, the performance shown in the figure needs to be improved before adopting the approach in real deployment scenarios. This means that the ML output should not be used raw, but extra scrutiny must be applied.

10.3.3 Window-based Attack Detection (WAD)

Rather than using the output of an ML model for attack detection (UL or SSL) directly as is for triggering alarms, Window-based Attack Detection (WAD) applies an additional inspection mechanism [9]. In it, an observation window of size δ is defined, along with a threshold τ on the number of positive attack samples that need to be detected before raising an alarm.

The WAD model provides a theoretical framework that allows for mathematically analyzing the probability of raising false and accurate alarms, as well as investigating the trade-off between the detection speed and the false alarm likelihood as a function of time. When these probabilities and trade-offs are assessed for different values of the observation window size δ and the threshold τ, on operator can select the working regime best suited to the considered threat scenario, taking into account also the cost and complexity of remediation actions.

Fig. 10.9 depicts the likelihood of raising an intrusion alarm over time (measured in terms of the number of samples reported to the network management system γ) for a family of different δ and τ settings. The time $\gamma = 0$ denotes the moment of an attack occurrence. All alarms raised before $\gamma = 0$ are false and attributed to the false positives generated by the ML model. When $\tau = 1$, detection of alarms is faster, but comes with a trade-off of a very high rate of false alarms. When τ is large, e.g. 7, the false alarm likelihood detection is driven to zero, but attack detection is slow. For example, when $\delta = 40$ and $\tau = 7$, around 12 samples should be received before the alarm probability approaches 1. Thus, a good operating regime may be around τ equal to 3 or 5, depending on the operator's priorities.

10.4. Runtime complexity of ML models

The runtime complexity of ML models has been gaining attention over the last few years. The reason is twofold. Firstly, the amount of data has been growing, and this directly increases the cost of training and running ML models. This also has a direct environmental impact, especially if the electricity sources used in datacenters are not renewable. The data growth in optical networks in particular is fueled by the increasing scope of monitoring embedded in network devices. Secondly, many applications using ML models require a short response time, which increases the need for models that can run faster. Network security qualifies as an important application scenario where the

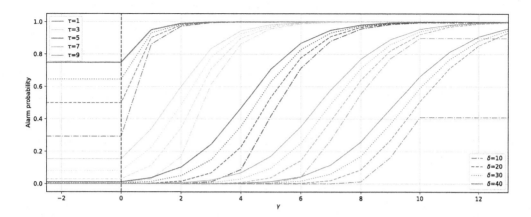

Figure 10.9 False positive and false negative rates when using WAD.

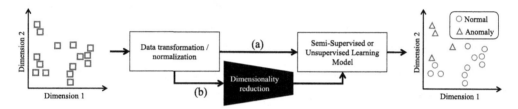

Figure 10.10 The sequence of procedures to apply dimensionality reduction to a dataset, thus modifying the regular workflow (a) into workflow (b).

quick detection and identification of an attack can play a significant role in effective remediation.

Because today's off-the-shelf coherent receivers can provide a very rich OPM dataset whose processing can be demanding and may impact the speed of attack detection, it is important to reduce the processing load by using the most relevant dataset features. To this end, dimensionality reduction techniques can be combined with ML models, especially semi- and unsupervised learning approaches. The sequence of actions taken when an ML model is applied directly to the data compared to the case when dimensionality reduction is incorporated in the workflow is depicted in Fig. 10.10. We consider three dimensionality reduction techniques:

- **Principal Component Analysis (PCA)**

  PCA reduces dataset dimensionality by encoding as much variance from the original dataset in the extracted features as possible [15], as illustrated in Fig. 10.11a. This is achieved by selecting a given number of orthogonal components from the dataset based on their variance. For each component to be extracted from the dataset, PCA finds the linear combination of the original features that results in the highest variance of the projected data points. Once the components are

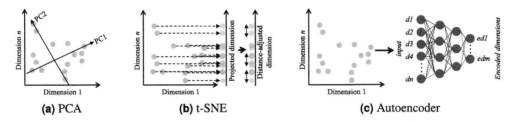

Figure 10.11 Three examples of dimensionality reduction techniques: (a) Principal Component Analysis (PCA); (b) t-distributed Stochastic Neighbor Embedding (t-SNE); and (c) Autoencoder.

selected, high-dimensional data can be projected onto a lower number of components using a linear combination of the original features. As PCA works with projected-dimension-wise steps (instead of, for instance, sample-wise steps), it works efficiently over large number of samples with large number of dimensions in the dataset.

- **t-Distributed Stochastic Neighbor Embedding (t-SNE)**
  t-SNE builds upon the Stochastic Neighbor Embedding with the capability of retaining local and global relations between data points [16], as illustrated in Fig. 10.11b. It starts by computing the pair-wise distances among all samples in the original dataset and then randomly projects the data points over a given number of dimensions. The method then iteratively moves the projected data points so that the distance in the projected dimensions is minimized between points adjacent in the original dataset, and maximized between distant points.

  The direction and amplitude of adjusting the position of each data point (known as gradient) can be computed using different algorithms such as stochastic gradient descent. As t-SNE works with sample-wise steps (as opposed to projected-dimensions-wise, for instance), much higher complexity (and therefore run time) can be expected compared to other dimensionality reduction methods such as PCA and autoencoder. There are also a few simplification methods that reduce the complexity of the t-SNE, making it more suitable for very large datasets with many features [17].

- **Autoencoder (AE)**
  Autoencoders are a type of a neural network, illustrated in Fig. 10.11c, that can be used for several different purposes, including denoising, one-shot learning and dimensionality reduction [18,19]. For the autoencoder, a neural network is used where input and output layers have the same dimension as the dataset. Among the hidden layers, the central layer is responsible for containing the encoded representation of the data. The objective of the autoencoder is for the neural network to

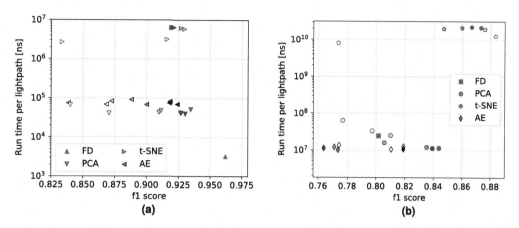

Figure 10.12 Tradeoff between accuracy and runtime obtained by the developed dimensionality reduction models for (a) OCSVM and (b) DBSCAN.

reproduce output values same as the input. The training is performed in a supervised manner, where the same samples are used as input and as the ground truth values at the output.

In an autoencoder built for dimensionality reduction, the feature extraction part is done by a central layer that represents the number of features to be extracted from the dataset. Once the autoencoder is trained, the neural network is split into an encoder and a decoder part, and only the encoder part is used for dimensionality reduction. The values output by the encoder can then be directly input to an SSL or a UL model to perform anomaly detection. The AE needs to be trained only once for each lightpath, at the beginning of its operation. Since the inference of the AE is quite efficient (and can be further assisted by specific-purpose hardware), its overhead in terms of additional run time is expected to be lower than other methods such as t-SNE.

The impact of the three dimensionality reduction techniques on the accuracy and runtime performance of DBSCAN and OCSVM is illustrated in Fig. 10.12. The red (mid gray in print version) icons denote the cases when DBSCAN and OCSVM are applied to the full dataset (FD) without transforming it.

For OCSVM (Fig. 10.12a), there are no observable benefits in applying dimensionality reduction methods, as the FD-OCSVM case yields the highest $f1$ score and runs the fastest. A different trend emerges for DBSCAN (Fig. 10.12b). Combined with either PCA or t-SNE, the accuracy of DBSCAN improves, while the PCA further offers a runtime reduction.

In general, OCSVM performs comparably better than DBSCAN, i.e. achieves higher accuracy at a lower runtime. This indicates that SSL represents a more accurate and more time efficient option for physical-layer attack detection in optical networks.

However, in spite of the obvious disadvantages in terms of lower accuracy and higher running times, the UL technique DBSCAN should not be discarded due to the following. Firstly, the inaccuracies can be mitigated by applying the window-based detection scheme. Secondly, the deployment of SSL in production environments comes at an added complexity of training the model for each new lightpath. Depending on the specific use case, an operator might give preference to simplicity of usage at a slight trade-off with accuracy and runtime.

10.5. Interpretability of ML models

One of the main aspects that hinders the use of ML schemes in real-world deployments is the difficulty in interpreting and explaining the outputs of the ML model. In the physical layer security domain, when an ML model flags an OPM sample as an attack, the first immediate action is to compute and apply countermeasures to the detected attack. However, the countermeasures are only the first step towards completely mitigating the attack. There are several other processes that need to be triggered by the network operation to understand and completely eliminate the threat. These are usually human-dependent (due to the need of specialized knowledge about the inner workings of the network) and time consuming (due to the amount of data to be analyzed).

Root Cause Analysis (RCA) is one of the techniques that can be applied to extract further insights from the ML models. While the term might give the idea that it is meant to find the ultimate root cause of the problem, in the context of ML that is not always possible. There, RCA is meant to identify the different degrees that a given feature has over the output. This means that RCA is not meant to be the final step of an analysis but rather to offer insight into which of the OPM parameters has the greatest impact on the ML output. It also means that RCA only considers the OPM parameters that are collected and used in the ML model. Therefore, if an attack is caused by a soft failure on an optical device that causes it to overheat, the RCA will not be able to precisely discover that unless the device temperature is being monitored and used in the ML model.

Having these limitations in mind, RCA can still provide significant value to the network operation by giving insight into what are the possible causes for the ML model to 'think' an attack is being launched in the network. For instance, in a state-of-the-art optical network deployment, the OPM can collect tens of parameters from each connection. Analyzing or even visualizing such high dimensional data is quite challenging. It becomes even more challenging in the security assessment task due to the fact that these parameters need to be analyzed over time in order to determine what was the state of the connection before and after the attack started. RCA can be used to rank OPM parameters according to their impact on the ML output, which can guide the manual analysis that is usually still required.

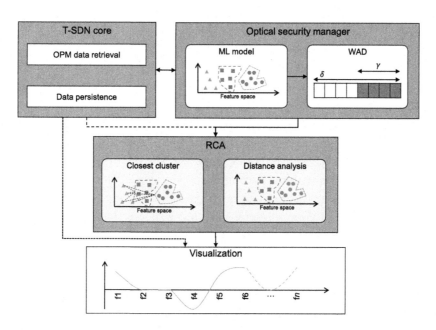

Figure 10.13 General overview of the Root Cause Analysis (RCA) framework applied to physical layer security management.

Fig. 10.13 illustrates the workflow of including RCA into the physical layer security assessment. The RCA module uses the output of the optical security manager and the OPM data collected by the network controller to perform its analysis. In the figure, two approaches are represented, but there are numerous other approaches that can be used [20]. These RCA approaches can also be connected with the specific ML model being used by the optical security manager. Finally, the data output by the RCA module can be combined with the OPM data from the network controller to compose a comprehensive visualization of the attack condition.

10.6. Open challenges

In many aspects, adoption of ML into the management of optical network security is still in its infancy and several important challenges are yet to be tackled before achieving truly autonomous, multi-layer network security management. A set of these challenges is related to the human–ML interaction. While the refinements in ML performance may reduce the level of human interventions, they are highly unlikely to be eliminated. Automation of suitable data processing tasks and attack remediation strategies, development of visualization platforms and improving explainability of ML outputs will be key to unburdening the security experts and allowing them to create advanced solutions for newly emerging threats.

Resolving security breaches might not always be possible through localized action, but may require network-wide collaboration. Multi-domain security management will need to rely on the exchange of relevant data and knowledge on security incidents among multiple stakeholders across different domains without violating confidential or proprietary information. Privacy-preserving federated learning models may provide a useful mechanism for collaborative training such that the cohorts of security managers may secure their networks without sharing possibly sensitive data. However, federated learning is vulnerable to attacks that target specific client or data (e.g. data poisoning, where the models are fed compromised data) and untargeted Byzantine attacks (e.g. model poisoning, aimed at tampering with the learning process) [21]. Should an attacker deceive an ML algorithm to falsely detect attacks, provoking unnecessary response actions can inflict substantial damage to network operators in terms of e.g. enlarged operating expenditures. Possible fraudulent and colluding agents can be thwarted using e.g. blockchain to ensure exchange of verified local model updates.

Tight integration of optical-layer security with existing attack management frameworks will require approaches that are capable of dealing with uncertainties stemming from both the ML models and the observed environment. The time scale of tasks and actions impacted by these uncertainties needs to be taken into account when deciding on the most appropriate form of modeling. For example, less time-critical tasks with a lower level of uncertainty can benefit from adaptive automation of converting learning models to algorithms, processes and workflows, while more time-critical ones with greater sensitivity require acceleration techniques and continuous incremental learning [22].

One of the still largely unaddressed problems is attack remediation, which should encompass definition of protocols and strategies to reconfigure the optical network upon the detection/identification of an attack. Here too ML techniques can play an important role in e.g. long-term tracking of reputation of different network elements and services, as well as adaptation of initially deterministic remediation strategies to cope with an intelligent and adaptive adversary.

The adoption of software-defined control plane in optical networks is another essential component of increasing the dynamicity in optical networks. However, this makes the security of the control plane a major concern as a breach of the 'network brain' may impose severe disruptions on the network connectivity. With ML models proliferating into control plane tasks, once compromised, there is also the risk of adversarial attacks targeting the ML models. Therefore, it is important to ensure the security of the control plane.

10.7. Conclusion

This chapter focused on the challenges related to the security of optical networks, fundamental component of the global communication network infrastructure. In spite of their immense importance underpinning the Internet and the ICT sector, optical networks are almost ubiquitously exposed to adversaries that can gain illegitimate access to the carried information or disrupt services. Moreover, the network operators are lacking tools to detect network breaches and counteract the threats. In this chapter, we discussed how machine learning can help secure the optical network through detection of physical-layer attacks. We examined appropriate supervised, semi-supervised and unsupervised learning techniques and analyzed their accuracy, runtime and interpretability of their outputs. Techniques to improve the performance of ML approaches were assessed: Window-based Attack Detection to improve the accuracy of semi- and unsupervised learning, dimensionality reduction to improve their runtime, and a Root Cause Analysis framework to improve output interpretability. Finally, several open challenges were listed along with possible directions for addressing them.

Acknowledgments

The authors gratefully acknowledge Marco Schiano and Andrea Di Giglio for their help with the attack experiments and Infinera for providing the Groove G30 transponders used therein. This work was supported by the Swedish Research Council (2019-05008), VINNOVA (AI-NET-PROTECT), and the European Commission 5GPPP TeraFlow (101015857).

References

[1] Marsh, Marsh Microsoft 2019 global cyber risk perception survey, 2019.
[2] Cisco, Annual cybersecurity report, 2018.
[3] ETSI, Tackling the challenges of cybersecurity, White Paper, 2016.
[4] Nexus-Net, Reasons why we still need to be careful about fiber optical security, 2017.
[5] N. Skorin-Kapov, M. Furdek, S. Zsigmond, L. Wosinska, Physical-layer security in evolving optical networks, IEEE Communications Magazine 54 (8) (2016) 110–117, https://doi.org/10.1109/MCOM.2016.7537185.
[6] S.K. Miller, Fiber optic network security a necessity, TechTarget, https://searchnetworking.techtarget.com/news/1263785/Fiber-optic-network-security-a-necessity, 2007.
[7] T. Uematsu, H. Hirota, T. Kawano, T. Kiyokura, T. Manabe, Design of a temporary optical coupler using fiber bending for traffic monitoring, IEEE Photonics Journal 9 (6) (2017) 1–13, https://doi.org/10.1109/JPHOT.2017.2762662.
[8] D. FitzGerald, Attacks on fiber networks in California baffle FBI, Wall Street Journal (2015), https://www.wsj.com/articles/attacks-on-fiber-networks-in-california-baffle-fbi-1439417515.
[9] M. Furdek, C. Natalino, F. Lipp, D. Hock, A. Di Giglio, M. Schiano, Machine learning for optical network security monitoring: a practical perspective, Journal of Lightwave Technology 38 (11) (2020) 2860–2871, https://doi.org/10.1109/JLT.2020.2987032.
[10] N. Skorin-Kapov, A. Jirattigalachote, L. Wosinska, An integer linear programming formulation for power equalization placement to limit jamming attack propagation in transparent optical networks, Security and Communication Networks 7 (12) (2014) 2463–2468, https://doi.org/10.1002/sec.958.

[11] N. Skorin-Kapov, J. Chen, L. Wosinska, A new approach to optical networks security: attack-aware routing and wavelength assignment, IEEE Transactions on Networking 18 (3) (2010) 750–760, https://doi.org/10.1109/TNET.2009.2031555.

[12] J. Zhu, B. Zhao, Z. Zhu, Leveraging game theory to achieve efficient attack-aware service provisioning in EONs, IEEE/OSA Journal of Lightwave Technology 35 (10) (2017) 1785–1796, https://doi.org/10.1109/JLT.2017.2656892.

[13] G. Savva, K. Manousakis, G. Ellinas, Eavesdropping-aware routing and spectrum/code allocation in OFDM-based EONs using spread spectrum techniques, Journal of Optical Communications and Networking 11 (7) (2019) 409–421, https://doi.org/10.1364/JOCN.11.000409.

[14] X. Glorot, Y. Bengio, Understanding the difficulty of training deep feedforward neural networks, in: Proceedings of Machine Learning Research, JMLR Workshop and Conference Proceedings, Chia Laguna Resort, Sardinia, Italy, vol. 9, 2010, pp. 249–256, http://proceedings.mlr.press/v9/glorot10a.html.

[15] R. Vidal, Yi Ma, S. Sastry, Generalized principal component analysis (GPCA), IEEE Transactions on Pattern Analysis and Machine Intelligence 27 (12) (2005) 1945–1959, https://doi.org/10.1109/TPAMI.2005.244.

[16] L.v.d. Maaten, G. Hinton, Visualizing data using t-SNE, Journal of Machine Learning Research 9 (Nov) (2008) 2579–2605.

[17] L. Van Der Maaten, Accelerating t-SNE using tree-based algorithms, Journal of Machine Learning Research 15 (1) (2014) 3221–3245.

[18] G.E. Hinton, S. Osindero, Y.-W. Teh, A fast learning algorithm for deep belief nets, Neural Computation 18 (7) (2006) 1527–1554, https://doi.org/10.1162/neco.2006.18.7.1527.

[19] Y. Wang, H. Yao, S. Zhao, Auto-encoder based dimensionality reduction, Neurocomputing 184 (2016) 232–242, https://doi.org/10.1016/j.neucom.2015.08.104.

[20] C. Natalino, A. Di Giglio, M. Schiano, M. Furdek, Root cause analysis for autonomous optical networks: a physical layer security use case, in: European Conference on Optical Communications (ECOC), 2020.

[21] O.A. Wahab, A. Mourad, H. Otrok, T. Taleb, Federated machine learning: survey, multi-level classification, desirable criteria and future directions in communication and networking systems, IEEE Communications Surveys and Tutorials (2021) 1–49, https://doi.org/10.1109/COMST.2021.3058573.

[22] G. Cirincione, D. Verma, Federated machine learning for multi-domain operations at the tactical edge, in: T. Pham (Ed.), Artificial Intelligence and Machine Learning for Multi-Domain Operations Applications, vol. 11006, SPIE, 2019, pp. 29–48.

CHAPTER ELEVEN

Machine learning for design and optimization of photonic devices

Keisuke Kojima[a], **Toshiaki Koike-Akino**[a], **Yingheng Tang**[a,b], **Ye Wang**[a], and **Matthew Brand**[a]

[a]Mitsubishi Electric Research Laboratories (MERL), Cambridge, MA, United States
[b]Electrical and Computer Engineering Dept., Purdue University, West Lafayette, IN, United States

11.1. Introduction

Subwavelength nanostructured materials can be used to control incident electromagnetic (EM) fields into specific transmitted and reflected wavefronts. Recent nanophotonic devices have used such complex structures to enable novel applications in optics, integrated photonics, sensing, and computational metamaterials in a compact and energy-efficient form. These include waveguide-type nanophotonic devices [1–9] and metasurfaces including plasmonics [10–17]. Optimizing nanostructures with a large number of possible combinations of parameters is a challenging task in practice.

EM simulations, including finite-difference time-domain (FDTD) methods and finite element methods (FEM), often require long simulation time, several minutes to hours depending on the size and the mesh of the photonic device, in order to estimate the optical transmission and/or reflection response. For designing nanostructures to achieve a target transmission/reflection profile, a large number of EM simulations needs to be performed, e.g., in a meta-heuristic manner. To resolve the issue, some optimization methods such as direct binary search (DBS) [4] and adjoint method [1,3,18,19] have been implemented. More recently, artificial intelligence using neural networks (NN) has been integrated in an optimization process that can accelerate optimization by reducing the required number of numerical simulations. For example, in [20], we demonstrated how NNs can help to streamline the design process, although the prediction accuracy was limited, due to the shallow network structure.

Deep learning methods are representation-learning techniques obtained by composition of non-linear models that transform the representation at the previous level into a higher and slightly more abstract level in a hierarchical manner [21]. The main idea is that by cascading a large number of such transformations, nearly arbitrary complex functions can be learned in a data-driven fashion using deep neural networks (DNN) [22]. The huge success of deep learning in modeling complex input-output relationship has attracted attention from several scientific communities such as material

Machine Learning for Future Fiber-Optic Communication Systems
https://doi.org/10.1016/B978-0-32-385227-2.00018-8

discovery and design [23–26], high energy physics [27], single molecule imaging [28], medical diagnosis [29], and particle physics [30]. The optical community also started working on signal processing and network automation of optical fiber communications [31–36], and inverse modeling for design of nanostructured optical components using DNN [37–42] Also, there have been reports on optical implementation of artificial neural networks [43–48] as well as characterization of optical pulses [49].

More recently, Liu *et al.* used a tandem NN architecture to learn non-unique EM scattering of alternating dielectric thin films with varying thickness [37]. Peurifoy *et al.* demonstrated DNNs to approximate light scattering of multilayer shell nanoparticles of SiO_2 and TiO_2 using a fully connected DNNs [40]. Asano and Noda provided an NN for prediction of the quality factor in two dimensional photonic crystals [42]. Hammond *et al.* used DNNs for forward and inverse modeling of strip waveguides and Bragg gratings [50]. Hedge paired DNN with evolutionary algorithms to accelerate antireflection coating designs [51]. Banerji *et al.* used reinforcement learning to design nanophotonic power splitters [52]. The design space for integrated photonic devices is considerably larger than previously demonstrated optical scattering applications, and calls for robust deeper networks such as Convolutional Neural Networks (CNN) [22].

Integrated photonic beam splitters have been widely used to equally divide the power into the output ports. Although an arbitrary split ratio can be applied in various applications such as signal monitoring, feedback circuits, or optical quantization [53], the design space is hardly explored due to design complexity. To design photonic power splitters with an arbitrary splitting ratio, the designer often begins with an overall structure based on analytical models and fine tunes the structure using parameter sweeps in numerical simulations. DNN can dramatically change the design process.

Metasurfaces and plasmonics are expected to play a major role in ultracompact and multi-functional lenses and various other applications. Possible applications for fiber-optics include quarter-wave plates, vortex plates to create orbital angular momentum (OAM), and plasmonic collimators [54]. However, designing metasurfaces and plasmonic devices also has the same issue of vast design space and time consuming EM simulations. Including DNNs in the design process will significantly advance the field.

DNNs can be used to predict an optical response of a topology (Forward Modeling) as well as to design a topology given a desired optical response (Inverse Modeling). Another class of DNNs for designing devices and materials is a generative DNN model (Generative Modeling). This paper first overviews these three categories of DNNs applied to designing/optimizing nanophotonic power splitters. We demonstrate that by using deep learning methods, we could efficiently learn the design space of a broadband integrated photonic power splitter to realize a compact device. We also review parallel activities for metasurfaces and plasmonics. The general concept of using DNNs for

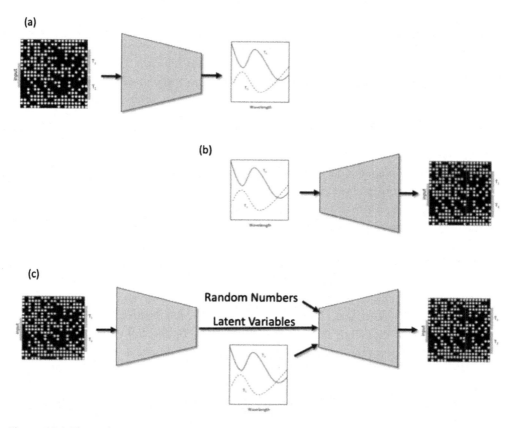

Figure 11.1 Three distinct categories of network models: (a) forward modeling, where the device topology is the input and the optical response is the output, (b) inverse modeling, where the optical response is the input and the device topology is the output, and (c) generative modeling, where the network is first trained with pairs of the device topology and the optical response, and later a new device topology is generated with the optical response and random numbers.

designing/optimizing photonic devices will be readily applicable to a broader area of photonics.

11.2. Deep neural network (DNN) models

Fig. 11.1 shows the three categories or methodologies of the DNN models for designing photonic devices; i.e., forward, inverse, and generative modeling.

The first category is forward modeling, where the DNN takes the device topology as an input and the optical response is the output, as shown in Fig. 11.1(a). In this case, for inverse design, we would use the trained DNN in surrogate optimization, e.g., optimization metric calculation in meta-heuristic optimizers such as DBS and evolutionary strategies (ES). The forward-modeling DNN can skip time-consuming EM simulations to predict the optical response. Let $x \in \mathbb{R}^n$ and $y \in \mathbb{R}^m$ be the topology parameters to design and the corresponding optical response, respectively. Given a device parameter x, the corresponding response y can be analyzed through the EM simulation. Let $f : \mathbb{R}^n \to \mathbb{R}^m$ be such a nonlinear function, transforming the geometry parameter x to the response y as $y = f(x)$. The forward-modeling DNN approximates the function f with a parameterized function f_θ such that $\hat{y} = f_\theta(x)$ is close to the true y. Because of the universal approximation theorem of DNN [55], the forward modeling can predict the response arbitrarily well as the size of DNN and training samples increases. Nevertheless, this forward-modeling DNN does not directly provide an optimized topology and, thus, it is less convenient for inverse design in general.

The second category is inverse modeling, where the optical response is the input and the device topology is the output, as shown in Fig. 11.1 (b). Letting $f^{-1} : \mathbb{R}^m \to \mathbb{R}^n$ be the inverse function of f, the inverse-modeling DNN tries to approximate f^{-1} such that its parameterized function $g_\phi : \mathbb{R}^m \to \mathbb{R}^n$ produces the output of $\hat{x} = g_\phi(y)$ close enough to the original geometry parameter x given the optical response y. In this case, using the trained network, when the target optical response is given, the desired device topology is directly generated. This inverse modeling can be more useful for inverse design than forward modeling as we can directly obtain device topology given a target profile. However, it typically requires more diverse training data as the design space is usually much larger than the diversity of the optical profile (specifically, $\mathbb{H}(x) \gg \mathbb{H}(y)$, where $\mathbb{H}$ denotes entropy, i.e., amount of information) and there is no guarantee that target profile is feasible. In addition, the inverse-modeling DNN usually provides only one topology candidate per target response, which limits the capability to explore better geometries. In fact, the f-function through EM simulations is not always bijective (even non-injective and non-surjective) and, hence, inverse modeling for any target profile y to obtain a single x is challenging (potentially infeasible in a strict sense).

The third category is generative modeling such as conditional autoencoders (CAE) and generative adversarial networks (GAN), wherein the network is trained using the device topology and optical response. Once the network is trained, we give a target optical response as well as random numbers, and the network generates a series of improved topologies. Fig. 11.1 (c) illustrates conditional variational autoencoder framework, where the encoder summarizes the geometry x into a latent representation and the decoder takes as input the randomized variables of the latent representation along with the target profile y as a condition to generate the geometry $\hat{x}$. Omitting the encoder to produce the latent variables, the decoder can be used alone with random latent variable inputs in a manner similar to sample generation with the generator block of a conditional GAN. When we further simplify it by excluding randomness and adversarial training, this model reduces to the inverse modeling. The generative modeling methods can resolve the issues of the forward and inverse modeling approaches as it can directly generate a series of inversely designed geometries.

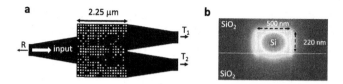

Figure 11.2 Schematic of the SOI-based power splitter. a) Top view, where T_1 and T_2 denote the modal transmissions of output ports 1 and 2, and R denotes the reflection at input port. b) Cross-section of the input/output waveguide [59]. By optimizing the binary sequence of positions of etch hole it is possible to adjust light propagation into either of the ports. In Section 11.3.5, continuous variables are used to represent the variable hole sizes.

11.3. Nanophotonic power splitter

11.3.1 Device structure

Integrated photonic beam splitters have been widely used to equally divide the power into the output ports. Although an arbitrary split ratio can be applied in various applications, the design space is hardly explored due to design complexity. Tian *et al.* [56] demonstrated a silicon-on-insulator (SOI)-based 1×3 coupler with variable splitting ratio in a 15×15 μm^2 device footprint with 60 nm wavelength range and 80% transmission efficiency. Xu *et al.* [57] optimized positioning of squared etched pixels to achieve 80% efficiency for arbitrary ratio power dividers in a 3.6×3.6 μm^2 device footprint. To design photonic power splitters with an arbitrary splitting ratio, the designer often begins with an overall structure based on analytical models and fine tunes the structure using parameter sweeps in numerical simulations. In this section, we consider a nanostructured power splitter with an arbitrary and fixed splitting ratio towards two output ports, targeting flat response with low insertion loss. Such a power splitter can be a building block of many types of photonic integrated circuits for optical communications and various other applications.

We use an SOI-based structure with one input and two output ports having 0.5 μm wide waveguides connected using a taper to the 2.25 μm wide square power splitter design as shown in Fig. 11.2. Each hole is a circle with a maximum diameter of 72 nm that is easily fabricable using well-established lithography methods [19,58]. In order to comply with the fabrication limit, we also put a constraint that the minimum hole diameter is 40 nm, when continous variables are used for the design process as in Section 11.3.5.

11.3.2 Simulation and DNN modeling procedures

Lumerical FDTD simulations are used to generate labeled data for training, where Lumerical script language is used to modify the device topology when training data is generated on a cluster. On the other hand, Python or Matlab® automation is used to modify the device topology for FDTD simulation, when the DNN is included in the optimization loop or active learning. This is only because it is difficult to run Python or Matlab script to automate FDTD simulation on a cluster in our environment. The fundamental transverse electric (TE) mode is used at input source and TE mode output power is recorded for transmission and reflection. It is verified that mode conversion from TE mode to transverse magnetic (TM) mode is negligible ($< 10^{-5}$).

The data for each structure consists of a hole vector (HV) of size 20×20 and labels of its spectral response (SPEC) at the two output ports and reflection at the input port whose size is 63 (3 data sets $\times$ 21 frequency points). In the HV, each hole can be represented by a binary state of 1 for etched ($n_{Silicon}$) and 0 for not etched (n_{Silica}) for the experiments described in Section 11.3.3 and 11.3.4, or continuous variables associated with hole area relative to the maximum size of 72 nm for the experiments described in Section 11.3.5. In order to comply with the fabrication limit, we also put a constraint that the minimum hole diameter is 40 nm, corresponding to the HV value of $40^2/72^2 = 0.31$. Hence, if the HV value is below 0.31, there is no hole. We use random patterned initial HVs and optimize them using heuristic optimization approaches for various optimization metrics to collect a diverse set of labeled training data for supervised learning. Even though each DBS optimization process itself is sequential, multiple DBS processes are run in parallel in a computer cluster, to accelerate the training data collection.

We use the open source machine learning framework of Keras using Tensorflow as a backend in Section 11.3.3, and PyTorch in Sections 11.3.4 and 11.3.5 in the Python language to build and test our DNNs. For the forward modeling case, the DNN needs to be invoked for training and multiple design outputs prior to each FDTD run (> 1000 times for one optimization run), and Tensorflow/Keras has the advantage of shorter start-up time. On the other hand, for inverse and generative modeling, only a few start-ups are needed for the whole design process, and the choice of the framework is more like a historical or personal choice. The training data are generated by FDTD simulations using a high-performance computing cluster with more than 100 processors. The DNN training, testing, and subsequent FDTD simulations are conducted on a computer with a graphics processing unit (GPU) board Nvidia GeForce GTX Titan Z (12 GB memory).

11.3.3 Deep learning for forward modeling to predict optical response

We first describe the forward regression model, using a DNN to predict the transmission and reflection spectra. Given the two-dimensional array (20×20), which is the binary image for the hole locations, we train a DNN using the corresponding transmission and reflection spectra vector which is 63-dimensional, consisting of spectral data for transmission at both port 1, port 2, and reflection at the input port each at 21 discrete wavelengths from 1300 to 1800 nm. Once the DNN is trained, it is used as the predicted (also called "surrogate") spectra within a DBS optimization loop.

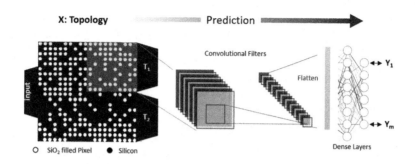

Figure 11.3 Convolutional neural network (CNN) architecture [61].

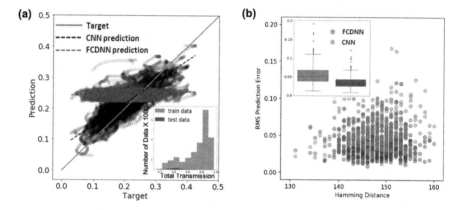

Figure 11.4 (a) Target and predicted total transmittance of CNN and FCDNN, and (b) RMS prediction error vs. hamming distance of CNN and FCDNN [61].

We initially used a fully-connected DNN with multiple layers where each layer has 100 neurons. The number of layers was considered as a hyperparameter which was optimized during the numerical experiments. However, we found that increasing the depth of the fully-connected DNNs did not improve the performance of the network. To achieve better results for deeper networks, we used a residual network (ResNet) to train both the forward and inverse problems as a next step [60].

Since the nanophotonic and photonic crystal designs are analogous to image processing and recognition problems, CNNs [22] are proposed to improve the forward prediction accuracy [42,61]. The 2D CNN architecture is shown in Fig. 11.3, where we treat the input HV data as two-dimensional images.

In our earlier work, we compared the generalization capability of CNN and fully-connected deep neural network (FCDNN) [61]. Fig. 11.4(a) shows the target and predicted total transmittance of CNN and FCDNN, when the tested total transmission is lower than 0.4, i.e., training data and the testing data are generally very far. It is shown that the predicted total transmittance using CNN is much better correlated with the target values. Fig. 11.4(b) shows that the root mean square (RMS) error of predicted total transmittance as a function of Hamming distance (number of holes with different binary values), and CNN shows much smaller RMS error compared to FCDNN. These indicate that 2D CNN achieves superior performance when the input data possess image-like properties.

Our forward modeling network consists of three 2D convolutional layers followed by one fully connected layer. The three convolutional layers have 64, 128, and 256 filters each with kernel sizes of 10×10, 5×5, and 4×4, and strides of 1×1, 2×2, and 2×2, respectively, with the Rectified Linear Unit (ReLU) activation function. The fully connected layer has 256 neurons with the sigmoid activation function [62].

In the network training process, we minimized the mean-square error (MSE) as follows:

$$\text{MSE} = \frac{1}{N} \sum_{\lambda=\lambda_{\min}}^{\lambda_{\max}} \left[\left| T_1(\lambda) - T_1^\star(\lambda) \right|^2 + \left| T_2(\lambda) - T_2^\star(\lambda) \right|^2 + \left| R(\lambda) - R^\star(\lambda) \right|^2 \right], \quad (11.1)$$

where $T_1(\lambda)$, $T_2(\lambda)$ and $R(\lambda)$ denote the transmission at the output ports 1 and 2 and reflection at the input port at a wavelength of λ, respectively, and $N = 21$ is the total number of spectral points. We let $[\cdot]^\star$ denote the corresponding target values. Here, we take a sum across uniformly sampled wavelengths λ from a minimum wavelength $\lambda_{\min}$ to a maximum $\lambda_{\max}$. In the DBS optimization process, we minimized the following loss function:

$$\text{Metric} = \left| T_1 - T_1^\star \right|^2 + \left| T_2 - T_2^\star \right|^2 + \alpha \left| R - R^\star \right|^2, \quad (11.2)$$

where T_1 and T_2 are the lowest transmitted power within the spectral range of 1300 nm and 1800 nm, while R is the largest reflection power. We chose $\alpha = 10$ as a weighting factor. We start with random patterns of 20×20 binary HVs, and generated about 11,000 training data by using the standard DBS, with target splitting ratios of 0 (0 : 10), 0.2 (2 : 8), 0.35 (3.5 : 6.5), and 0.5 (5 : 5). Blue (dark gray in print version) points in Fig. 11.5 show the training data, where the total transmission ($T_1 + T_2$) is plotted against the splitting ratio $T_1/(T_1 + T_2)$. Due to the symmetry, the actual number of distinct data used in the training is about 22,000. We then try to design a power splitter with $T_1^\star = 0.27$ and $T_2^\star = 0.73$, starting from the three initial conditions indicated by the red (mid gray in print version) circles in Fig. 11.5.

In the conventional DBS process, starting with the lower left corner hole, we flip the binary value to evaluate the metric using the FDTD simulation. When the new metric is lower, we choose the new binary value, or if the metric is higher, we revert back to the original binary value. Then, we continue with a sequential application of the same procedure to the other holes.

We use the forward DNN modeling to accelerate the DBS process. In the DNN-assisted DBS, we first train the DNN with 300 epochs using the initial training data, which takes about an hour on the computer with the GPU board. Next we virtually flip each of the 400 holes, using the output of the DNN as a predicted SPEC value. We then select the new pattern with a flipped hole corresponding to the lowest metric and verify the spectra via an FDTD simulation. When the actual FDTD result is better as expected, the flipped pattern is retained, unless otherwise, we try another hole corresponding to the next best metric and verify with an FDTD simulation. We train with one epoch of the original training data and 15 epochs of the accumulated newly acquired data. This is essentially an accelerated DBS using metric values predicted by the DNN.

Fig. 11.6 shows a comparison between a conventional DBS and a DNN-assisted DBS (denoted as DL-DBS), plotting the metric as a function of the number of FDTD runs. It is confirmed that the DNN-assisted DBS optimizes the device structure faster than the conventional DBS and leads to better device designs overall. Note that each FDTD run takes about two minutes, while the additional training (active learning) for each FDTD run takes 20 seconds. Hence, there is an overhead of about 20% per FDTD. Fig. 11.7 shows an example of the optimized device via the DNN-assisted DBS, where the spectral response is very flat across a wide range of wavelengths at least from 1300 to 1800 nm. The total transmittance greater than 87.5% is achieved with a low reflection below 0.5%.

In this section, we verified that the forward modeling is effective to accelerate the optimization process to design high-performance devices. The key is to take advantage of the very fast forward model computation to predict performance of numerous candidates, which can save time-consuming EM simulations.

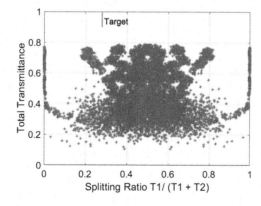

Figure 11.5 Total transmittance of 22,000 training data as a function of the splitting ratio. The red (mid gray in print version) circles indicate the three initial conditions used for training the forward regression model, and the line shows the target splitting.

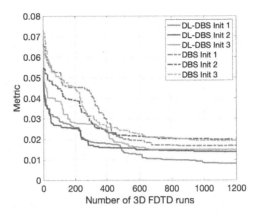

Figure 11.6 Metric as a function of the number of 3D FDTD runs, for the conventional DBS and the DNN-assisted DBS methods. Three different initial conditions are used [62].

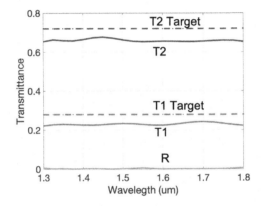

Figure 11.7 Example of a 1 : 3 power splitter designed by the forward regression model. For the range of 1300 nm and 1800 nm, the sum of the worst case T_1 and T_2 was 87.5%, while the worst case reflection was less than 0.5%.

11.3.4 Deep learning for inverse modeling to construct device topology

The forward modeling described in the previous section is generally good enough to predict the performance given the device topology. However, it needs to be combined with an external optimization method such as DBS or other meta-heuristic algorithms to explore different device topologies. In contrast, the inverse modeling takes the spectral response as the input and directly generates the device topology as the output, as shown in Fig. 11.1 (b).

We here consider an inverse modeling DNN consisting of three fully connected layers and two 2D deconvolution layers. The deconvolutional layers, not included in our original inverse modeling paper [60], improve the performance. Note that if we use very small number of performance metrics such as only transmitted power at a single wavelength, we will face a severe issue of non-bijectivity for the f-function, where multiple geometries $\boldsymbol{x}$ may end up having the identical performance $\boldsymbol{y}$, making the inverse design challenging as we discussed. Therefore, it is desirable to have a relatively large number of performance metrics (in our case $m = 63$) to increase the entropy of $\mathbb{H}(\boldsymbol{y})$ in comparison to $\mathbb{H}(\boldsymbol{x})$ for the HV dimension of $n = 400$.

The DNN is trained to predict HV in favor of minimizing the binary cross-entropy (BCE) loss as follows:

$$\text{BCE} = -\sum_{i=1}^{n}\left[x_i\log\hat{x}_i + (1-x_i)\log(1-\hat{x}_i)\right], \tag{11.3}$$

where n is the maximum number of holes, x_i denotes the ith HV value of the training data, and $\hat{x}_i$ is the output of DNN as the corresponding estimate of x_i. The predicted HVs $\hat{x}_i$ can take any value from 0 to 1 from a Bernoulli distribution classifier. The classification tends to converge to either 0 or 1 as the loss reduces by increasing the number of training epochs. We train the network using the same data as used in Section 11.3.3.

To test the generalization capabilities of the network, we investigate the design performance on arbitrary and unseen target cases. We try to optimize the device at around a target splitting ratio of 0.27 as an example, where there are no good samples in the training dataset, as shown in Fig. 11.8. In the design stage, the target splitting ratios are chosen to be $0.23, 0.24, \ldots, 0.32$, and the total transmittances are $0.80, 0.82, 0.84$, and 0.86, and hence we obtain 40 combinations of input spectrum to feed in the inverse-modeling DNN. The DNN output is quantized with a threshold of 0.5, and the binary sequence is then fed back into the FDTD solver. The results are shown as the red (mid gray in print version) circles in Fig. 11.8. In this first round, some data points fill the gap, while many are overlapping with the original training data clouds.

In the second run, we add these 40 new data points to the training data, retrain the network, and repeat this active learning process. The results are shown as the dark blue (dark gray in print version) triangles, and those from the third round are indicated by the green (mid gray in print version) squares in Fig. 11.8. These results show that the inverse modeling has the capability of generating the unseen results out of the training data, and the results further improve with active learning.

In this process, the training takes about half an hour at each round on a high-performance computer with a GPU board. Once trained, the inverse design process is instantaneous (less than 1 s) for 40 devices.

11.3.5 Deep learning for generative modeling to produce device topology candidates

The inverse modeling has a potential to generate an unseen good device structure achieving near the target SPEC. However, the generated topology is usually deterministic given the desired SPEC input. Although we can still employ a stochastic variational

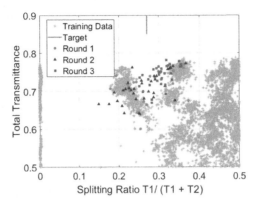

Figure 11.8 Demonstration of inverse design for a power splitter to fill gap around a splitting ratio of 0.27. The light blue (light gray in print version) asterisks, red (mid gray in print version) circles, dark blue (dark gray in print version) triangles, and green (mid gray in print version) squares the training data, the first, the second, and the third round results, where the total transmittance is plotted against the splitting ratio. The red (mid gray in print version) line denotes the target splitting ratio of 0.27 [62].

sampling at the output nodes according to the Bernoulli distribution, the variation of the topology candidates tends to be limited. This can significantly constrain the capability to explore different candidates of topology to achieve improved devices. Besides the forward and inverse modeling, another methodology based on generative modeling has been proposed for photonic devices [37,38,59,63–66]. The generative network can produce a series of improved designs from the training data, based on random number sampling, in a more explicit and systematic way. We have applied generative deep learning models based on an improved version of CVAE to generate integrated photonics devices [66], and later added a concept of active learning to further improve the performance [59].

The CVAE network models the distribution of the device topologies associated with target spectrum characteristics. In our power splitter application, we define the topology pattern using variable-size holes (rather than binary holes). Such pattern has high probability to perform better at light guiding and more stable spectrum response. Our device footprints are 2.25×2.25 μm^2 as previously described in Fig. 11.2.

We first constructed a conventional CVAE [67] as shown in Fig. 11.9(a). The original HVs (20×20) are passed to two convolutional layers [61] and reduced to two sets of intermediate parameters of a probability density function (PDF), representing the means $\mu := (\mu_1, \ldots, \mu_J)$ and standard deviations $\sigma := (\sigma_1, \ldots, \sigma_J)$. In order to have a complete back propagation flow for the network, the reparameterization trick is applied, described by the following equation:

$$z_i = \mu_i + \sigma_i \varepsilon_i, \qquad (11.4)$$

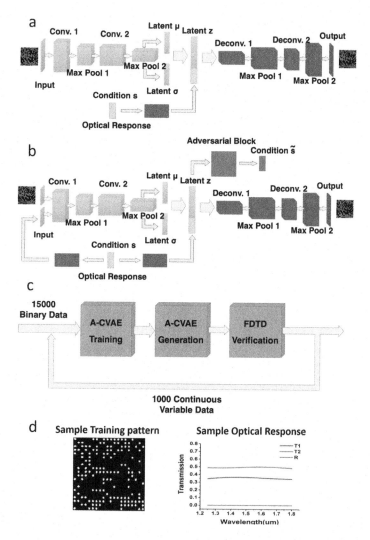

Figure 11.9 a) The CVAE model structure, wherein the input is the 20 × 20 hole vector. The first convolutional layer has 16 channels with a kernel size of 3. The second convolutional layer has 32 channels with a kernel of 3. The condition *s* is the transmission and reflection spectra obtained from the FDTD simulation to form a 3 × 21 matrix, and is fed to the latent variable through a fully connected (FC) layer. b) The A-CVAE model structure, wherein the input right now becomes the two channels of 20 × 20 hole vector, where the first channel is the hole vector of the device and the second channel is the decoded spectra data. The main difference from a) is that one adversarial block is added, which is composed by two FC layers. c) The active learning method added for the A-CVAE method, where 1000 newly generated continuous variable data are added to the original 15,000 binary training data. d) One sample training data, wherein the left figure is the pattern (hole vector) and the right figure shows the three spectra of the training device (the transmission for the two ports and the reflection). Each spectra response has 21 data points, which will be fed into the network as the optical response.

where ε_i are independent and identical distributed samples drawn from the standard Gaussian distribution. After reparameterization, the latent variable $z := (z_1, \ldots, z_J)$ is concatenated with the encoded condition parameter s (the dimension reduced from the original 60 to 9 through one fully connected layer) to deconvolute back to the HV. The loss function for this conventional CVAE is constructed by two parts: a) a cross-entropy loss between the original HV $x := (x_1, \ldots, x_n)$ and the decoded HV $y := (y_1, \ldots, y_n)$, and the Kullback–Leibler (KL) divergence between the latent variables and the Gaussian prior. The first term is to reconstruct the original structure as accurate as possible. The second term is to make the latent variable as close to the standard normal distribution($\mathcal{N}(0, I)$) as possible. The equation of loss function is shown as follows:

$$\text{Loss} = -\sum_{i=1}^{n}\left[y_i \log x_i + (1 - y_i)\log(1 - x_i)\right]$$
$$+ \frac{1}{2}\sum_{j=1}^{J}\left[\mu_j^2 + \sigma_j^2 - \log(\sigma_j^2) - 1\right]. \tag{11.5}$$

For device generation, the trained decoder of the CVAE model is used with the desired condition along with a latent variable sampled from the normal distribution $\mathcal{N}(0, I)$, by which a series of HV topology candidates are generated. The HV will be expressed as hole structure on the MMI with different radius. However, the power splitters generated by CVAE have a total transmission of $\sim 80\%$, which still has quite some room for improvement. We believe the reason is that the latent variable in CVAE tends to be correlated with the condition SPEC s, which will result in degradation of the device performance for the pattern generation because random latent space sampling can adversely impact the target spectra.

To address such a problem of the conventional CVAE model, an Adversarial CVAE (A-CVAE) is introduced as shown in Fig. 11.9(c), where a separate branch to the adversary block is used for isolating the latent variable z from the nuisance variations s (the target SPEC) [68–70]. Our A-CVAE model also has two convolutional layers both for the encoder and the decoder networks (similar to the CVAE model). The encoder is followed by one fully connected layer to obtain the latent variable. The number of channels for the two layers are 16 and 32, and the max pooling stride is 2, after that there is one fully connected layer to reduce the number of the latent variable to 63. The latent variables are then concatenated with the SPEC performance data and are fed into the decoder to generate the output HV. The validations are calculated by using the FDTD simulation to verify a figure of merit (FOM) of generated patterns, where the FOM is calculated by:

$$\text{FOM} = 1 - 10\sum_{\lambda=\lambda_{\min}}^{\lambda_{\max}}\left[\left|T_1(\lambda) - T_1^{\star}(\lambda)\right|^2 + \left|T_2(\lambda) - T_2^{\star}(\lambda)\right|^2 + \alpha\left|R(\lambda) - R^{\star}(\lambda)\right|^2\right], \tag{11.6}$$

where $\alpha = 4$ is used as a weighting factor to balance between the contributions from transmission and the reflection. We take the average of **FOM** over the FDTD simulation spectral range. As the SPEC performance approaches the target, the **FOM** increases towards 1, in which case we obtain an ideal power splitter without excess loss for $R^{\star}(\lambda) = 0$ and $T_1^{\star}(\lambda) + T_2^{\star}(\lambda) = 1$.

We use an encoder structure feeding the performance SPEC s projected on a 20×20 matrix which is combine with the original 20×20 hole vector to form a 2-channel input, then process it through two convolution layers. In A-CVAE, the latent z variable will also be fed into an adversarial block to estimate the SPEC $\bar{s} := (\bar{s}_1, \ldots, \bar{s}_n)$. The loss function for the A-CVAE model is shown as follows:

$$\mathsf{Loss} = - \sum_{i=1}^{n} \Big[y_i \log x_i + (1 - y_i) \log(1 - x_i) \Big]$$
$$+ \frac{1}{2} \sum_{j=1}^{J} \Big[\mu_{zj}^2 + \sigma_{zj}^2 - \log(\sigma_{zj}^2) - 1 \Big] - \frac{\beta}{n} \sum_{i=1}^{n} (s_i - \bar{s}_i)^2. \tag{11.7}$$

The loss function has three parts. The first loss function term is the VAE reconstruction loss in the BCE criterion and the second term is the Kullback–Leibler (KL) divergence. These two terms are the same as in Eq. (11.5). The last term is a regularization term which is the MSE loss of the adversarial block. Since the condition information contained in the latent variable z shall be minimized, the MSE loss between s and $\bar{s}$ needs to be maximized. A complete update of the network generally requires alternating updates in two iteration cycles. The first iteration updates the CVAE model based on the loss function in (11.7). The second iteration updates the adversarial block solely based on the MSE loss between s and $\bar{s}$. The total training time using a computer with a GPU board is around 5 minutes.

Fig. 11.10 shows the training loss and the validation result as a function of the training iterations. The solid black line represents the training loss and the solid blue (dark gray in print version) line represents the trend for validations for the CVAE model. The plot shows that the training result approaches its optimal point when the epoch number is between 10 to 15. The validation results further decreased beyond the epoch number of 30. Here, we used about 15,000 binary training data, including power splitters semi-optimized by DBS with the target splitting ratios of $5:5$, $6.5:3.5$, and $1:0$. The validation result is an FOM calculated over 20 devices of each type. This result clearly indicates that A-CVAE outperforms CVAE with a large margin.

Fig. 11.11(a) shows the latent variable distribution of the CVAE model for groups of devices with four different splitting ratios. The original latent variables are in dimension of 63 and the t-distributed Stochastic Neighbor Embedding (t-SNE) method is used to reduce the dimension to 2 for better visualization. This clearly shows that the four groups are clustered, and their centroids are widely distributed. Fig. 11.11(b) shows the

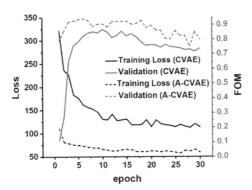

Figure 11.10 The general model performance (training loss and validation result) with different epoch number, for both CVAE and A-CVAE model. Regarding the validation, the FOM is calculated after running FDTD simulation for different generated HV patterns. It shows that the model has generally good performance when the epoch number is between 5 and 15.

similar plot for the A-CVAE model. The figure clearly shows that with adversarial censoring, all the latent variables are distributed similarly, with centroids almost overlapping. They obey the normal distribution $N(0, 1)$, which is expected. It implies that A-CVAE offers more degrees of freedom to generate different potential devices achieving the target performance by sampling normal latent variables concatenated with desired spectra in the conditional decoder.

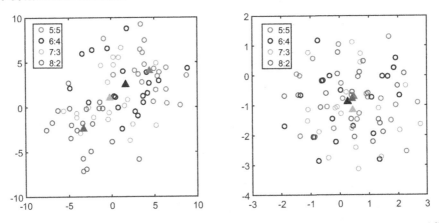

Figure 11.11 t-SNE output of the latent variables. The latent variables are obtained from 4 different group of devices with different splitting ratios. We use the t-SNE to reduce the dimension from 63 to 2 for better visualization. The filled triangle markers are the centroids for each device group. a) Latent variables for the CVAE model, which shows clear clustering. b) Latent variables for the A-CVAE model, where the centroids of the four groups overlap with each other.

The initial training data do not contain any data close to the splitting ratios of 6 : 4, 7 : 3, and 8 : 2, and it is difficult to generate very good designs around these splitting ratios.

So we employ an active learning as shown in Fig. 11.9(c). Here, for the second round of training, we use the new data generated from the first trained model. When we train the second model, the original BCE is replaced with the MSE loss since the training data now contains non-binary data. After training the proposed machine learning model, we test the decoder output of the final A-CVAE model by sampling random latent variable z combined with different splitting conditions to generate nanopatterned power splitter devices according to the generated HVs. Fig. 11.12 shows the comparison of the performance among the devices generated by the CVAE model, by the A-CVAE model without active learning, and by the A-CVAE model with active learning, for four different splitting ratios (5 : 5, 6 : 4, 7 : 3, 8 : 2). The FOM is calculated for 20 randomly generated devices from the trained CVAE, A-CVAE models, and A-CVAE models with active learning. This figure shows that the adversarial censoring and active learning generates devices with much better performance across a very broad bandwidth (from 1250 µm to 1800 µm), compared to the conventional CVAE model. The devices generated by our final A-CVAE model with active learning can fit the target splitting ratio better with excellent total transmission. The average FOMs for the CVAE model, A-CVAE model, and A-CVAE model with active learning are 0.771, 0.888, and 0.9009, respectively.

As the above results show, generative deep learning models can generate a series of improved results based on the statistical characteristics of the training data. In particular, it is that the adversarial censoring and especially with active learning further improves the performance of the model with high stability. Note that part of the good performances of CVAE and A-CVAE comes from the fact that variable hole sizes are adopted, which could not be done in the case of DBS.

11.3.6 Nanophotonic power splitter experiment

In order to verify the validity of the simulations our DNN-based design method, power splitters are prototyped using a commercial multi project wafer (MPW) service by Applied Nanotools Inc. The wafer is processed through a standard 220 nm SOI process with SiO_2 cladding. Direct electron beam writing is used for lithography. In this particular design, the size of the square region is 2.6×2.6 µm^2 and the binary hole size is 90 nm in diameter, from a historical reason.

The scanning electron microscope (SEM) image of a 1 : 3 beam splitter is shown in Fig. 11.13. The device is designed by the inverse model as described in Section 11.3.4 using binary variables, and the holes are clearly defined.

The measurement is conducted with grating couplers, using amplified spontaneous emission (ASE) from an Erbium-doped fiber amplifier as a light source. The transmittance is defined by the ratio of the transmitted power of the device under test and a reference device having the same designs of grating couplers. The measurement wavelength range is limited by the grating couplers and the ASE source. Solid lines in

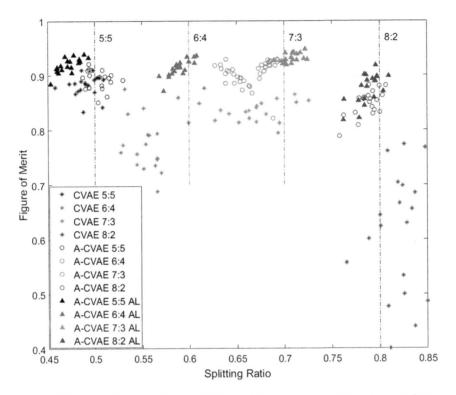

Figure 11.12 FOM comparison for different CVAE models: conventional CVAE (star marker), A-CVAE (round marker) and A-CVAE with active learning (triangle marker). Four different splitting ratios are used as a target value to test the model performance (marked with dashed lines). The devices generated by the active learning assisted CVAE model can fit the target splitting ratio better with excellent total transmission. The average FOM for the three models are: 0.771, 0.888, 0.901, respectively.

Fig. 11.14 show the measured transmittance, and the dashed lines show the transmittance obtained from simulations. The measured values agree well with the simulated values, validating our simulation and optimization results. The cause of the small ripples in the measured transmittance is not clear at this moment.

11.3.7 Comparison with other optimization methods

The adjoint method is widely used in the inverse design of nanophotonic devices [1, 3,71]. Given an ideal initial condition for parameters, the optimization process can be done in a small number of iterations (tens of EM simulations in some cases). However, the initial condition needs to be carefully chosen in order to obtain the optimal result. A DNN method is very different, in that it is trained from a library (training data) of EM simulation results. The training data may come from previous imperfect optimization results with multiple target conditions (splitting ratio in the case of Section 11.3). Then

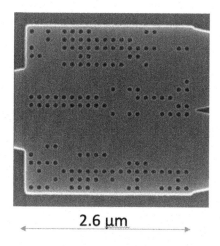

Figure 11.13 An SEM image of a prototyped 1 : 3 power splitter.

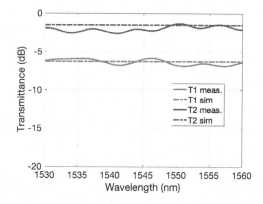

Figure 11.14 Solid and dashed lines show the measured and the simulated transmittance, respectively, of a 1 : 3 power splitter. The red (mid gray in print version) and blue (dark gray in print version) lines show transmittance to output ports 1 and 2, respectively.

the generative model will try to learn/generalize from the library, and generate a series of improved results for a given condition. Once the model is trained, inverse designs for multiple conditions can be generated in almost no time. Further FDTD simulations are not required, and can be used only for verification purposes.

Table 11.1 provides a comparison of power splitters using different optimization methods, including a more conventional y-junction device optimized by classical particle swarm optimization (PSO) [72], and nanophotonic splitters designed by the adjoint method [1,71] and the fast search method (FDM) [57]. Our work has the most broad bandwidth with very small device footprint and low insertion loss. Although, the set-up

Table 11.1 Comparison of simulation results for photonics power splitters using different optimization methods.

Split ratio	Insertion loss	Bandwidth	Footprint	Method	Reference	Setup time	Computational time
1 : 1	0.09 dB	100 nm	$2 \times 2\ \mu m^2$	Adjoint	Lalau-Keraly et al. [1]		NA
1 : 1	0.32 dB	40 nm	$2.6 \times 2.6\ \mu m^2$	Adjoint	Wang et al. [71]		1.2 hr
1 : 1	0.13 dB	80 nm	$1.2 \times 2\ \mu m^2$	PSO	Zhang et al. [72]		NA
4 : 6	1 dB (measured)	30 nm	$3.6 \times 3.6\ \mu m^2$	FSM	Xu et al. [57]		120 hr
4 : 6	0.65 dB	550 nm	$2.25 \times 2.25\ \mu m^2$	A-CVAE	Our work [59]	~ 89 hr*	~ 5 min**
3 : 7	0.51 dB	550 nm	$2.25 \times 2.25\ \mu m^2$	A-CVAE	Our work [59]	~ 89 hr*	~ 5 min**

* Here the time is mostly for the data collection. The data collection is ~ 20 s $\times$ 16,000 = 89 hr.
** The training time is ~ 5 min and the new design generation time is ~ 5 s.

time is long, it is a one-time process, and the actual computational time is much shorter for each new condition, compared to other optimization methods.

11.4. Metasurfaces and plasmonics

Metasurfaces including plasmonics are in principle ultrathin layer of metamaterials which control the wavefront of optical beams [54,73]. There is so much freedom in the design of the nanostructures, and the wavefront can be made to be dependent on the wavelength, polarization, or the combination of both. These are expected to create ultra-thin, high-performance, and multi-functionality lenses. However, there are so many design parameters, so it can be very challenging to design metasurfaces. In this section, we overview the application of DNNs for the design of metasurfaces [38,63–65, 74–79]. Just like we discussed three distinct types of DNN models for the inverse design of nanophotonic power splitters, inverse design on metasurfaces has attracted many research groups working on various types of DNNs. One difference is that in the former group as discussed in Section 11.3, the device structure is represented by $n = 400$ (20 × 20) vectors, while the latter group uses either very small number $(3 − 8)$ of parameters to represent nanodisks or nanopillars, or very large number (as much as 64 × 64 or 45 × 45 × 10) of pixels. Also, in cases where electric field components are used, DNNs typically treat the real and imaginary part of the electric field separately. Nonetheless, the overall concept of connecting the structure parameters to optical responses using DNNs for the inverse design is very similar as we will see in this section.

11.4.1 Deep learning for forward modeling

An *et al.* [75] and Nadell *et al.* [74] used DNNs for forward modeling of the all-dielectric metasurfaces in the terahertz region. They used four cylinders within a unit cell, wherein each cylinder radius is denoted as r_1, r_2, r_3, and r_4, and each cylinder height is denoted as h_1, h_2, h_3, and h_4. In addition to these eight geometrical parameters, the ratio of these parameters, like r_1/h_1, r_1/h_2,... and r_4/h_4 are used as the inputs of the DNN. These ratios, motivated by the knowledge of the underlining physics, minimize the need for network parameters. This is analogous to the "tensor layer" [38], which essentially pre-computes products of input variables. The DNN comprises of nine fully connected layers, essentially upsampling the length 25 input vector to a length 165 output vector, followed by three transposed-convolution layers and a convolution layer, further upsampling to a spectrum of 300 points. The first hidden layer is called "neural tensor network (NTN)", which takes a tensor of input parameters as well as linear inputs. This layer can be related the two inputs multiplicatively instead of only implicitly through nonlinearity. This layer significantly accelerates the training process. An illustration of the DNN is shown in Fig. 11.15. Using 18,000 pairs of structural parameters and spectra data, this forward network is trained and the typical mean square error (MSE) is

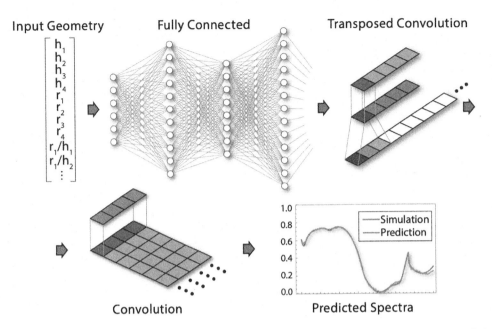

Figure 11.15 Forward DNN architecture for predicting the metasurface optical spectra. Nine fully connected layers feed a set of 24 geometric inputs (four radii, four heights, and 16 ratios of these parameters). Data is then smoothed and upsampled in a learnable manner via transpose convolution layers (top row). The final layer (bottom row) is a convolutional layer, which produces a predicted spectrum of 300 points, shown as the blue (mid gray in print version) curve, compared to the ground truth (red (dark gray in print version) curve) [74].

less than 3.4×10^{03}. The inverse design problem is performed using this trained forward DNN and a full search algorithm, where 814 million (13^8) forward modeling can be done very quickly (9400 spectra/s) using a GPU board.

Wiecha and Muskens used a so-called U-Net to model the electric field in the 3D space [76]. U-Net, a variant of encoder-decoder network comprising of CNNs and shortcut connections, is known to be especially efficient in reconstructing spatial images [80]. As shown in Fig. 11.16, the 3D geometric data (0 for no silicon, 1 for silicon) are used as inputs, and six channels of electric fields (real and imaginary parts of the x, y, and z components of the electric field at 3D grid points) are used as outputs. The direct connection between the corresponding CNN layers in the input side and the output side, called "short cut connection", helps to accurately reconstruct spatial information from the strongly compressed center layers of the DNN. After the training, given the input geometry, the electric fields are reconstructed, which can be used for calculating the near-field and far-field of scattering (reflection and transmission) and various other properties.

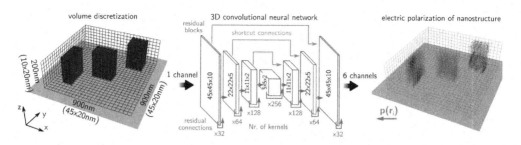

Figure 11.16 U-Net for the example of the silicon nanostructure model. The volume discretization of the three-dimensional geometry is fed into the neural network. The 3D convolutional network follows an encoder-decoder architecture and is organized in a sequence of residual blocks. The six output channels of the network contain the real and imaginary parts of the *x, y,* and *z* components of the complex electric field inside the nanostructure [76].

11.4.2 Deep learning for inverse modeling

An *et al.* [75] proposed to use an inverse modeling network called "a meta-filter model generator". It has four hidden layers with 500, 500, 500, and 50 neurons, respectively, and takes target spectra of 31 spectral points as inputs, and four device structure parameters, index, gap, thickness, and radius of nanocylinders, as outputs as shown in Fig. 11.17 [75]. Then it is connected to a pretrained forward modeling network, called "predicting neural network" (PNN), consisting of four hidden layers with 50, 500, 500, and 200 neurons, respectively. The PNN predicts the real and imaginary parts of the actual reflection spectra. For the training of the inverse network, the difference between the target spectra and the output spectra of the PNN are minimized, while the weights of the PNN are fixed. This cascaded network solves the nonconvergence problem resulting from non-bijective solutions. Once the network is trained, inverse designs can be generated in 22 seconds.

Gao *et al.* [77] proposed to use a tandem (forward and inverse) DNN for modeling silicon metasurfaces for color filter applications. The metasurface involves a unit cell of four equally-spaced silicon nanodisks represented by four parameters, i.e., the diameter (D) and height (H) of the nanodisks, the gap (G) between the two nanodisks within a unit cell, and the period (P) of repeating units, as shown in Fig. 11.18. The reflectance spectrum of the metasurface can be represented by the three values *x, y,* and *Y*. For the forward design, they use a DNN with four hidden layers with 320 neurons in each layer, with the four structural parameters as inputs, and the three color values as outputs. For the inverse design, they use three color parameters as inputs and four structural parameters as outputs. The optimal structure consists of four hidden layers each with 300 neurons. If the inverse network is used alone, the validation error is fairly large. In a tandem DNN as shown in Fig. 11.19, a pretrained network is added at the output of the inverse network. The inverse network is trained such that the differences between the input values *x, y,* and *Y,* and the output values of the pretrained forward network *x′, y′,*

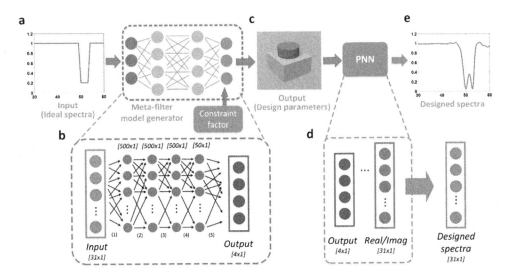

Figure 11.17 Architecture of the meta-filter design network and design examples. (a) Target spectra designated as input. (b) The model generator of the meta-filter design network, which is constructed using DNNs. (c) Output of the model generator, which is a combination of design parameters such as permittivity and meta-atom dimensions. (d) These parameters are then fed into the DNN to yield the complex transmission coefficient. (e) The refined amplitude response of the generated design as derived from the complex transmission coefficient [75].

Figure 11.18 Schematic illustration of the silicon nanostructure and the generated colors. a) The studied geometry parameters include the diameter (D) and height (H) of nanodisks, the gap (G) between nanodisks in a unit and the period (P) of repeating units. b) The coverage of training data generated colors on the CIE 1931 chromatic diagram [77].

and Y' are minimized. In the tandem DNN case, the validation error is significantly lower than the inverse network is used alone, indicating that the non–bijective issue between color and structure is overcome.

11.4.3 Deep learning for generative modeling

There are two mainstreams of generative modeling, i.e., CVAE and GAN. Generative adversarial networks (GANs) are algorithmic architectures that use two neural networks,

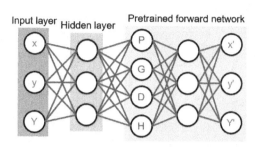

Figure 11.19 Tandem DNNs architecture with an input layer of color values x, y, and Y, hidden layers, and a connected pretrained forward modeling network. The tandem network is trained such that the differences between the input values x, y, and Y, and the output values of the pretrained forward network x', y', and Y' are minimized [77].

pitting one against the other (thus the "adversarial") in order to generate new, synthetic instances of data that can pass for real data, and are used widely in image, video, and voice generation. For photonic device designs, both CVAE and GAN are similar in that they generate new device topology candidates based on random numbers and input conditions.

Ma *et al.* proposed a variant of CVAE for the inverse design of plasmonic metamaterials [64]. The design data consists of 64×64 binary image for the metal patterns, while the optical response consists of three polarization dependent reflection spectra each with 31 frequency points. Fig. 11.20 shows the CVAE model for the inverse design of plasmonic metamaterials, comprising of three submodels, the recognition model, the prediction model, and the generation model. The recognition model encodes the metamaterial pattern with its optical response into a low-dimensional latent space. The prediction model outputs a deterministic prediction of the optical response given the metamaterial design. The generation model accepts the optical response and the sampled latent variable to produce feasible metamaterial designs according to specific requirements. The trained network can predict optical responses when a plasmonic metamaterial structure is given. Also, given target optical responses, a series of metamaterial designs can be generated with sampling of latent variables.

Liu *et al.* proposed using GAN for the inverse design of metasurfaces [63]. As shown in Fig. 11.21, the proposed network consists of three networks, the generator (G), the simulator (S), and the critic (D, also called a discriminator). The generator (an inverse network) accepts the spectra T (four transmission polarization components comprising of 32 elements) and noise z (50 element vector) and produces possible patterns. The simulator is a pretrained forward network that approximates the transmittance spectrum $\hat{T}$ for a given pattern (64×64 binary image) at its input, and the critic evaluates the distance of the distributions between the geometric data and the patterns from the generator. While training the generator, the produced patterns vary according to the

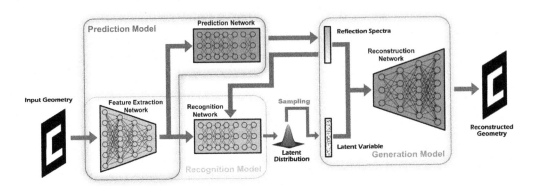

Figure 11.20 CVAE model for inverse design of metamaterials. Three submodels, the recognition model, the prediction model, and the generation model, constitute the complete architecture, which is implemented by four neural networks with deliberately designed structures for different purposes [64].

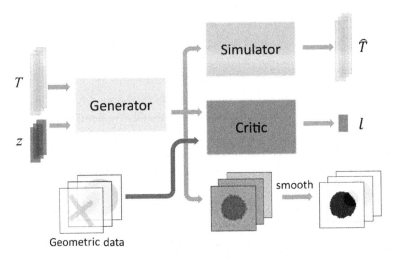

Figure 11.21 Architecture of the proposed GAN. Three networks, the generator (G), the simulator (S), and the critic (D) constitute the complete architecture. The generator accepts the spectra T and noise z and produces possible patterns [63].

feedback obtained from S and D. Valid patterns are documented during the training process, and are smoothed to qualify as candidate structures.

There are also proposals of hybrid approaches to combine the generative model with other optimization technologies. Liu *et al.*, the same group as above, first trained a variational autoencoders (VAE) network as shown in Fig. 11.22(a). The geometric data are 64×64 binary vectors, and the spectral data are T_{xx}, T_{xy}, T_{yx}, and T_{yy}, each with 32 spectral points between 170 THz and 600 THz. Random variables μ, σ, and υ have the same size of 10. The trained decoder works as an inverse network, with randomly sam-

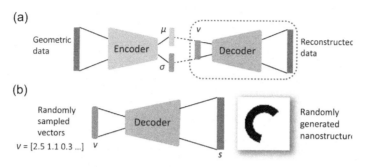

Figure 11.22 Basic architecture of a VAE implemented in the framework. (a) The illustration of the vanilla VAE. The encoder accepts the geometric data and produces two parametric vectors μ and σ. Random vectors υ are samples from the normal distribution with the mean and the standard deviation defined by μ and σ. The decoder then reconstructs vectors υ to images of photonic structures. The VAE encodes the geometric data into a compact latent space where the optimization algorithms can be applied efficiently. (b) After the training, the decoder encircled in (a) can be separated and treated as a generator of geometric data. It transforms randomly sampled vectors v to their correspondent structures [78].

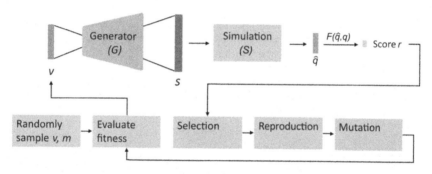

Figure 11.23 Flowchart of the VAE-ES framework. The generator (G) is utilized to reconstruct the structure of each υ. Simulation process (S) is then followed to produce the simulation results. The simulation can be carried out by either a neural network approximation or a real physical simulation such as FEM. The yellow (light gray in print version) boxes show the functionalities of a traditional ES algorithm [78].

pled vectors as inputs, and randomly generated nanostructures as outputs (Fig. 11.23). It takes an input vector of length 10, and outputs 64×64 images, with four transposed convolutional layers. Fig. 11.23 shows the flowchart of framework combining both VAE and an evolutional strategy (ES) [78]. Here, the generator (G) is the trained decoder, and the Simulation (S) can be carried out by either a forward DNN, or a real EM simulation.

Wen *et al.* used progressive growth of GANs (PGGANs) to learn spatially fine features from high–resolution training data. PGGANs are often used in the computer vision community and they support improved training stability and the ability to capture spa-

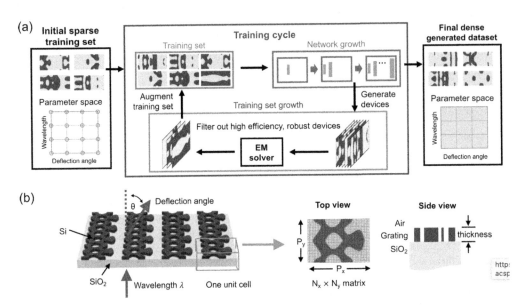

Figure 11.24 Overview of progressively growing GAN (PGGAN) network architectures and training sets. (a) Schematic of the training protocol. The initial training set consists of high-performing devices sampling the coarse grid of the device parameter space. The training process consists of multiple training cycles, each of which involves training the PGGAN from scratch, evaluating/validating the generated devices, and augmenting the training set with the best generated devices. The final trained PGGAN can generate device structures that span the full device parameter space. (b) Freeform silicon metagratings deflect normally incident light to the angle θ [81].

tially fine features from a high-resolution training set. Here, the generative model, the discriminative model, and network resolution progressively grow over the course of training.

Fig. 11.24(a) shows the overview of a PGGAN training protocol. The initial training set consists of high-performing devices sampling the coarse grid of the device parameter space as shown in the left box. The training process includes multiple training cycles, each of which PGGAN is trained from scratch, evaluating/validating the generated devices once the training is complete, and augmenting the training set with the best generated devices (active learning). More specifically, in each training cycle, the training starts with a training set comprising of 8 × 16 pixels, downsampled from the 64 × 128 pixels. The training data sets having 16 × 32, 32 × 64, and 64 × 128 are used as the training progresses, and layer structures of the generator and the discriminator grow accordingly. The final trained PGGAN can output device layouts that continuously span the full device parameter space. The PGGAN is shown to significantly outperform the conventional GAN.

11.5. Other types of optical devices

In this section, we review activities of using DNN for designing optical devices other than nanophotonic power splitters and metasurfaces. The examples below are forward modeling methods and generative modeling methods for waveguide type devices, dielectric films, and photonic crystals.

11.5.1 Deep learning for forward modeling

Hammond and Camacho used DNNs for forward and inverse modeling of strip waveguides and Bragg gratings [50]. In the case of silicon photonic sidewall-corrugated linearly-chirped Bragg gratings, the structural parameters are the width of the first part of the waveguide w_0, thickness (t), waveguide corrugation with difference ($w = w_1 - w_0$), the length of the first grating period (a_0), and the last grating period (a_1). The output properties are the reflection spectrum and group delay response. They generated 104,000 training grating structures with 250 wavelength points, using a layered dielectric media transfer matrix method (LDMTMM). The DNN consists of 10 hidden layers and 128 neurons for each layer. Inverse design is performed using gradient-based methods by directly evaluating the Jacobian and Hessian tensors from the DNN without any extra sampling or discretization.

In order to accelerate design antireflection coating, Hedge paired DNN with evolutionary algorithms [51]. The structure is an air-clad 16-layered thin-film made of alternating layers of silica and titania on a semi-infinite substrate of refractive index 1.52. The input parameters are the thicknesses of each layer. The output parameters are reflection spectra with 64 points between 400 nm and 800 nm. The DNN comprises of fully connected 3 hidden layers each with 128 neurons. Once the DNN is trained with > 50,000 training data, differential evolution (DE) algorithm, which is one of the evolutionary strategy variants, uses the output of the DNN, instead of the actual calculated reflection spectra by "oracle", which is an actual optical simulator based on the transfer matrix method (TMM) [82]. This DNN-assisted DE significantly accelerates the optimization process.

Asano and Noda used a DNN as a forward regression model to predict the Q factor, to optimize the microcavity structure [42,83]. Fig. 11.25 shows the schematic of the photonic crystal structure with a three-missing-air-holes (L3) cavity as the base structure. The displacement of the 50 air holes is the structural parameters to be optimized to maximize Q. The network structure is shown in Fig. 11.26, where the input nodes are two-channel 2D tables with each channel corresponds to the x and y components of the displacement vector with a length of 50. The CNN layer consists of 50 filters with a size of 3 holes × 5 rows × 2 channels. After training a CNN using 1000 initial training data calculated from 3D FDTD, a gradient method is used to obtain semi-optimal nanocavity structures. 70 new candidates are selected and 3D FDTD verifies the Q factor. The new data are added to the training data to update the CNN (active learning), and repeat the process. A record Q factor of 1.1×10^7 is obtained. They found that exploring structure not only near the present highest Q factor, but also those distant from it, accelerate the optimization process. Effects of fabrication errors on the Q factor are also analyzed using the CNN model.

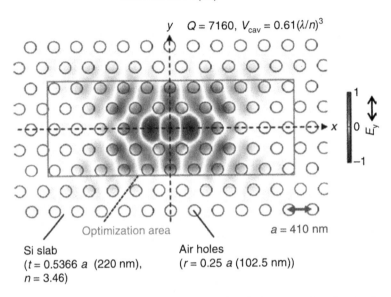

Figure 11.25 A three-missing-air-holes (L3) cavity is used as the base structure for structural optimization. The lattice constant a is 410 nm. The distribution of the y component of the electric field (E_y) of the fundamental resonant mode is plotted in color. The theoretical Q factor of the base structure determined by FDTD is 7160. The displacements of the 50 air holes inside the red (mid gray in print version) square are the structural parameters that are used to optimize the cavity design with respect to Q [83].

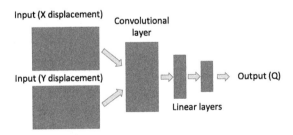

Figure 11.26 Configuration of the neural network prepared to learn the relationship between displacements of air holes and Q factors.

11.5.2 Deep learning for generative modeling

Christensen *et al.* [84] used DNNs for predicting and generating photonic crystals as shown in Fig. 11.27. They first used a CNN-based encoder and a decoder network to predict the band gap of photonic crystals as shown in Fig. 11.28. The encoder consists of three CNN layers and two fully connected layers, mapping 32×32 input space into a linear 64-dimensional feature space (latent variable). The decoder was implemented with six feed-forward networks, each consisting of five fully-connected layers that were separately optimized for each band. The relative error for the predicted bandgap is typically very small ($< 2\%$). As for data generation, they adapted a standard off-the-shelf implementation of GAN [85]. The input data are 64×64 unit cell permittivity profiles, in which TM band gap is greater than 5%. The discriminator network outputs real when the bandgap is greater than 5% as shown in Fig. 11.29. With enough training, the GAN generates new data satisfying the bandgap greater than 5%. They compared a conventional GAN, a least squares GAN (LSGAN), and Deep Regret Analytic GAN (DRAGAN), and the fidelity (fraction of the unit cells hosting a band gap greater than 5% is slightly higher with the LSGAN than the conventional GAN, while it was lowest with the DRAGAN in this particular problem.

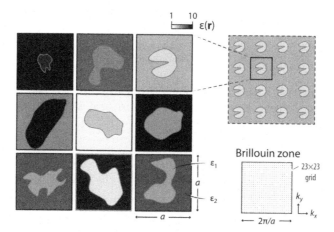

Figure 11.27 Several representative unit cells and the Brillouine grid-sampling used in the calculation of band structures [84].

11.6. Discussion

DNNs can be used to take device structure data (shape, depth, and permittivity) to predict the optical response of the nanostructure in the forward modeling framework. In this case DNNs can be used as a viable counterpart for fast approximation of the optical response, in comparison to the use of computationally heavy FDTD or FEM

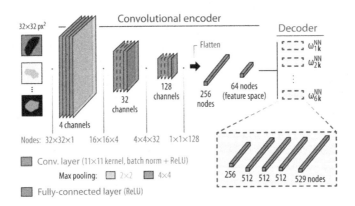

Figure 11.28 Predictive network architecture showing the convolutional encoder and fully-connected decoder [84].

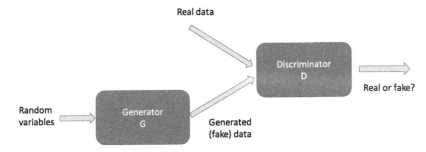

Figure 11.29 GAN consists of a generator (G) and a discriminator (D), and new synthetic examples (fake) of 2D unit cells can be generated from a genuine data set (real) with a TM bandgap greater than 5%.

simulations. In contrast, inverse-modeling DNNs take an optical response to provide the user with an approximate solution of nanostructure in the inverse modeling framework, which does not rely on external meta-heuristic optimizers unlike forward modeling. The generative model, on the other hand, implicitly integrates forward and inverse models. Once the network is trained, the generation of new designs takes practically little time. In the case of CVAE, the use of adversarial censoring further improves the design capability. Overall, DNNs have the capability of learning the training data with a high generalizability.

Although the DNN initially needs a sufficient amount of data for the training purpose, it is possible to process several heuristic optimization metrics in parallel on a computing cluster to speed up generating the training data. We can design the nanostructured geometry in a fraction of a second once the network is trained to represent the topology as optical response and vice versa. There is an overhead of generating large amount of training data upfront, however, the data can be accumulated from multiple optimization runs, such as DBS, which did not necessarily generate optimal results, or targeting different splitting ratios or other optical performances.

We have seen that there are multiple ways to generate new device structures, and also some of them use hybrid optimization algorithms. At this moment, there is no comprehensive study to compare all the variants of networks architectures using a common data set. Therefore, we do not know if one network architecture works best for most of the device categories, or different device categories call for different network architectures. This will be the task for the future.

There are also many more future research items in this field. We can incorporate many of the deep learning techniques developed in other fields and apply them to photonic devices. Some of the examples are as follows.

1) In most DNN applications, a small number of hyperparameters (e.g., the number of hidden nodes and hidden layers) are manually optimized, however, there is no guarantee that they are optimal. Determining such hyperparameters may often require human effort and insight to design the DNN architecture. Automation of hyperparameter and architecture exploration in the context of AutoML [86–94] will facilitate DNN design suited for photonic device optimization.

2) Transfer learning and fine tuning consist of learning features on one problem, and transferring them to a new, similar problem [95]. They are usually used for tasks where the dataset is too limited in size for training a full-scale model from scratch. For the inverse design of photonic devices, we may use the transfer learning and/or fine tuning by pre-training a network using a larger set of training data, for example, using the 2D FDTD simulations using effective refractive index. Then we generate a smaller set of new training data in a different domain, for example, 3D FDTD simulation data, experimental data, or slightly different device or material parameters such as silicon layer thickness. For the case of transfer learning, we take layers from the pre-trained network and make them frozen in order to avoid destroying any important features, and add some trainable layers on top of them, which are then trained using the new training data set for the generation of devices in the new domain. The fine-tuning is similar but is different in that the whole network is unfrozen and can be trained to the new domain.

11.7. Conclusion

We gave an overview of how photonic devices can be designed using deep learning, and how they are different from conventional optimization processes. Especially, to design complicated nanophotonic devices and metasurfaces with hundreds or thousands of parameters, a sophisticated design algorithm is necessary, and deep learning offers a promising solution. Most of the problems are characterized as structural parameters as inputs and transmission/reflection spectral characteristics as outputs.

We demonstrated three different types of DNN methodologies, i.e., forward modeling, inverse modeling, and generative modeling, to design nanophotonic power splitters.

The forward modeling uses a DNN to predict the spectral performances given the device topology. The DNN is integrated with an optimization method, to reduce the requirement of computationally intensive FDTD simulations. As more data are accumulated for online training, the prediction accuracy further improves. The inverse modeling uses a DNN to directly generate the device topology given a target performance. By supplying a series of modified SPEC performance, a series of good device structure candidates is generated. This can avoid the use of external optimizer methods. For the generative model, we reviewed a CVAE with adversarial censoring. Once trained, the CVAE can generate a series of improved device structures given spectral performance data as a condition along with random variables sampled from the normal distribution. We confirmed that the adversarial censoring significantly improves the design capability.

We also gave an overview of parallel activities of these DNN modeling approaches for metasurfaces and plasmonics and other photonic devices.

Among all the efforts and activities for the design of photonic devices, we can find some common themes and techniques. 1) The current general trend is moving towards generative modeling. 2) When the structure is represented in periodic parameters such as grids, 2D/3D CNN is a very powerful tool. 3) Use large number of output values (such as fine spectrums) instead of a few numbers, to minimize non-bijectivity issue. 4) If complex numbers are used, such as electric fields, real and imaginary part are handled separately. 5) When small number of input structural parameters are used, if the physics allows, using multiplications or ratios of these parameters in addition to linear terms, can accelerate training or improve performance. 6) Active learning usually helps to improve the DNN for iterative design/optimization processes.

The area of deep learning-assisted photonic device design is quickly expanding, and these modeling methods will play important roles in the advancement of photonic device design activities. The future task will involve determining if there is a single network architecture or framework which work for almost all the device types, or each device category has one specific framework which work best for it.

References

[1] C.M. Lalau-Keraly, S. Bhargava, O.D. Miller, E. Yablonovitch, Adjoint shape optimization applied to electromagnetic design, Optics Express 21 (18) (2013) 21693–21701.

[2] A. Silva, F. Monticone, G. Castaldi, V. Galdi, A. Alù, N. Engheta, Performing mathematical operations with metamaterials, Science 343 (6167) (2014) 160–163.

[3] A.Y. Piggott, J. Lu, K.G. Lagoudakis, J. Petykiewicz, T.M. Babinec, J. Vučković, Inverse design and demonstration of a compact and broadband on-chip wavelength demultiplexer, Nature Photonics 9 (6) (2015) 374–377.

[4] B. Shen, P. Wang, R. Polson, R. Menon, An integrated-nanophotonics polarization beamsplitter with 2.4×2.4 μm 2 footprint, Nature Photonics 9 (6) (2015) 378–382.

[5] M. Teng, K. Kojima, T. Koike-Akino, B. Wang, C. Lin, K. Parsons, Broadband SOI mode order converter based on topology optimization, in: 2018 Optical Fiber Communications Conference and Exposition (OFC), 2018, pp. 1–3.

[6] A. Motayed, G. Aluri, A.V. Davydov, M.V. Rao, V.P. Oleshko, R. Bajpai, M.E. Zaghloul, B. Thomson, B. Wen, T. Xie, et al., Highly selective nanostructure sensors and methods of detecting target analytes, US Patent 9,983,183, May 29 2018.

[7] Z. Chu, Y. Liu, J. Sheng, L. Wang, J. Du, K. Xu, On-chip optical attenuators designed by artificial neural networks, in: 2018 Asia Communications and Photonics Conference (ACP), IEEE, 2018, pp. 1–3.

[8] Z. Liu, X. Liu, Z. Xiao, C. Lu, H.-Q. Wang, Y. Wu, X. Hu, Y.-C. Liu, H. Zhang, X. Zhang, Integrated nanophotonic wavelength router based on an intelligent algorithm, Optica 6 (10) (2019) 1367–1373.

[9] M. Teng, A. Honardoost, Y. Alahmadi, S.S. Polkoo, K. Kojima, H. Wen, C.K. Renshaw, P. LiKamWa, G. Li, S. Fathpour, et al., Miniaturized silicon photonics devices for integrated optical signal processors, Journal of Lightwave Technology (2019).

[10] X. Ni, Z.J. Wong, M. Mrejen, Y. Wang, X. Zhang, An ultrathin invisibility skin cloak for visible light, Science 349 (6254) (2015) 1310–1314.

[11] A. Alù, N. Engheta, Achieving transparency with plasmonic and metamaterial coatings, Physical Review E 72 (1) (2005) 016623.

[12] F. Monticone, N.M. Estakhri, A. Alù, Full control of nanoscale optical transmission with a composite metascreen, Physical Review Letters 110 (2013) 203903, https://doi.org/10.1103/PhysRevLett.110.203903.

[13] E. Arbabi, A. Arbabi, S.M. Kamali, Y. Horie, A. Faraon, Multiwavelength polarization-insensitive lenses based on dielectric metasurfaces with meta-molecules, Optica 3 (6) (2016) 628–633.

[14] A.K. Azad, W.J. Kort-Kamp, M. Sykora, N.R. Weisse-Bernstein, T.S. Luk, A.J. Taylor, D.A. Dalvit, H.-T. Chen, Metasurface broadband solar absorber, Scientific Reports 6 (2016) 20347.

[15] M. Khorasaninejad, W.T. Chen, R.C. Devlin, J. Oh, A.Y. Zhu, F. Capasso, Metalenses at visible wavelengths: diffraction-limited focusing and subwavelength resolution imaging, Science 352 (6290) (2016) 1190–1194.

[16] A. Krasnok, M. Tymchenko, A. Alù, Nonlinear metasurfaces: a paradigm shift in nonlinear optics, Materials Today 21 (1) (2018) 8–21, https://doi.org/10.1016/j.mattod.2017.06.007.

[17] R. Pestourie, C. Pérez-Arancibia, Z. Lin, W. Shin, F. Capasso, S.G. Johnson, Inverse design of large-area metasurfaces, Optics Express 26 (26) (2018) 33732–33747.

[18] L.H. Frandsen, O. Sigmund, Inverse design engineering of all-silicon polarization beam splitters, in: Photonic and Phononic Properties of Engineered Nanostructures VI, vol. 9756, International Society for Optics and Photonics, 2016, p. 97560Y.

[19] A.Y. Piggott, J. Petykiewicz, L. Su, J. Vučković, Fabrication-constrained nanophotonic inverse design, Scientific Reports 7 (1) (2017) 1786.

[20] K. Kojima, B. Wang, U. Kamilov, T. Koike-Akino, K. Parsons, Acceleration of FDTD-based inverse design using a neural network approach, in: Integrated Photonics Research, Silicon and Nanophotonics, Optical Society of America, 2017, p. ITu1A–4.

[21] Y. LeCun, Y. Bengio, G. Hinton, Deep learning, Nature 521 (7553) (2015) 436–444.

[22] A. Krizhevsky, I. Sutskever, G.E. Hinton, ImageNet classification with deep convolutional neural networks, Communications of the ACM 60 (6) (2017) 84–90, https://doi.org/10.1145/3065386.

[23] J. Ghaboussi, J. Garrett Jr, X. Wu, Knowledge-based modeling of material behavior with neural networks, Journal of Engineering Mechanics 117 (1) (1991) 132–153.

[24] R. Gómez-Bombarelli, J.N. Wei, D. Duvenaud, J.M. Hernández-Lobato, B. Sánchez-Lengeling, D. Sheberla, J. Aguilera-Iparraguirre, T.D. Hirzel, R.P. Adams, A. Aspuru-Guzik, Automatic chemical design using a data-driven continuous representation of molecules, ACS Central Science 4 (2) (2018) 268–276.

[25] B. Sanchez-Lengeling, A. Aspuru-Guzik, Inverse molecular design using machine learning: generative models for matter engineering, Science 361 (6400) (2018) 360–365.

[26] Z. Shi, E. Tsymbalov, M. Dao, S. Suresh, A. Shapeev, J. Li, Deep elastic strain engineering of bandgap through machine learning, Proceedings of the National Academy of Sciences 116 (10) (2019) 4117–4122.

[27] P. Baldi, P. Sadowski, D. Whiteson, Searching for exotic particles in high-energy physics with deep learning, Nature Communications 5 (2014) 4308.

[28] M. Yasui, M. Hiroshima, J. Kozuka, Y. Sako, M. Ueda, Automated single-molecule imaging in living cells, Nature Communications 9 (1) (2018) 3061.

[29] Y. Jun, T. Eo, T. Kim, H. Shin, D. Hwang, S.H. Bae, Y.W. Park, H.-J. Lee, B.W. Choi, S.S. Ahn, Deep-learned 3D black-blood imaging using automatic labelling technique and 3D convolutional neural networks for detecting metastatic brain tumors, Scientific Reports 8 (1) (2018).

[30] A. Radovic, M. Williams, D. Rousseau, M. Kagan, D. Bonacorsi, A. Himmel, A. Aurisano, K. Terao, T. Wongjirad, Machine learning at the energy and intensity frontiers of particle physics, Nature 560 (7716) (2018).

[31] F.N. Khan, Y. Zhou, A.P.T. Lau, C. Lu, Modulation format identification in heterogeneous fiber-optic networks using artificial neural networks, Optics Express 20 (11) (2012) 12422–12431.

[32] T. Koike-Akino, Perspective of statistical learning for nonlinear equalization in coherent optical communications, in: Advanced Photonics for Communications, Optical Society of America, 2014, p. ST2D.2.

[33] D. Zibar, M. Piels, R. Jones, C.G. Schäeffer, Machine learning techniques in optical communication, Journal of Lightwave Technology 34 (6) (2015) 1442–1452.

[34] D. Rafique, L. Velasco, Machine learning for network automation: overview, architecture, and applications (invited tutorial), Journal of Optical Communications and Networking 10 (10) (2018) D126–D143, https://doi.org/10.1364/JOCN.10.00D126.

[35] B. Karanov, M. Chagnon, F. Thouin, T.A. Eriksson, H. Bülow, D. Lavery, P. Bayvel, L. Schmalen, End-to-end deep learning of optical fiber communications, Journal of Lightwave Technology 36 (20) (2018) 4843–4855.

[36] T. Koike-Akino, Y. Wang, D.S. Millar, K. Kojima, K. Parsons, Neural turbo equalization to mitigate fiber nonlinearity, in: European Conference on Optical Communication (ECOC), 2019, p. Tu.1.B.1.

[37] D. Liu, Y. Tan, E. Khoram, Z. Yu, Training deep neural networks for the inverse design of nanophotonic structures, ACS Photonics 5 (4) (2018) 1365–1369.

[38] W. Ma, F. Cheng, Y. Liu, Deep-learning enabled on-demand design of chiral metamaterials, ACS Nano 12 (6) (2018) 6326–6334.

[39] I. Malkiel, M. Mrejen, A. Nagler, U. Arieli, L. Wolf, H. Suchowski, Deep learning for the design of nano-photonic structures, in: 2018 IEEE International Conference on Computational Photography (ICCP), 2018, pp. 1–14.

[40] J. Peurifoy, Y. Shen, L. Jing, Y. Yang, F. Cano-Renteria, B.G. DeLacy, J.D. Joannopoulos, M. Tegmark, M. Soljačić, Nanophotonic particle simulation and inverse design using artificial neural networks, Science Advances 4 (6) (2018), https://doi.org/10.1126/sciadv.aar4206, http://advances.sciencemag.org/content/4/6/eaar4206.full.pdf.

[41] Y. Sun, Z. Xia, U.S. Kamilov, Efficient and accurate inversion of multiple scattering with deep learning, Optics Express 26 (11) (2018) 14678–14688.

[42] T. Asano, S. Noda, Optimization of photonic crystal nanocavities based on deep learning, Optics Express 26 (25) (2018) 32704–32717.

[43] J. Ohta, K. Kojima, N. Y, T. Shuichi, K. Kazuo, Optical neurochip based on a three-layered feedforward model, Optics Letters 15 (23) (1990) 1362–1364.

[44] A.N. Tait, T.F. Lima, E. Zhou, A.X. Wu, M.A. Nahmias, B.J. Shastri, P.R. Prucnal, Neuromorphic photonic networks using silicon photonic weight banks, Scientific Reports 7 (1) (2017).

[45] A. Mehrabian, Y. Al-Kabani, V.J. Sorger, T. El-Ghazawi, PCNNA: a photonic convolutional neural network accelerator, arXiv preprint, arXiv:1807.08792, 2018.

[46] X. Lin, Y. Rivenson, N.T. Yardimci, M. Veli, Y. Luo, M. Jarrahi, A. Ozcan, All-optical machine learning using diffractive deep neural networks, Science (2018), https://doi.org/10.1126/science.aat8084.

[47] J. Chiles, S.M. Buckley, S.W. Nam, R.P. Mirin, J.M. Shainline, Design, fabrication, and metrology of 10×100 multi-planar integrated photonic routing manifolds for neural networks, APL Photonics 3 (10) (2018).

[48] T.W. Hughes, M. Minkov, Y. Shi, S. Fan, Training of photonic neural networks through in situ backpropagation and gradient measurement, Optica 5 (7) (2018) 864–871.

[49] W. Xiong, B. Redding, S. Gertler, Y. Bromberg, H.D. Tagare, H. Cao, Deep learning of ultrafast pulses with a multimode fiber, APL Photonics 5 (9) (2020) 096106.

[50] A.M. Hammond, R.M. Camacho, Designing integrated photonic devices using artificial neural networks, Optics Express 27 (21) (2019) 29620–29638.

[51] R.S. Hegde, Photonics inverse design: pairing deep neural networks with evolutionary algorithms, IEEE Journal of Selected Topics in Quantum Electronics 26 (1) (2019) 1–8.

[52] S. Banerji, A. Majumder, A. Hamrick, R. Menon, B. Sensale-Rodriguez, Machine learning enables design of on-chip integrated silicon T-junctions with footprint of 1.2 μm × 1.2 μm, arXiv preprint, arXiv:2004.11134, 2020.

[53] Z. Kang, X. Zhang, J. Yuan, X. Sang, Q. Wu, G. Farrell, C. Yu, Resolution-enhanced all-optical analog-to-digital converter employing cascade optical quantization operation, Optics Express 22 (18) (2014) 21441–21453.

[54] N. Yu, F. Capasso, Optical metasurfaces and prospect of their applications including fiber optics, Journal of Lightwave Technology 33 (12) (2015) 2344–2358, http://jlt.osa.org/abstract.cfm?URI= jlt-33-12-2344.

[55] Y. Lu, J. Lu, A universal approximation theorem of deep neural networks for expressing probability distributions, Advances in Neural Information Processing Systems 33 (2020).

[56] Y. Tian, J. Qiu, M. Yu, Z. Huang, Y. Qiao, Z. Dong, J. Wu, Broadband 1×3 couplers with variable splitting ratio using cascaded step-size MMI, IEEE Photonics Journal 10 (3) (2018) 1–8.

[57] K. Xu, L. Liu, X. Wen, W. Sun, N. Zhang, N. Yi, S. Sun, S. Xiao, Q. Song, Integrated photonic power divider with arbitrary power ratios, Optics Letters 42 (4) (2017) 855–858.

[58] L. Lu, M. Zhang, F. Zhou, D. Liu, An ultra-compact colorless 50: 50 coupler based on PhC-like metamaterial structure, in: Optical Fiber Communications Conference and Exhibition (OFC), 2016, IEEE, 2016, pp. 1–3.

[59] Y. Tang, K. Kojima, T. Koike-Akino, Y. Wang, P. Wu, Y. Xie, M.H. Tahersima, D.K. Jha, K. Parsons, M. Qi, Generative deep learning model for inverse design of integrated nanophotonic devices, Laser & Photonics Reviews (2020) 2000287.

[60] M.H. Tahersima, K. Kojima, T. Koike-Akino, D. Jha, B. Wang, C. Lin, K. Parsons, Deep neural network inverse design of integrated photonic power splitters, Scientific Reports 9 (1) (2019) 1368.

[61] M.H. Tahersima, K. Kojima, T. Koike-Akino, D. Jha, B. Wang, C. Lin, K. Parsons, Nanostructured photonic power splitter design via convolutional neural networks, in: CLEO: Science and Innovations, Optical Society of America, 2019, p. SW4J-6.

[62] K. Kojima, M.H. Tahersima, T. Koike-Akino, D. Jha, Y. Tang, Y. Wang, K. Parsons, Deep neural networks for inverse design of nanophotonic devices (invited), Journal of Lightwave Technology 39 (4) (2021) 1010–1019.

[63] Z. Liu, D. Zhu, S.P. Rodrigues, K.-T. Lee, W. Cai, Generative model for the inverse design of metasurfaces, Nano Letters 18 (10) (2018) 6570–6576.

[64] W. Ma, F. Cheng, Y. Xu, Q. Wen, Y. Liu, Probabilistic representation and inverse design of metamaterials based on a deep generative model with semi-supervised learning strategy, Advanced Materials 31 (35) (2019) 1901111.

[65] S. An, B. Zheng, H. Tang, M.Y. Shalaginov, L. Zhou, H. Li, T. Gu, J. Hu, C. Fowler, H. Zhang, Multifunctional metasurface design with a generative adversarial network, arXiv preprint, arXiv:1908.04851, 2019.

[66] Y. Tang, K. Kojima, T. Koike-Akino, Y. Wang, P. Wu, M.H. Tahersima, D. Jha, K. Parsons, M. Qi, Generative deep learning model for a multi-level nano-optic broadband power splitter, in: 2020 Optical Fiber Communications Conference and Exhibition (OFC), 2020, p. Th1A.1.

[67] K. Sohn, H. Lee, X. Yan, Learning structured output representation using deep conditional generative models, in: Advances in Neural Information Processing Systems, 2015, pp. 3483–3491.

[68] G. Lample, N. Zeghidour, N. Usunier, A. Bordes, L. Denoyer, M. Ranzato, Fader networks: manipulating images by sliding attributes, in: Advances in Neural Information Processing Systems, 2017, pp. 5967–5976.

[69] Y. Wang, T. Koike-Akino, D. Erdogmus, Invariant representations from adversarially censored autoencoders, arXiv preprint, arXiv:1805.08097, 2018.

[70] O. Özdenizci, Y. Wang, T. Koike-Akino, D. Erdoğmuş, Transfer learning in brain-computer interfaces with adversarial variational autoencoders, in: 2019 9th International IEEE/EMBS Conference on Neural Engineering (NER), IEEE, 2019, pp. 207–210.

[71] K. Wang, X. Ren, W. Chang, L. Liu, D. Liu, M. Zhang, Inverse design of digital nanophotonic devices using the adjoint method, Photonics Research 8 (2020) 528–533.

[72] Y. Zhang, S. Yang, A.E.-J. Lim, G.-Q. Lo, C. Galland, T. Baehr-Jones, M. Hochberg, A compact and low loss y-junction for submicron silicon waveguide, Optics Express 21 (2013) 1310–1316.

[73] N. Yu, F. Capasso, Flat optics with designer metasurfaces, Nature Materials 13 (2) (2014) 139–150.

[74] C.C. Nadell, B. Huang, J.M. Malof, W.J. Padilla, Deep learning for accelerated all-dielectric meta-surface design, Optics Express 27 (20) (2019) 27523–27535.

[75] S. An, C. Fowler, B. Zheng, M.Y. Shalaginov, H. Tang, H. Li, L. Zhou, J. Ding, A.M. Agarwal, C. Rivero-Baleine, et al., A deep learning approach for objective-driven all-dielectric metasurface design, ACS Photonics 6 (12) (2019) 3196–3207.

[76] P.R. Wiecha, O.L. Muskens, Deep learning meets nanophotonics: a generalized accurate predictor for near fields and far fields of arbitrary 3D nanostructures, Nano Letters 20 (1) (2019) 329–338.

[77] L. Gao, X. Li, D. Liu, L. Wang, Z. Yu, A bidirectional deep neural network for accurate silicon color design, Advanced Materials 31 (51) (2019) 1905467, https://doi.org/10.1002/adma.201905467.

[78] Z. Liu, L. Raju, D. Zhu, W. Cai, A hybrid strategy for the discovery and design of photonic structures, IEEE Journal on Emerging and Selected Topics in Circuits and Systems 10 (1) (2020) 126–135.

[79] J. Jiang, M. Chen, J.A. Fan, Deep neural networks for the evaluation and design of photonic devices, Nature Reviews Materials (2020) 1–22.

[80] O. Ronneberger, P. Fischer, T. Brox, U-net: convolutional networks for biomedical image segmentation, in: International Conference on Medical Image Computing and Computer-Assisted Intervention, Springer, 2015, pp. 234–241.

[81] F. Wen, J. Jiang, J.A. Fan, Robust freeform metasurface design based on progressively growing generative networks, ACS Photonics 7 (8) (2020) 2098–2104.

[82] S.J. Byrnes, Multilayer optical calculations, arXiv:1603.02720, 2020.

[83] T. Asano, S. Noda, Iterative optimization of photonic crystal nanocavity designs by using deep neural networks, Nanophotonics 8 (12) (2019) 2243–2256.

[84] T. Christensen, C. Loh, S. Picek, D. Jakobović, L. Jing, S. Fisher, V. Ceperic, J.D. Joannopoulos, M. Soljačić, Predictive and generative machine learning models for photonic crystals, Nanophotonics 9 (13) (2020) 4183–4192.

[85] H. Kang, T. Wang, Pytorch generative model collections, https://github.com/znxlwm/pytorch-generative-model-collections, 2017.

[86] A. Ashok, N. Rhinehart, F. Beainy, K.M. Kitani, N2N learning: network to network compression via policy gradient reinforcement learning, arXiv preprint, arXiv:1709.06030, 2017.

[87] A. Brock, T. Lim, J.M. Ritchie, N. Weston, Smash: one-shot model architecture search through hypernetworks, arXiv preprint, arXiv:1708.05344, 2017.

[88] H. Cai, T. Chen, W. Zhang, Y. Yu, J. Wang, Reinforcement learning for architecture search by network transformation, arXiv preprint, arXiv:1707.04873, 2017.

[89] R. Miikkulainen, J. Liang, E. Meyerson, A. Rawal, D. Fink, O. Francon, B. Raju, H. Shahrzad, A. Navruzyan, N. Duffy, et al., Evolving deep neural networks, in: Artificial Intelligence in the Age of Neural Networks and Brain Computing, Elsevier, 2019, pp. 293–312.

[90] Y. He, J. Lin, Z. Liu, H. Wang, L.-J. Li, S. Han, AMC: AutoML for model compression and acceleration on mobile devices, in: Proceedings of the European Conference on Computer Vision (ECCV), 2018, pp. 784–800.

[91] E. Real, S. Moore, A. Selle, S. Saxena, Y.L. Suematsu, J. Tan, Q.V. Le, A. Kurakin, Large-scale evolution of image classifiers, in: Proceedings of the 34th International Conference on Machine Learning-Volume 70, JMLR.org, 2017, pp. 2902–2911.

[92] E. Real, C. Liang, D.R. So, Q.V. Le AutoML-Zero, Evolving machine learning algorithms from scratch, arXiv preprint, arXiv:2003.03384, 2020.

[93] B. Zoph, V. Vasudevan, J. Shlens, Q.V. Le, Learning transferable architectures for scalable image recognition, in: Proceedings of the IEEE Conference on Computer Vision and Pattern Recognition, 2018, pp. 8697–8710.

[94] K.O. Stanley, R. Miikkulainen, Evolving neural networks through augmenting topologies, Evolutionary Computation 10 (2) (2002) 99–127.

[95] S.J. Pan, Q. Yang, A survey on transfer learning, IEEE Transactions on Knowledge and Data Engineering 22 (10) (2009) 1345–1359.

Index

Printed in the United States
by Baker & Taylor Publisher Services